Springer Series in Statistics

Advisors:
J. Berger, S. Fienberg, J. Gani,
K. Krickeberg, I. Olkin, B. Singer

Springer Series in Statistics

(continued after index)

Peter Hall

The Bootstrap and Edgeworth Expansion

Springer-Verlag

New York Berlin Heidelberg London Paris
Tokyo Hong Kong Barcelona Budapest

Peter Hall
Centre for Mathematics and its Applications
Australian National University
Canberra, Australia
and
CSIRO Division of Mathematics and Statistics
Sydney, Australia

Mathematical Subject Classification: 62E20, 62G09, 62E25

With six figures.

Library of Congress Cataloging-in-Publication Data
Hall, P.
    The bootstrap and Edgeworth expansion / Peter Hall.
        p.   cm. — (Springer series in statistics)
    Includes bibliographical references and index.
    ISBN 0-387-97720-1. — ISBN 3-540-97720-1
    1. Bootstrap (Statistics)   2. Edgeworth expansions.   I. Title.
    II. Series.
    QA276.8.H34   1992
    519.5 — dc20   91-34951

Printed on acid-free paper.

Production managed by Hal Henglein; manufacturing supervised by Jacqui Ashri.
Printed and bound by R. R. Donnelley & Sons, Harrisonburg, VA.
Printed in the United States of America.

9 8 7 6 5 4 3 2 1

ISBN 0-387-97720-1 Springer-Verlag New York Berlin Heidelberg
ISBN 3-540-97720-1 Springer-Verlag Berlin Heidelberg New York

Preface

This monograph addresses two quite different topics in the belief that each can shed light on the other. First, we lay the foundation for a particular view, even a personal view, of theory for Brad Efron's bootstrap. This topic is of course relatively new, its history extending over little more than a decade. Second, we give an account of theory for Edgeworth expansion, which has recently received a revival of interest owing to its usefulness for exploring properties of contemporary statistical methods. The theory of Edgeworth expansion is now a century old, if we date it from Chebyshev's paper in 1890.

Methods based on Edgeworth expansion can help explain the performance of bootstrap methods, and on the other hand, the bootstrap provides strong motivation for reexamining the theory of Edgeworth expansion. We do not claim to provide a definitive account of the totality of either subject. This is particularly so in the case of Edgeworth expansion, where our description of the theory is almost entirely restricted to results that are useful in exploring properties of the bootstrap. On the part of the bootstrap, our theoretical study is largely confined to a collection of problems that may be treated using classical theory for sums of independent random variables — the so-called "smooth function model." The corresponding class of statistics is wide (it includes sample means, variances, correlations, ratios and differences of means and variances, etc.) but is by no means comprehensive. It does not include, for example, the sample median or other sample quantiles, whose treatment is relegated to an appendix. Our study of the bootstrap is directed mainly at problems involving distribution estimation, since the theory of Edgeworth expansions is most relevant there. (However, our introduction to bootstrap methods in Chapter 1 is more wide-ranging.) We consider linear regression and nonparametric regression, but not general linear models or functional regression. We treat some but by no means all possible approaches to using the bootstrap in density estimation. We do not treat bootstrap methods for dependent data. There are many other examples of our selectivity, which is designed to present a coherent and unified account of theory for the bootstrap.

Chapter 1 is about the bootstrap, with almost no mention of Edgeworth expansion; Chapter 2 is about Edgeworth expansion, with scarcely a word about the bootstrap; and Chapter 3 brings these two themes together, using Edgeworth expansion to explore and develop properties of the bootstrap. Chapter 4 continues this approach along less conventional lines, and Chapter 5 sketches details of mathematical rigour that are missing from earlier chapters. Finally, five appendices present material that complements the main theoretical themes of the monograph. For example, Appendix II on Monte Carlo simulation summarizes theory behind the numerical techniques that are essential to practical implementation of the bootstrap methods described in Chapters 1, 3, and 4. Each chapter concludes with a bibliographical note; citations are kept to a minimum elsewhere in the chapters.

I am grateful to D.R. Cox, D.J. Daley, A.C. Davison, B. Efron, N.I. Fisher, W. Härdle, P. Kabaila, J.S. Marron, M.A. Martin, Y.E. Pittelkow, T.P. Speed and T.E. Wehrly for helpful comments and corrections of some errors. J., N. and S. provided much-needed moral support throughout the project. Miss N. Chin typed all drafts with extraordinary accuracy and promptness, not to say good cheer. My thanks go particularly to her. Both of us are grateful to Graeme Bailey for his help, at all hours of the day and night, in getting the final TEX manuscript to the production phase.

Contents

Common Notation

Alphabetical Notation

He_j : jth Hermite polynomial; the functions He_j, $j \geq 0$, are orthogonal with respect to the weight function $e^{-x^2/2}$. (See Section 2.2.)

$\widehat{I}$: percentile bootstrap confidence interval.

$\widehat{J}$: percentile-t bootstrap confidence interval.

I, J : nonbootstrap version of $\widehat{I}, \widehat{J}$, obtained by reference to the distributions of $(\hat{\theta} - \theta_0)/\sigma$, $(\hat{\theta} - \theta_0)/\hat{\sigma}$ respectively.

$\mathcal{I}, \mathcal{J}$: generic confidence intervals.

   (If numeric subscripts are appended to $\widehat{I}$, $\widehat{J}$, I, J, $\mathcal{I}$, or $\mathcal{J}$, then the first subscript (1 or 2) indicates the number of sides to the interval.)

$I(\cdot)$: indicator function.

n : sample size.

p, q (denoting functions; with or without subscripts) : polynomials in Edgeworth or Cornish-Fisher expansions. We always reserve q for the case of a Studentized statistic.

$\mathbb{R}$: real line.

$\mathbb{R}^d$: d-dimensional Euclidean space.

$\mathcal{R}$: subset of $\mathbb{R}^d$.

t_α : α-level quantile of Student's t distribution with $n - \nu$ degrees of freedom, for a specified integer $\nu \geq 1$.

u_α, v_α : α-level quantiles of distributions of $n^{1/2}(\hat{\theta} - \theta_0)/\sigma$, $n^{1/2}(\hat{\theta} - \theta_0)/\hat{\sigma}$ respectively. That is, u_α and v_α solve the equations

$$P\{n^{1/2}(\hat{\theta} - \theta_0)/\sigma \leq u_\alpha\} \ = \ P\{n^{1/2}(\hat{\theta} - \theta_0)/\hat{\sigma} \leq v_\alpha\} \ = \ \alpha.$$

$\hat{u}_\alpha, \hat{v}_\alpha$: bootstrap estimates of u_α, v_α, respectively.

$\mathcal{X} = \{X_1, \dots, X_n\}$: random sample from population. (A generic X_i is denoted by X.)

$\mathcal{X}^* = \{X_1^*, \dots, X_n^*\}$: resample obtained by sampling from $\mathcal{X}$ with replacement. (A generic X_i^* is denoted by X^*.)

$\bar{X}, \bar{X}^*$: means of $\mathcal{X}, \mathcal{X}^*$ respectively.

z_α : α-level quantile of Standard Normal distribution; that is, $z_\alpha = \Phi^{-1}(\alpha)$.

α : nominal coverage level of confidence interval; $\xi = \frac{1}{2}(1 + \alpha)$.

xi

γ : skewness of a distribution. (Generally, $\gamma = \kappa_3$.)

θ_0 : true value of unknown parameter θ.

$\hat{\theta}$: estimate of θ, computed from $\mathcal{X}$.

$\hat{\theta}^*$: bootstrap version of $\hat{\theta}$, computed from $\mathcal{X}^*$. (If bootstrap quantities are computed by Monte Carlo simulation, involving calculation of a sequence $\hat{\theta}_1^*, \ldots, \hat{\theta}_B^*$, then $\hat{\theta}^*$ denotes a generic $\hat{\theta}_b^*$.)

κ_j : jth cumulant of a distribution.

κ (without subscript) : kurtosis of a distribution. (Generally, $\kappa = \kappa_4$.)

μ : population mean.

ξ : see α.

σ^2 : asymptotic variance of $n^{1/2}\,\hat{\theta}$.

$\hat{\sigma}^2$: estimate of σ^2, computed from $\mathcal{X}$.

$\hat{\sigma}^{*2}$: bootstrap version of $\hat{\sigma}^2$, computed from $\mathcal{X}^*$.

ϕ, Φ (without subscripts) : Standard Normal density, distribution functions respectively.

χ : characteristic function.

χ^* : bootstrap version of χ.

Nonalphabetical Notation, and Notes

The transpose of a vector or matrix $\mathbf{v}$ is indicated by $\mathbf{v}^{\mathrm{T}}$. If $\mathbf{v}$ is a vector, its ith element is denoted by $v^{(i)} = (\mathbf{v})^{(i)}$. If $\mathbf{x}, \mathbf{y}$ are two vectors of length d, then $\mathbf{x} \geq \mathbf{y}$ means $x^{(i)} \geq y^{(i)}$ for $1 \leq i \leq d$.

Complementation is denoted by $\sim$. For example, if $\mathcal{R} \subseteq \mathbb{R}^d$ then

$$\tilde{\mathcal{R}} \;=\; \mathbb{R}^d \backslash \mathcal{R} \;=\; \{\mathbf{x} \in \mathbb{R}^d : \mathbf{x} \notin \mathcal{R}\}.$$

Given a sample $\mathcal{X} = \{X_1, \ldots, X_n\}$, the empirical measure is the measure that assigns to a set $\mathcal{R}$ a value equal to the proportion of sample values contained within $\mathcal{R}$. The empirical distribution function (d.f.) and empirical characteristic function (ch.f.) are respectively the d.f. and ch.f. of the empirical measure.

If $\{\Delta_n\}$ is a sequence of random variables and $\{\delta_n\}$ is a sequence of positive constants, we write $\Delta_n = O_p(\delta_n)$ ("Δ_n is of order δ_n in probability") to mean that

$$\lim_{\lambda \to \infty} \limsup_{n \to \infty} P(|\Delta_n/\delta_n| > \lambda) \;=\; 0.$$

If in addition

$$\lim_{\epsilon \to 0} \limsup_{n \to \infty} P(|\Delta_n/\delta_n| > \epsilon) > 0,$$

we say that Δ_n is of precise order or precise size δ_n, in probability.

The "first term" in an Edgeworth expansion is the term of largest order that converges to zero as $n \to \infty$. The term in the expansion that does not depend on n is called the "0th term."

Theorems are stated under explicit regularity conditions; *Propositions* are not given under explicit conditions.

Wherever possible we use the term *estimate*, rather than estimator, to indicate a function of the data that approximates an unknown quantity. For example, $\hat{\theta}$ is an estimate of θ. However, if θ is function-valued (e.g., if θ is a density) then we call $\hat{\theta}$ an estimator. It is sometimes difficult to be absolutely consistent in this notation.

1

Principles of Bootstrap Methodology

1.1 Introduction

The idea behind the bootstrap is an old one. If we wish to estimate a functional of a population distribution function F, such as a population mean

$$\mu = \int x \, dF(x) \,,$$

consider employing the same functional of the sample (or empirical) distribution function $\widehat{F}$, which in this instance is the sample mean

$$\bar{X} = \int x \, d\widehat{F}(x) \,.$$

(The empirical distribution is that probability measure that assigns to a set a measure equal to the proportion of sample values that lie in that set.) This argument is not always practicable — a probability density is just one example of a functional of F that is not directly amenable to this treatment. Nevertheless, the scope of the bootstrap idea was much broader than most of us realized before Efron (1979) drew our attention to its considerable promise and virtue and gave it its name. With modification and elaboration it applies even more widely, to probability densities and beyond.

Efron also showed that in many complex situations, where bootstrap statistics are awkward to compute, they may be approximated by Monte Carlo "resampling." That is, same-size resamples may be drawn repeatedly from the original sample, the value of a statistic computed for each individual resample, and the bootstrap statistic approximated by taking an average of an appropriate function of these numbers. (See Appendix II.) The approximation improves as the number of resamples increases. This approach works because, in the majority of applications of the bootstrap, the bootstrap quantity may be expressed as an expectation conditional on the sample, or equivalently, as an integral with respect to the sample distribution function. Our development of bootstrap principles focuses on that property. Later chapters will use Edgeworth expansions to explore deeper and more detailed aspects of the principles. For the sake of connectivity,

1

some of that work will be cited in this chapter, but Edgeworth expansions will not be needed for the present development.

Somewhat unfortunately, the name "bootstrap" conveys the impression of "something for nothing" — of statisticians idly resampling from their samples, presumably having about as much success as they would if they tried to pull themselves up by their bootstraps. This perception still exists in some quarters. One of the aims of this monograph is to dispel such mistaken impressions by presenting the bootstrap as a technique with a sound and promising theoretical basis.

We show in Section 1.2 that many statistical problems may be posed in terms of a solution to a "population equation," involving integration with respect to the population distribution function. The bootstrap estimate of the solution to this equation comes from solving an equation in which integration is with respect to the sample distribution function; the equation is now a "sample equation." By posing problems in this manner we concentrate attention on the extent to which bootstrap methods achieve perfect accuracy in a well-defined sense, such as error-free coverage accuracy of confidence intervals or unbiasedness of point estimates. This approach goes well beyond a mere definition of bootstrap estimators, as we stress in the last paragraph of Section 1.2. It is particularly relevant to bootstrap iteration, studied in Section 1.4. There, it is shown that the bootstrap principle can be repeated successively, to improve the accuracy of the solution to the population equation. Examples of the bootstrap principle and bootstrap iteration are detailed in Sections 1.3 and 1.5, respectively. Example 1.2 in Section 1.3 points out the importance of "pivotalness" in confidence interval construction, and this concept is germane to work in future chapters.

Notation will be introduced at various places in this chapter, as new ideas are developed. However, some notation deserves special mention here. A *sample* $\mathcal{X} = \{X_1, \ldots, X_n\}$ is a collection of n numbers (usually scalars in this chapter, but often vectors in later chapters), without regard to order, drawn at random from the population. By "at random" we mean that the X_i's are independent and identically distributed random variables, each having the population distribution function F. In so-called nonparametric problems, a *resample* $\mathcal{X}^* = \{X_1^*, \ldots, X_n^*\}$ is an unordered collection of n items drawn randomly from $\mathcal{X}$ with replacement, so that each X_i^* has probability n^{-1} of being equal to any given one of the X_j's,

$$P(X_i^* = X_j \mid \mathcal{X}) = n^{-1}, \quad 1 \leq i, j \leq n.$$

That is, the X_i^*'s are independent and identically distributed, conditional on $\mathcal{X}$, with this distribution. Of course, this means that $\mathcal{X}^*$ is likely to contain repeats, all of which must be listed in the collection $\mathcal{X}^*$. In this sense our use of set notation for $\mathcal{X}^*$ is a trifle nonstandard. For example, when $n = 4$ we do not intend the *collection* $\mathcal{X}^* = \{1.5, 1.7, 1.7, 1.8\}$ to be mistaken for the *set* $\{1.5, 1.7, 1.8\}$ and, of course, $\mathcal{X}^*$ is the same as $\{1.5, 1.7, 1.8, 1.7\}$, $\{1.7, 1.5, 1.8, 1.7\}$, etc.

In parametric problems $\mathcal{X}^*$ denotes a sample drawn at random from a population depending on a parameter whose value has been estimated; Section 1.2 will give details. If that population is continuous then with probability one, all values in $\mathcal{X}^*$ are different, and so the issue of ties discussed in the preceding paragraph does not arise. In both parametric and nonparametric problems, $\widehat{F}$ denotes the distribution function of the "population" from which $\mathcal{X}^*$ was drawn.

The abbreviation "d.f." stands for "distribution function." The pair $(F, \widehat{F})$, representing (population d.f. , sample d.f.), will often be written as (F_0, F_1) during this chapter, hence making it easier to introduce bootstrap iteration, where for general $i \geq 1$, F_i denotes the d.f. of a sample drawn from F_{i-1} conditional on F_{i-1}. For the same reason, the pair $(\mathcal{X}, \mathcal{X}^*)$ will often be written as $(\mathcal{X}_1, \mathcal{X}_2)$.

The ith application of the bootstrap to a particular problem is termed the ith iteration, not the $(i - 1)$th iteration. Given a sample $\mathcal{X} = \{X_1, \ldots, X_n\}$,

$$\hat{\sigma}^2 = n^{-1} \sum_{i=1}^{n} (X_i - \bar{X})^2$$

denotes the *sample variance* and

$$\tilde{\sigma}^2 = (n - 1)^{-1} \sum_{i=1}^{n} (X_i - \bar{X})^2$$

is the *unbiased variance estimate*. (In subsequent chapters it will be convenient to use $\hat{\sigma}^2$ to represent a more general variance estimate.)

An estimate $\hat{\theta}$ is a function of the data and may also be regarded as a functional of the sample distribution function $\widehat{F}$. While these two functions are numerically equivalent, it is helpful to distinguish between them. This we do by using square brackets for the former and round brackets for the latter:

$$\hat{\theta} = \theta[\mathcal{X}] = \theta(\widehat{F}).$$

For example, if $\theta = \theta(F) = \int x \, dF(x)$ is population mean and $\widehat{F}$ is the empirical distribution function (which assigns weight n^{-1} to each data point X_i), then

$$\hat{\theta} = \theta[\mathcal{X}] = n^{-1} \sum_{i=1}^{n} X_i$$

$$= \theta(\widehat{F}) = \int x \, d\widehat{F}(x),$$

the sample mean.

1.2 The Main Principle

We begin by describing a convenient physical analogy, and later (following equation (1.5)) give an explicit statement of the principle.

A Russian "matryoshka" doll is a nest of wooden figures, usually with slightly different features painted on each. Call the outer figure "doll 0," the next figure "doll 1," and so on; see Figure 1.1. Suppose we are not allowed to observe doll 0 — it represents the population in a sampling scheme. We wish to estimate the number n_0 of freckles on its face. Let n_i denote the number of freckles on the face of doll i. Since doll 1 is smaller than doll 0, n_1 is likely to be an underestimate of n_0, but it seems reasonable to suppose that the ratio of n_1 to n_2 should be close to the ratio of n_0 to n_1. That is, $n_1/n_2 \simeq n_0/n_1$, so that $\hat{n}_0 = n_1^2/n_2$ might be a reasonable estimate of n_0.

The key feature of this argument is our hypothesis that the relationship between n_2 and n_1 should closely resemble that between n_1 and the unknown n_0. Under the (fictitious) assumption that the relationships are *identical*, we equated the two ratios and obtained our estimate $\hat{n}_0$. We could refine the argument by delving more deeply into the nest of dolls, adding correction terms to $\hat{n}_0$ so as to take account of the relationship between doll i and doll $i + 1$ for $i \geq 2$. We shall have more to say about that in Section 1.4. But for the time being we study the more obvious implications of the "Russian doll principle."

Much of statistical inference amounts to describing the relationship between a sample and the population from which the sample was drawn. Formally, given a functional f_t from a class $\{f_t : t \in T\}$, we wish to determine that value t_0 of t that solves an equation

$$ E\left\{f_t(F_0, F_1) \mid F_0\right\} = 0, \tag{1.1} $$

where F_0 denotes the population distribution function and F_1 the distribution function "of the sample." An explicit definition of F_1 will be given shortly. Conditioning on F_0 in (1.1) serves to stress that the expectation is taken with respect to the distribution F_0. We call (1.1) the *population equation* because we need properties of the population if we are to solve this equation exactly.

For example, let $\theta_0 = \theta(F_0)$ denote a true parameter value, such as the rth power of a mean,

$$ \theta_0 = \left\{\int x \, dF_0(x)\right\}^r . $$

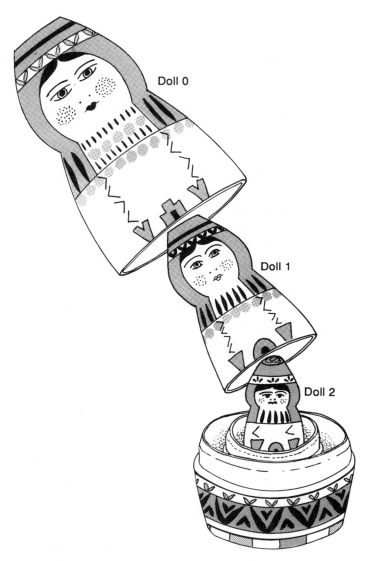

Doll 0

Doll 1

Doll 2

FIGURE 1.1. A Russian matryoshka doll.

Here $\theta(G)$ is the functional $\{\int x\, dG(x)\}^r$. (The term "parameter" is not entirely satisfactory since it conveys the impression that we are working under a parametric model, which is not necessarily the case. However, there does not appear to be a more appropriate term.) Let $\hat{\theta} = \theta(F_1)$ be our bootstrap estimate of θ_0, such as the rth power of a sample mean,

$$\hat{\theta} = \left\{ \int x\, dF_1(x) \right\}^r = \bar{X}^r,$$

where F_1 is the empirical distribution function of the sample. Correcting $\hat{\theta}$ for bias is equivalent to finding that value t_0 that solves (1.1) when

$$f_t(F_0, F_1) = \theta(F_1) - \theta(F_0) + t. \tag{1.2}$$

Our bias-corrected estimate would be $\hat{\theta} + t_0$. On the other hand, to construct a symmetric, 95% confidence interval for θ_0, solve (1.1) when

$$f_t(F_0, F_1) = I\{\theta(F_1) - t \le \theta(F_0) \le \theta(F_1) + t\} - 0.95, \tag{1.3}$$

where the indicator function $I(\mathcal{E})$ is defined to equal 1 if event $\mathcal{E}$ holds and 0 otherwise. The confidence interval is $(\hat{\theta} - t_0,\ \hat{\theta} + t_0)$, where $\hat{\theta} = \theta(F_1)$.

Equation (1.1) provides an explicit description of the relationship between F_0 and F_1 that we are trying to determine. Its analogue in our "find the number of freckles on the face of the Russian doll" problem would be

$$n_0 - t n_1 = 0, \tag{1.4}$$

where (as before) n_i denotes the number of freckles on doll i. If we could solve (1.4) for $t = t_0$, then the value of n_0 would be $n_0 = t_0 n_1$. Recall from the first paragraph of this section that our approach to estimating t_0 is to replace the pair (n_0, n_1) in (1.4) by (n_1, n_2), which we know, thereby transforming (1.4) to

$$n_1 - t n_2 = 0.$$

We obtain the solution $\hat{t}_0$ of this equation, and thereby our estimate $\hat{n}_0 = \hat{t}_0 n_1 = n_1^2/n_2$ of n_0.

To obtain an approximate solution of the population equation (1.1), all we do is apply the same argument. The analogue of n_2 is F_2, the distribution function of a sample drawn from F_1 (conditional on F_1). Simply replace the pair (F_0, F_1) in (1.1) by (F_1, F_2), thereby transforming (1.1) to

$$E\{f_t(F_1, F_2) \mid F_1\} = 0. \tag{1.5}$$

We call this the *sample equation* because we know (or can find out) everything about it once we know the sample distribution function F_1. In particular, its solution $\hat{t}_0$ is a function of the sample values. The idea, of course, is that the solution of the sample equation should be a good approximation of the solution of the population equation, the latter being unobtainable in practice.

We shall refer to this as "the bootstrap principle".

We call $\hat{t}_0$ and $E\{f_t(F_1, F_2) \mid F_1\}$ "the bootstrap estimates" of t_0 and $E\{f_t(F_0, F_1) \mid F_0\}$, respectively. They are obtained by replacing F_0 by F_1 in formulae for t_0 and $E\{f_t(F_0, F_1) \mid F_0\}$. In the bias correction problem, where f_t is given by (1.2), the bootstrap version of our bias-corrected estimate is $\hat{\theta} + \hat{t}_0$. In the confidence interval problem where (1.3) describes f_t, our bootstrap confidence interval is $(\hat{\theta} - \hat{t}_0, \ \hat{\theta} + \hat{t}_0)$. The latter is commonly called a (symmetric) percentile-method confidence interval for θ_0.

It is appropriate now to give detailed definitions of F_1 and F_2. There are at least two approaches, suitable for nonparametric and parametric problems respectively. In both, inference is based on a sample $\mathcal{X}$ of n random (independent and identically distributed) observations of the population. In the nonparametric case, F_1 is simply the empirical distribution function of $\mathcal{X}$; that is, the distribution function of the distribution that assigns mass n^{-1} to each point in $\mathcal{X}$. The associated empirical probability measure assigns to a region $\mathcal{R}$ a value equal to the proportion of the sample that lies within $\mathcal{R}$. Similarly, F_2 is the empirical distribution function of a sample drawn at random from the population with distribution function F_1; that is, the empiric of a sample $\mathcal{X}^*$ drawn randomly, with replacement, from $\mathcal{X}$. If we denote the population by $\mathcal{X}_0$ then we have a nest of sampling operations, like our nest of dolls: $\mathcal{X}$ is drawn at random from $\mathcal{X}_0$ and $\mathcal{X}^*$ is drawn at random from $\mathcal{X}$.

In the parametric case, F_0 is assumed completely known up to a finite vector $\boldsymbol{\lambda}_0$ of unknown parameters. To indicate this dependence we write $F_0 = F_{(\lambda_0)}$, an element of a class $\{F_{(\lambda)}, \ \boldsymbol{\lambda} \in \Lambda\}$ of possible distributions. Let $\widehat{\boldsymbol{\lambda}}$ be an estimate of $\boldsymbol{\lambda}_0$ computed from $\mathcal{X}$, often (but not necessarily) the maximum likelihood estimate. It will be a function of sample values, so we may write it as $\boldsymbol{\lambda}[\mathcal{X}]$. Then $F_1 = F_{(\widehat{\lambda})}$, the distribution function obtained on replacing "true" parameter values by their sample estimates. Let $\mathcal{X}^*$ denote the sample drawn at random from the distribution with distribution function $F_{(\widehat{\lambda})}$ (*not* simply drawn from $\mathcal{X}$ with replacement), and let $\widehat{\boldsymbol{\lambda}}^* = \boldsymbol{\lambda}[\mathcal{X}^*]$ denote the version of $\widehat{\boldsymbol{\lambda}}$ computed for $\mathcal{X}^*$ instead of $\mathcal{X}$. Then $F_2 = F_{(\widehat{\lambda}^*)}$.

In both parametric and nonparametric cases, $\mathcal{X}^*$ is found by resampling from a distribution determined by the original sample $\mathcal{X}$. Thus, the "Russian doll principle" has led us to a *bootstrap principle* for conducting statistical inference.

The reader will appreciate that the bootstrap principle is not needed in all circumstances. As pointed out in the opening paragraph of the Introduction, the key idea behind the bootstrap (in nonparametric problems) is that of replacing the true distribution function $F = F_0$ by the empiric $\widehat{F} = F_1$ in a formula that expresses a parameter as a functional of F. This entails replacing the pair (F_0, F_1) by (F_1, F_2), and may be applied very generally to construct estimates of means, quantiles, etc. For example, mean squared error

$$\tau^2 = E(\hat{\theta} - \theta_0)^2 = E\left[\{\theta(F_1) - \theta(F_0)\}^2 \mid F_0\right]$$

has bootstrap estimate

$$\hat{\tau}^2 = E\{(\hat{\theta}^* - \hat{\theta})^2 \mid \mathcal{X}\} = E\left[\{\theta(F_2) - \theta(F_1)\}^2 \mid F_1\right],$$

where $\hat{\theta}^* = \theta[\mathcal{X}^*]$ is the version of $\hat{\theta}$ computed for $\mathcal{X}^*$ rather than $\mathcal{X}$. In this circumstance the facility offered by the resampling principle, of focusing attention on a particular characteristic (such as bias or coverage error) that should ideally be vanishingly small, is not required.

We are now in a position to illustrate the resampling principle in detail.

1.3 Four Examples

Example 1.1: Bias Reduction

Here the function f_t is given by (1.2), and the sample equation (1.5) assumes the form

$$E\{\theta(F_2) - \theta(F_1) + t \mid F_1\} = 0,$$

whose solution is

$$t = \hat{t}_0 = \theta(F_1) - E\{\theta(F_2) \mid F_1\}.$$

The bootstrap bias-reduced estimate is thus

$$\hat{\theta}_1 = \hat{\theta} + \hat{t}_0 = \theta(F_1) + \hat{t}_0 = 2\theta(F_1) - E\{\theta(F_2) \mid F_1\}. \tag{1.6}$$

Note that our basic estimate $\hat{\theta} = \theta(F_1)$ is also a bootstrap estimate since it is obtained by substituting F_1 for F_0 in the functional formula $\theta_0 = \theta(F_0)$.

The expectation $E\{\theta(F_2) \mid F_1\}$ may always be computed (or approximated) by Monte Carlo simulation, as follows. Conditional on F_1, draw B resamples $\{\mathcal{X}_b^*, 1 \leq b \leq B\}$ independently from the distribution with distribution function F_1. In the nonparametric case, where F_1 is the empirical distribution function of the sample $\mathcal{X}$, let F_{2b} denote the empirical distribution function of $\mathcal{X}_b^*$. In the parametric case, let $\widehat{\lambda}_b^* = \lambda(\mathcal{X}_b^*)$ be that estimate of λ_0 computed from resample $\mathcal{X}_b^*$, and put $F_{2b} = F_{(\widehat{\lambda}_b^*)}$.

Define $\hat{\theta}_b^* = \theta(F_{2b})$ and $\hat{\theta} = \theta(F_1)$. Then in both parametric and nonparametric circumstances,

$$\hat{u}_B = B^{-1} \sum_{b=1}^{B} \theta(F_{2b}) = B^{-1} \sum_{b=1}^{B} \hat{\theta}_b^*$$

converges to $\hat{u} = E\{\theta(F_2) \mid F_1\} = E(\hat{\theta}^* \mid \mathcal{X})$ (with probability one, conditional on F_1) as $B \to \infty$. If the expectation cannot be computed exactly then $\hat{u}_B$, for a suitably large value of B, may be used as a practical and accurate approximation. See Appendix II for an account of efficient Monte Carlo simulation.

In some simple cases we may compute the expectation explicitly. For example, let $\mu = \int x \, dF_0(x)$ denote the mean of a population, assumed to be a scalar, and put $\theta_0 = \theta(F_0) = \mu^3$, the third power of the mean. Write $\mathcal{X}_1 = \{X_1, \dots, X_n\}$ for the sample and $\bar{X} = n^{-1} \sum X_i$ for its mean. In nonparametric problems, at least, $\hat{\theta} = \theta(F_1) = \bar{X}^3$. Elementary calculations show that

$$E\{\theta(F_1) \mid F_0\} = E\left\{\mu + n^{-1} \sum_{i=1}^{n} (X_i - \mu)\right\}^3$$
$$= \mu^3 + n^{-1} 3\mu\sigma^2 + n^{-2}\gamma, \qquad (1.7)$$

where $\sigma^2 = E(X_1 - \mu)^2$ and $\gamma = E(X_1 - \mu)^3$ denote population variance and skewness respectively. In the nonparametric case we obtain in direct analogy to (1.7),

$$E\{\theta(F_2) \mid F_1\} = \bar{X}^3 + n^{-1} 3\bar{X}\hat{\sigma}^2 + n^{-2}\hat{\gamma},$$

where $\hat{\sigma}^2 = n^{-1} \sum(X_i - \bar{X})^2$ and $\hat{\gamma} = n^{-1} \sum(X_i - \bar{X})^3$ denote sample variance and sample skewness respectively. (We have defined $\hat{\sigma}^2$ using divisor n rather than $(n-1)$, because in the nonparametric case the variance of the distribution with distribution function F_1 is $n^{-1} \sum(X_i - \bar{X})^2$, not $(n-1)^{-1} \sum(X_i - \bar{X})^2$.) Therefore the bootstrap bias-reduced estimate is, by (1.6),

$$\hat{\theta}_1 = 2\theta(F_1) - E\{\theta(F_2) \mid F_1\}$$
$$= 2\bar{X}^3 - \left(\bar{X}^3 + n^{-1} 3\bar{X}\hat{\sigma}^2 + n^{-2}\hat{\gamma}\right)$$
$$= \bar{X}^3 - n^{-1} 3\bar{X}\hat{\sigma}^2 - n^{-2}\hat{\gamma}. \qquad (1.8)$$

In parametric contexts, the estimate $\hat{\theta}_1$ can have a formula different from this. We shall consider two cases, Normal and Exponential distributions. If the population is Normal $N(\mu, \sigma^2)$ then $\gamma = 0$ and (1.7) becomes

$$E\{\theta(F_1) \mid F_0\} = \mu^3 + n^{-1} 3\mu\sigma^2. \qquad (1.9)$$

We could use the maximum likelihood estimate $\widehat{\boldsymbol{\lambda}} = (\bar{X}, \hat{\sigma}^2)$ to estimate the parameter vector $\boldsymbol{\lambda}_0 = (\mu, \sigma^2)$. Then, since $\theta(F_2)$ denotes the statistic $\hat{\theta}$ computed for a sample from a Normal $N(\bar{X}, \hat{\sigma}^2)$ distribution, we have in direct analogy to (1.9),

$$E\left\{\theta(F_2) \mid F_1\right\} = \bar{X}^3 + n^{-1} 3 \bar{X} \hat{\sigma}^2.$$

Therefore, by (1.6),

$$\hat{\theta}_1 = 2\theta(F_1) - E\left\{\theta(F_2) \mid F_1\right\} = \bar{X}^3 - n^{-1} 3 \bar{X} \hat{\sigma}^2. \qquad (1.10)$$

Alternatively, we might take $\widehat{\boldsymbol{\lambda}} = (\bar{X}, \tilde{\sigma}^2)$ where $\tilde{\sigma}^2 = (n-1)^{-1} \sum (X_i - \bar{X})^2$ is the unbiased estimate of variance, in which case

$$\hat{\theta}_1 = \bar{X}^3 - n^{-1} 3 \bar{X} \tilde{\sigma}^2. \qquad (1.11)$$

As our second parametric example, take the population to be Exponential with mean μ, having density $f_\mu(x) = \mu^{-1} e^{-x/\mu}$ for $x > 0$. Then $\sigma^2 = \mu^2$ and $\gamma = 2\mu^3$, so that (1.7) becomes

$$E\left\{\theta(F_1) \mid F_0\right\} = \mu^3 \left(1 + 3n^{-1} + 2n^{-2}\right). \qquad (1.12)$$

Taking our estimate of $\lambda_0 = \mu$ to be the maximum likelihood estimate $\hat{\lambda} = \bar{X}$, we obtain in direct analogy to (1.12),

$$E\left\{\theta(F_2) \mid F_1\right\} = \bar{X}^3 \left(1 + 3n^{-1} + 2n^{-2}\right).$$

Therefore

$$\hat{\theta}_1 = 2\theta(F_1) - E\left\{\theta(F_2) \mid F_1\right\} = \bar{X}^3 \left(1 - 3n^{-1} - 2n^{-2}\right). \qquad (1.13)$$

The estimates $\hat{\theta}_1$ defined by (1.8), (1.10), (1.11), and (1.13) each represent improvements, in the sense of bias reduction, on the basic bootstrap estimate $\hat{\theta} = \theta(F_1)$. To check that they really do reduce bias, observe that for general distributions with finite third moments,

$$E(\bar{X}^3) = \mu^3 + n^{-1} 3 \mu \sigma^2 + n^{-2} \gamma,$$
$$E(\bar{X} \hat{\sigma}^2) = \mu \sigma^2 + n^{-1} (\gamma - \mu \sigma^2) - n^{-2} \gamma,$$

and

$$E(\hat{\gamma}) = \gamma \left(1 - 3n^{-1} + 2n^{-2}\right).$$

It follows that

$$E(\hat{\theta}_1) - \theta_0 =$$

$$\begin{cases} n^{-2}\, 3\, (\mu\sigma^2 - \gamma) + n^{-3}\, 6\,\gamma - n^{-4}\, 2\,\gamma & \text{in the case of (1.8),} \\ & \text{for general populations;} \\[2mm] n^{-2}\, 3\,\mu\sigma^2 & \text{in the case of (1.10), for Normal populations;} \\[2mm] 0 & \text{in the case of (1.11), for Normal populations;} \\[2mm] -\mu^3\, (9n^{-2} + 12n^{-3} + 4n^{-4}) & \text{in the case of (1.13),} \\ & \text{for Exponential populations.} \end{cases}$$

Therefore bootstrap bias reduction has diminished bias to at most $O(n^{-2})$ in each case. Compare this with the bias of $\hat{\theta}$, which is of size n^{-1} unless $\mu = 0$ (see (1.7)). Furthermore, if the population has finite sixth-order moments then $\hat{\theta}$ and each version of $\hat{\theta}_1$ has variance asymptotic to a constant multiple of n^{-1}, the constant being the same for both $\hat{\theta}$ and $\hat{\theta}_1$. While the actual variance may have increased a little as a result of bootstrap bias reduction, the first-order asymptotic formula for variance has not changed.

More generally than these examples, bootstrap bias correction reduces the order of magnitude of bias by the factor n^{-1}; see Section 1.4 for a proof.

Example 1.5 in Section 1.5 treats the problem of bias-correcting an estimate of a squared mean.

Example 1.2: Confidence Interval

A symmetric confidence interval for $\theta_0 = \theta(F_0)$ may be constructed by applying the resampling principle using the function f_t given by (1.3). The sample equation then assumes the form

$$P\Big\{\theta(F_2) - t \le \theta(F_1) \le \theta(F_2) + t \;\Big|\; F_1\Big\} - 0.95 = 0. \qquad (1.14)$$

In a nonparametric context, $\theta(F_2)$ conditional on F_1 has a discrete distribution, and so it would seldom be possible to solve (1.14) exactly. However, any error in the solution of (1.14) will usually be very small, since the size of even the largest atom of the distribution of $\theta(F_2)$ decreases exponentially quickly with increasing n (see Appendix I). The largest atom is of size only 3.6×10^{-4} when $n = 10$. We could remove this minor difficulty by smoothing the distribution function F_1. See Section 4.1 for a discussion of that approach. In parametric cases, (1.14) may usually be solved exactly for t.

Since any error in solving (1.14) is usually so small, and decreases so rapidly with increasing n, there is no harm in saying that the quantity

$$\hat{t}_0 = \inf\left\{t \ : \ P[\theta(F_2) - t \leq \theta(F_1) \leq \theta(F_2) + t \mid F_1] - 0.95 \geq 0\right\}$$

actually solves (1.14). We shall indulge in this abuse of notation throughout the monograph. Now, $(\hat{\theta} - \hat{t}_0, \ \hat{\theta} + \hat{t}_0)$ is a bootstrap confidence interval for $\theta_0 = \theta(F_0)$, usually called a (two-sided, symmetric) *percentile* interval since $\hat{t}_0$ is a percentile of the distribution of $|\theta(F_2) - \theta(F_1)|$ conditional on F_1. Other nominal 95% percentile intervals include the two-sided, equal-tailed interval $(\hat{\theta} - \hat{t}_{01}, \ \hat{\theta} + \hat{t}_{02})$ and the one-sided interval $(-\infty, \ \hat{\theta} + \hat{t}_{03})$, where $\hat{t}_{01}$, $\hat{t}_{02}$, and $\hat{t}_{03}$ solve

$$P\{\theta(F_1) \leq \theta(F_2) - t \mid F_1\} - 0.025 = 0,$$
$$P\{\theta(F_1) \leq \theta(F_2) + t \mid F_1\} - 0.975 = 0,$$

and

$$P\{\theta(F_1) \leq \theta(F_2) + t \mid F_1\} - 0.95 = 0,$$

respectively. The former interval is called *equal-tailed* because it attempts to place equal probability in each tail:

$$P(\theta_0 \leq \hat{\theta} - \hat{t}_{01}) \approx P(\theta_0 \geq \hat{\theta} + \hat{t}_{02}) \approx 0.025.$$

The "ideal" form of this interval, obtained by solving the population equation rather than the sample equation, does place equal probability in each tail.

Still other 95% percentile intervals are $\hat{I}_2 = (\hat{\theta} - \hat{t}_{02}, \hat{\theta} + \hat{t}_{01})$ and $\hat{I}_1 = (-\infty, \hat{\theta} + \hat{t}_{04})$, where $\hat{t}_{04}$ is the solution of

$$P\{\theta(F_1) \leq \theta(F_2) - t \mid F_1\} - 0.05 = 0.$$

These do not fit naturally into a systematic development of bootstrap methods by frequentist arguments, and we find them a little contrived. They are sometimes motivated as follows. Define $\hat{\theta}^* = \theta(F_2)$, $\widehat{H}(x) = P(\hat{\theta}^* \leq x \mid \mathcal{X})$, and

$$\widehat{H}^{-1}(\alpha) = \inf\{x : \widehat{H}(x) \geq \alpha\}.$$

Then

$$\hat{I}_2 = \left(\widehat{H}^{-1}(0.025), \ \widehat{H}^{-1}(0.975)\right) \quad \text{and} \quad \hat{I}_1 = \left(-\infty, \ \widehat{H}^{-1}(0.95)\right). \quad (1.15)$$

This "other percentile method" will be discussed in more detail in Chapter 3.

All these intervals cover θ_0 with probability approximately 0.95, which we call the *nominal coverage*. *Coverage error* is defined to be true coverage minus nominal coverage; it generally converges to zero as sample size increases.

We now treat in more detail the construction of two-sided, symmetric percentile intervals in parametric problems. There, provided the distribution functions $F_{(\lambda)}$ (introduced in Section 1.2) are continuous, equation (1.14) may be solved exactly. We focus attention on the cases where $\theta_0 = \theta(F_0)$ is a population mean and the population is Normal or Exponential. Our main aim is to bring out the virtues of pivoting, which usually amounts to rescaling so that the distribution of a statistic depends less on unknown parameters.

If the population is Normal $N(\mu, \sigma^2)$ and we use the maximum likelihood estimate $\widehat{\boldsymbol{\lambda}} = (\bar{X}, \hat{\sigma}^2)$ to estimate $\lambda_0 = (\mu, \sigma^2)$, then the sample equation (1.14) may be rewritten as

$$P\big(|n^{-1/2}\,\hat{\sigma}\,N| \le t \mid F_1\big) \;=\; 0.95\,, \tag{1.16}$$

where N is Normal $N(0,1)$ independent of F_1. Therefore the solution of (1.14) is $t = \hat{t}_0 = x_{0.95}\,n^{-1/2}\,\hat{\sigma}$, where x_α is defined by

$$P\big(|N| \le x_\alpha\big) \;=\; \alpha\,.$$

The bootstrap confidence interval is therefore

$$\left(\bar{X} - n^{-1/2}\,x_{0.95}\,\hat{\sigma}\,,\ \bar{X} + n^{-1/2}\,x_{0.95}, \hat{\sigma}\right),$$

with coverage error

$$P\big(\bar{X} - n^{-1/2}\,x_{0.95}\,\hat{\sigma} \le \mu \le \bar{X} + n^{-1/2}\,x_{0.95}\,\hat{\sigma}\big)$$
$$= P\big\{\,|n^{1/2}(\bar{X} - \mu)/\hat{\sigma}| \le x_{0.95}\,\big\} \;-\; 0.95\,. \tag{1.17}$$

Of course, $n^{1/2}(\bar{X} - \mu)/\hat{\sigma}$ does not have a Normal distribution, but a rescaled Student's t distribution with $n - 1$ degrees of freedom. Therefore the coverage error is essentially that which results from approximating to Student's t distribution by a Normal distribution, and so is $O(n^{-1})$. (See Kendall and Stuart (1977, p. 404), and also Example 2.1 in Section 2.6.)

For our second parametric example, take the population to be Exponential with mean $\theta_0 = \mu$ and use the maximum likelihood estimator $\hat{\lambda} = \bar{X}$ to estimate $\lambda_0 = \mu$. Then equation (1.14) may be rewritten as

$$P\big(\bar{X}|n^{-1}Y - 1| \le t \mid F_1\big) \;=\; 0.95\,,$$

where Y has a gamma distribution with mean n and is independent of F_1. Therefore the solution of the sample equation is $\hat{t}_0 = y_{0.95}\,\bar{X}$, where $y_\alpha = y_\alpha(n)$ is defined by $P\big(|n^{-1}Y - 1| \le y_\alpha\big) = \alpha$. The symmetric percentile confidence interval is therefore

$$\left(\bar{X} - y_{0.95}\,\bar{X}\,,\ \bar{X} + y_{0.95}\,\bar{X}\right),$$

with coverage error

$$P\left(\bar{X} - y_{0.95}\,\bar{X} \le \mu \le \bar{X} + y_{0.95}\,\bar{X}\right) - 0.95$$
$$= P\left(|n^{-1}Y - 1| \le y_{0.95}\,Y\right) - 0.95$$
$$= O(n^{-1})\,.$$

Generally, a symmetric percentile confidence interval has coverage error $O(n^{-1})$; see Section 3.5 for a proof. Even the symmetric "Normal approximation" interval, which (in either parametric or nonparametric cases) is $\left(\bar{X} - n^{-1/2}\,x_{0.95}\,\hat{\sigma},\ \bar{X} + n^{-1/2}\,x_{0.95}\,\hat{\sigma}\right)$ when the unknown parameter θ_0 is a population mean, has coverage error $O(n^{-1})$. Therefore there is nothing particularly virtuous about the percentile interval.

To appreciate why the percentile interval has this inadequate performance, let us go back to our parametric example involving the Normal distribution. The root cause of the problem there is that $\hat{\sigma}$, and not σ, appears on the right-hand side in (1.17). This happens because the sample equation (1.14), equivalent here to (1.16), depends on $\hat{\sigma}$. Put another way, the population equation (1.1), equivalent to

$$P\left\{|\theta(F_1) - \theta(F_0)| \le t\right\} = 0.95\,,$$

depends on σ^2, the population variance. This occurs because the distribution of $|\theta(F_1) - \theta(F_0)|$ depends on the unknown σ. We should try to eliminate, or at least minimize, this dependence.

A function T of both the data and an unknown parameter is said to be (exactly) *pivotal* if it has the same distribution for all values of the unknowns. It is *asymptotically pivotal* if, for sequences of known constants $\{a_n\}$ and $\{b_n\}$, $a_n\,T + b_n$ has a proper nondegenerate limiting distribution not depending on unknowns. We may convert $\theta(F_1) - \theta(F_0)$ into a pivotal statistic by correcting for scale, changing it to $T = \{\theta(F_1) - \theta(F_0)\}/\hat{\tau}$ where $\hat{\tau} = \tau(F_1)$ is an appropriate scale estimate. In our example about the mean there are usually many different choices for $\hat{\tau}$, e.g., the sample standard deviation $\{n^{-1}\sum(X_i - \bar{X})^2\}^{1/2}$, the square root of the unbiased variance estimate, Gini's mean difference, and the interquartile range. In more complex problems, a jackknife standard deviation estimate is usually an option. Note that exactly the same confidence interval will be obtained if $\hat{\tau}$ is replaced by $c\hat{\tau}$, for any given $c \ne 0$, and so it is inessential that $\hat{\tau}$ be consistent for the asymptotic standard deviation of $\theta(F_1)$. What is important is *pivotalness* — exact pivotalness if we are to obtain a confidence interval with zero coverage error, asymptotic pivotalness if exact pivotalness is unattainable. If we change to a pivotal statistic then the function

f_t alters from the form given in (1.3) to

$$f_t(F_0, F_1)$$
$$= I\{\theta(F_1) - t\tau(F_1) \leq \theta(F_0) \leq \theta(F_1) + t\tau(F_1)\} - 0.95. \qquad (1.18)$$

In the case of our parametric Normal model, any reasonable scale estimate $\hat{\tau}$ will give exact pivotalness. We shall take $\hat{\tau} = \hat{\sigma}$, where $\hat{\sigma}^2 = \sigma^2(F_1)$ $= n^{-1}\sum(X_i - \bar{X})^2$ denotes sample variance. Then f_t becomes

$$f_t(F_0, F_1)$$
$$= I\{\theta(F_1) - t\sigma(F_1) \leq \theta(F_0) \leq \theta(F_1) + t\sigma(F_1)\} - 0.95. \qquad (1.19)$$

Using this functional in place of that at (1.3), but otherwise arguing exactly as before, equation (1.16) changes to

$$P\{(n-1)^{-1/2}|T_{n-1}| \leq t \mid F_1\} = 0.95, \qquad (1.20)$$

where T_{n-1} has Student's t distribution with $n-1$ degrees of freedom and is stochastically independent of F_1. (Therefore the conditioning on F_1 in (1.20) is irrelevant.) Thus, the solution of the sample equation is $\hat{t}_0 = (n-1)^{-1/2} w_{0.95}$, where $w_\alpha = w_\alpha(n)$ is given by $P(|T_{n-1}| \leq w_\alpha) = \alpha$. The bootstrap confidence interval is $(\bar{X} - \hat{t}_0\,\hat{\sigma},\ \bar{X} + \hat{t}_0\,\hat{\sigma})$, with *perfect* coverage accuracy,

$$P\{\bar{X} - (n-1)^{-1/2} w_{0.95}\,\hat{\sigma} \leq \mu \leq \bar{X} + (n-1)^{-1/2} w_{0.95}\,\hat{\sigma}\} = 0.95.$$

(Of course, the latter statement applies only to the parametric bootstrap under the assumption of a Normal model.)

Such confidence intervals are usually called *percentile-t intervals* since $\hat{t}_0$ is a percentile of the Student's t-like statistic $|\theta(F_2) - \theta(F_1)|/\tau(F_2)$.

We may construct confidence intervals in the same manner for Exponential populations. Define f_t by (1.19) and redefine w_α by

$$P\left[\left|n^{-1/2}\sum_{i=1}^{n}(Z_i - 1)\right|\left\{n^{-1}\sum_{i=1}^{n}(Z_i - \bar{Z})^2\right\}^{-1/2} \leq w_\alpha\right] = \alpha,$$

where $Z_1, \ldots, Z_n$ are independent Exponential variables with unit mean. The solution $\hat{t}_0$ of the sample equation is $\hat{t}_0 = w_{0.95}\, n^{-1/2}$, and the percentile-t bootstrap confidence interval $(\bar{X} - \hat{t}_0\,\hat{\sigma},\ \bar{X} + \hat{t}_0\,\hat{\sigma})$ again has perfect coverage accuracy.

This Exponential example illustrates a case where taking $\hat{\tau} = \hat{\sigma}$ (the sample standard deviation) as the scale estimate is not the most appropriate choice. For in this circumstance, $\sigma(F_0) = \theta(F_0)$, of which the maximum likelihood estimate is $\theta(F_1) = \bar{X}$. This suggests taking

$\hat{\tau} = \tau(F_1) = \theta(F_1) = \bar{X}$ in the definition (1.18) of f_t. Then if $w_\alpha = w_\alpha(n)$ is redefined by

$$P\left\{ |n^{-1}Y - 1| \, (n^{-1}Y)^{-1} \le w_\alpha \right\} = \alpha,$$

where Y is Gamma with mean n, the solution of the sample equation is $\hat{t}_0 = w_{0.95}$. The resulting bootstrap confidence interval is

$$\left(\bar{X} - w_{0.95}\,\bar{X}, \; \bar{X} + w_{0.95}\,\bar{X} \right),$$

again with perfect coverage accuracy.

Perfect coverage accuracy of percentile-t intervals usually holds only in certain parametric problems where the underlying statistic is exactly pivotal. More generally, if symmetric percentile-t intervals are constructed in parametric and nonparametric problems by solving the sample equation when f_t is defined by (1.18), where $\tau(F_1)$ is chosen so that $T = \{\theta(F_1) - \theta(F_0)\}/\tau(F_1)$ is asymptotically pivotal, then coverage error will usually be $O(n^{-2})$ rather than the $O(n^{-1})$ associated with ordinary percentile intervals. See Section 3.6 for details.

We have discussed only two-sided, symmetric percentile-t intervals. Of course, there are other percentile-t intervals, some of which may be defined as follows. Let $\hat{v}_\alpha$ denote the solution in t of the sample equation

$$P\left[\{\theta(F_2) - \theta(F_1)\}/\tau(F_2) \le t \mid F_1 \right] - \alpha = 0,$$

where $\hat{\tau}^* = \tau(F_2)$ is an estimate of scale computed for the bootstrap resample. Then $(-\infty, \hat{\theta} - \hat{\tau}\hat{v}_{1-\alpha})$ is a one-sided percentile-t interval, and $(\hat{\theta} - \hat{\tau}\hat{v}_{(1+\alpha)/2}, \hat{\theta} - \hat{\tau}\hat{v}_{(1-\alpha)/2})$ is a two-sided equal-tailed percentile-t interval, both having nominal coverage α. They will be discussed in more detail in Chapter 3. In Example 1.3 we describe short confidence intervals inspired by the percentile-t construction.

We conclude this example with remarks on the computation of critical points, such as $\hat{v}_\alpha$, by uniform Monte Carlo simulation. Further details, including an account of efficient Monte Carlo simulation, are given in Appendix II.

Assume we wish to compute the solution $\hat{v}_\alpha$ of the equation

$$P\left[\{\theta(F_2) - \theta(F_1)\} / \tau(F_2) \le \hat{v}_\alpha \mid F_1 \right] = \alpha, \qquad (1.21)$$

or, to be more precise, the value

$$\hat{v}_\alpha = \inf\left\{ x : P\left[\{\theta(F_2) - \theta(F_1)\}/\tau(F_2) \le x \mid F_1 \right] \ge \alpha \right\}.$$

Choose integers $B \ge 1$ and $1 \le \nu \le B$ such that $\nu/(B+1) = \alpha$. For example, if $\alpha = 0.95$ then we could take $(\nu, B) = (95, 99)$ or $(950, 999)$. Conditional on F_1, draw B resamples $\{\mathcal{X}_b^*, \, 1 \le b \le B\}$ independently from the distribution with distribution function F_1. In the nonparametric case, write $F_{2,b}$ for the empirical distribution function of $\mathcal{X}_b^*$. In the parametric case, where the population distribution function is $F_{(\lambda_0)}$ and λ_0 is a vector of unknown parameters, let $\widehat{\lambda}$ and $\widehat{\lambda}_b^*$ denote the estimates of λ_0 computed from the sample $\mathcal{X}$ and the resample $\mathcal{X}_b^*$, respectively, and put $F_{2,b} = F_{(\widehat{\lambda}_b^*)}$. For both parametric and nonparametric cases, define

$$T_b^* = \{\theta(F_{2,b}) - \theta(F_1)\}/\tau(F_{2,b}),$$

and write T^* for a generic T_b^*. In this notation, equation (1.21) is equivalent to $P(T^* \leq \hat{v}_\alpha \mid \mathcal{X}) = \alpha$. Let $\hat{v}_{\alpha,B}$ denote the νth largest value of T_b^*. Then $\hat{v}_{\alpha,B} \to \hat{v}_\alpha$ with probability one, conditional on $\mathcal{X}$, as $B \to \infty$. The value $\hat{v}_{\alpha,B}$ is a Monte Carlo approximation to $\hat{v}_\alpha$.

Example 1.3: Short Confidence Intervals

Let $\hat{\tau} = \tau(F_1)$ be proportional to an estimate of the scale of $\hat{\theta}$, and let v_α be the α-level critical point of the Studentized statistic $(\hat{\theta} - \theta)/\hat{\tau}$,

$$P\{(\hat{\theta} - \theta)/\hat{\tau} \leq v_\alpha\} = \alpha, \quad 0 < \alpha < 1.$$

Then for each $0 < a < 1 - \alpha$, the interval

$$J(a) = (\hat{\theta} - \hat{\tau}v_{\alpha+a}, \ \hat{\theta} - \hat{\tau}v_a)$$

is an α-level confidence interval for θ,

$$P\{\theta \in J(a)\} = \alpha, \quad 0 < a < 1 - \alpha.$$

The length of $J(a)$ equals

$$(\hat{\theta} - \hat{\tau}v_a) - (\hat{\theta} - \hat{\tau}v_{\alpha+a}) = \hat{\tau}(v_{\alpha+a} - v_a),$$

and so is proportional to $v_{\alpha+a} - v_a$. The shortest of the intervals $J(a)$ is that obtained by minimizing $v_{\alpha+a} - v_a$ with respect to a,

$$J_{\mathrm{SH}} = (\hat{\theta} - \hat{\tau}v_{\alpha+a_1}, \ \hat{\theta} - \hat{\tau}v_{a_1}),$$

where a_1 minimizes $v_{\alpha+a} - v_a$. If the distribution of $(\hat{\theta} - \theta)/\hat{\tau}$ is unimodal then the shortest confidence is *likelihood based*, in the sense that the "likelihood" of each parameter value inside J_{SH} is greater than or equal to that of any parameter outside J_{SH},

$$L(\theta_1) \geq L(\theta_2) \quad \text{whenever} \quad \theta_1 \in J_{\mathrm{SH}} \quad \text{and} \quad \theta_2 \notin J_{\mathrm{SH}},$$

where $L(\theta)$ denotes the probability density of $T = (\hat{\theta} - \theta)/\hat{\tau}$, evaluated at the observed value of T, when θ is the true parameter value. This result follows from a little calculus of variations and is essentially the Neyman-Pearson lemma. (Our use of the term "likelihood" here is nonstandard since it refers to the density of T rather than the likelihood of the data set.)

Of course, $v_\alpha = v_\alpha(F_0)$ is the solution of the population equation (1.1) when

$$f_t(F_0, F_1) = I\{\theta(F_1) - \theta(F_0) \le \tau(F_1)t\} - \alpha.$$

The bootstrap estimate of v_α, $\hat{v}_\alpha = v_\alpha(F_1)$, is obtained by solving the sample equation (1.5) using this f. And the bootstrap version of J_{SH} is

$$\widehat{J}_{\text{SH}} = \left(\hat{\theta} - \hat{\tau}\hat{v}_{\alpha+\hat{a}_1}, \ \hat{\theta} - \hat{\tau}\hat{v}_{\hat{a}_1}\right),$$

where $\hat{a}_1$ minimizes $\hat{v}_{\alpha+a} - \hat{v}_a$. This is an interval of the percentile-t type.

Shortness of the confidence interval is a virtuous property in the sense that overly long confidence intervals convey relatively imprecise information about the position of the unknown parameter θ. Furthermore, the likelihood-based method of construction produces an interval that utilizes important information about the distribution of $(\hat{\theta} - \theta)/\hat{\tau}$, ignored by other types of intervals. However, in principle, intervals that are shortest in the sense described above could tend to be short when they fail to cover θ and long when they cover θ. Lehmann (1959, p. 182) pointed out that this is an undesirable characteristic: "short confidence intervals are desirable when they cover the parameter value, but not necessarily otherwise." See also Lehmann (1986, p. 223). From this viewpoint, an ideal α-level interval might be

$$J_{\text{LS}} = \left(\hat{\theta} - \hat{\tau}v_{\alpha+a_2}, \ \hat{\theta} - \hat{\tau}v_{a_2}\right),$$

where a_2 minimizes mean length *conditional on coverage*,

$$l(a) = E\{\hat{\tau}(v_{\alpha+a} - v_a) \mid \theta \in J(a)\}$$
$$= \alpha^{-1}(v_{\alpha+a} - v_a) E\{\hat{\tau}I (\hat{\theta} - \hat{\tau}v_{\alpha+a} \le \theta \le \hat{\theta} - \hat{\tau}v_a)\}.$$

The bootstrap estimate $\hat{a}_2$ of a_2 is obtained by minimizing

$$\hat{\lambda}(a) = \alpha^{-1}(\hat{v}_{\alpha+a} - \hat{v}_a) E\left[\tau(F_2) I\{\theta(F_2) - \tau(F_2)\hat{v}_{\alpha+a}\right.$$
$$\left. \le \theta(F_1) \le \theta(F_2) - \tau(F_2)\hat{v}_a\} \mid F_1\right].$$

The bootstrap version of J_{LS} is

$$\widehat{J}_{\text{LS}} = \left(\hat{\theta} - \hat{\tau}\hat{v}_{\alpha+\hat{a}_2}, \ \hat{\theta} - \hat{\tau}\hat{v}_{\hat{a}_2}\right).$$

We should mention two caveats concerning the percentile-t argument, which do not emerge clearly from our discussion in Examples 1.2 and 1.3 but are nevertheless important. First of all, the efficacy of percentile-t depends on how well the scale parameter τ can be estimated. In particular, it is important that the relative deviation of $\hat{\tau}$ (the ratio of the root mean squared error of $\hat{\tau}$ to the value of τ) not be too large. Certain problems, such as that where θ is a correlation coefficient or a ratio of two means, stand out as examples where it can be difficult to apply the percentile-t method satisfactorily without first employing an appropriate variance-stabilizing transformation, such as Fisher's z-transformation in the former case and taking logarithms in the latter. This potential for difficulties will be discussed at greater length in Chapter 3, particularly Section 3.10. Secondly, the percentile-t method is not transformation-respecting That is, if it produces a confidence interval $(\hat{a},\,\hat{b})$ for θ and if f is a monotone increasing function, then percentile-t does not, in general, give $(f(\hat{a}),\,f(\hat{b}))$ as a confidence interval for $f(\theta)$. The "other percentile method", introduced in Section 1.2, is transformation-respecting.

Example 1.4: L^p Shrinkage

An elementary version of the L^p shrinkage problem is that of choosing $t = t_0$ to minimize

$$l(t) \;=\; E\big|\,(1-t)\,\hat{\theta} + tc - \theta_0\,\big|^p\,,$$

where c is a given constant. The estimate $\hat{\theta}_s = (1 - t_0)\,\hat{\theta} + t_0\,c$ represents $\hat{\theta}$ shrunken towards c. Now, minimizing $l(t)$ is equivalent to solving the population equation (1.1) with

$$
\begin{aligned}
f_t(F_0,\,F_1) \;&=\; (\partial/\partial t)\,\big|\,(1-t)\,\theta(F_1) + tc - \theta(F_0)\,\big|^p \\
&=\; p\,\big\{c - \theta(F_1)\big\}\,\big|\,(1-t)\,\theta(F_1) + tc - \theta(F_0)\,\big|^{p-1} \\
&\qquad\qquad \times\, \mathrm{sgn}\big\{\,(1-t)\,\theta(F_1) + tc - \theta(F_0)\,\big\}\,, \qquad (1.22)
\end{aligned}
$$

where sgn denotes the sign function:

$$
\mathrm{sgn}(x) \;=\; \left\{
\begin{array}{ll}
1 & \text{if } x > 0, \\
-1 & \text{if } x < 0, \\
0 & \text{if } x = 0.
\end{array}
\right.
$$

Therefore L^p shrinkage problems may be treated by bootstrap methods.

Traditionally, L^p shrinkage problems have been solved only for $p = 2$, since a simple, explicit solution is available there. When $p = 2$,

$$
\begin{aligned}
l(t) \;&=\; E\,\big|\,(1-t)\,(\hat{\theta} - E\hat{\theta}) + \big\{(1-t)\,(E\hat{\theta} - \theta_0) + t\,(c - \theta_0)\big\}\,\big|^2 \\
&=\; (1-t)^2\,\mathrm{var}(\hat{\theta}) + \big\{\,(1-t)\,(E\hat{\theta} - \theta_0) + t\,(c - \theta_0)\big\}^2\,,
\end{aligned}
$$

so that

$$
l'(t) \;=\; 2\,(t-1)\,\mathrm{var}(\hat{\theta}) + 2\,\big\{\,(1-t)\,(E\hat{\theta} - \theta_0) + t\,(c - \theta_0)\,\big\}\,(c - E\hat{\theta})\,,
$$

which vanishes when

$$t = t_0 = \left\{ (c - E\hat{\theta})^2 + \text{var}(\hat{\theta}) \right\}^{-1} \left\{ (c - E\hat{\theta})(\theta_0 - E\hat{\theta}) + \text{var}(\hat{\theta}) \right\}.$$

Should $\hat{\theta}$ be unbiased for θ_0, this simplifies to

$$t_0 = \left\{ (c - \theta_0)^2 + \text{var}(\hat{\theta}) \right\}^{-1} \text{var}(\hat{\theta}).$$

If $\hat{s}^2$ estimates $\text{var}(\hat{\theta})$ then an adaptive version of t_0 is

$$\hat{t}_0 = \left\{ (c - \hat{\theta})^2 + \hat{s}^2 \right\}^{-1} \hat{s}^2.$$

An estimate of θ_0 shrunken towards c is $(1 - \hat{t}_0)\hat{\theta} + \hat{t}_0 c$.

The bootstrap allows shrinkage problems to be solved in any L^p metric, $1 \le p < \infty$. Simply solve the sample equation (1.5) for t, with f_t given by (1.22), obtaining $t = \hat{t}_0$ say. If $\text{var}(\hat{\theta}) = s^2(F_0)$, $\hat{s}^2 = s^2(F_1)$, and $E(\hat{\theta}) = \theta_0$, then in the case $p = 2$ the value $\hat{t}_0$ obtained by solving the sample equation is identical to that given in the previous paragraph.

Should $\hat{\theta}$ be $\sqrt{n}$-consistent for θ_0, so that $E(\hat{\theta} - \theta_0)^2 = O(n^{-1})$, then it is generally true for $1 \le p < \infty$ that $\hat{t}_0 = O_p(n^{-1})$. Therefore

$$(1 - \hat{t}_0)\hat{\theta} + \hat{t}_0 c = \hat{\theta} + O_p(n^{-1}),$$

implying that the shrunken estimate satisfies the same central limit theorem as $\hat{\theta}$.

1.4 Iterating the Principle

In Section 1.2 we introduced a general bootstrap resampling principle by arguing in analogy with the problem of estimating the number of freckles on the face of a Russian matryoshka doll. As in that work, let n_i denote the number of freckles on the ith doll in the nest, for $i \ge 0$. (Doll 0 is the outer doll, doll i the ith inner doll.) We wish to estimate n_0, representing a true parameter value, using n_i for $i \ge 1$. Our suggestion in Section 1.2 was that n_0 be thought of as a multiple of n_1; that is, $n_0 = tn_1$ for some $t > 0$, or

$$n_0 - tn_1 = 0,$$

which we called the population equation. Not knowing n_0 we could not solve this equation exactly. However, arguing that the relationship between n_0 and n_1 should be similar to that between n_1 and n_2 we were led to the sample equation,

$$n_1 - tn_2 = 0,$$

whose solution $t = \hat{t}_{01} = n_1/n_2$ gave us the estimate $\hat{n}_{01} = \hat{t}_{01} n_1 = n_1^2/n_2$. (We have appended the extra subscript 1 to $\hat{t}_0$ and $\hat{n}_0$ to indicate that this is the first iteration.)

In the present section we delve more deeply into our nest of dolls and iterate this idea to adjust the estimate $\hat{n}_{01}$. We now seek a correction factor t such that $n_0 = t\hat{n}_{01}$. That is, we wish to solve the new population equation

$$n_0 - t\,n_1^2\,n_2^{-1} \;=\; 0.$$

Once again, our lack of knowledge of n_0 prevents a practical solution, so we pass to the sample equation in which the triple (n_0, n_1, n_2) is replaced by (n_1, n_2, n_3),

$$n_1 - t\,n_2^2\,n_3^{-1} \;=\; 0.$$

Its solution is $t = \hat{t}_{02} = n_1 n_3/n_2^2$, which results in the new estimate

$$\hat{n}_{02} \;=\; \hat{t}_{02}\,\hat{n}_{01} \;=\; n_1^3\,n_3/n_2^3$$

of n_0. A third iteration involves correcting $\hat{n}_{02}$ by the factor t, resulting in a new population equation,

$$n_0 - t\,n_1^3\,n_3\,n_2^{-3} \;=\; 0,$$

which, on replacing (n_0, n_1, n_2, n_3) by (n_1, n_2, n_3, n_4), leads to a new sample equation,

$$n_1 - t\,n_2^3\,n_4\,n_3^{-3} \;=\; 0.$$

This has solution $t = \hat{t}_{03} = n_1 n_3^3/(n_2^3\,n_4)$, which gives us the thrice-iterated estimate

$$\hat{n}_{03} \;=\; \hat{t}_{03}\,\hat{n}_{02} \;=\; (n_1 n_3)^4/(n_2^6\,n_4)$$

of n_0. The ith iteration produces an estimate $\hat{n}_{0i}$ that depends on $n_1, \ldots, n_{i+1}$.

Bootstrap iteration proceeds in an entirely analogous manner. To begin, let us recap the argument in Section 1.2. There we suggested that many statistical problems consist of seeking a solution t to a population equation of the form

$$E\{f_t(F_0, F_1) \mid F_0\} \;=\; 0, \qquad (1.23)$$

where F_0 denotes the population distribution function, F_1 is the sample distribution function, and f_t is a functional given for example by (1.2) in the case of bias correction or by (1.3) in the case of confidence interval construction. We argued that the relationship between F_1 and the unknown F_0 should be similar to that between F_2 and F_1, where F_2 is the distribution function of a sample drawn from F_1 conditional on F_1. That led us to the sample equation,

$$E\{f_t(F_1, F_2) \mid F_1\} \;=\; 0. \qquad (1.24)$$

Our estimate of the value $t_0 = t_{01}$ that solves (1.23) was the value $\hat{t}_0 = \hat{t}_{01}$ that solves (1.24). Of course, t_{01} may be written as $T(F_0)$ for some functional T. Then $\hat{t}_{01}$ is just $T(F_1)$ for the same functional T, and we have an approximate solution to (1.23),

$$E\{f_{T(F_1)}(F_0,\, F_1) \mid F_0\} \;\approx\; 0\,. \tag{1.25}$$

In many instances we would like to improve on this approximation — for example, to further reduce bias in a bias correction problem, or to improve coverage accuracy in a confidence interval problem. Therefore we introduce a correction term t to the functional T, so that $T(\cdot)$ becomes $U(\cdot,\, t)$ with $U(\cdot,\, 0) \equiv T(\cdot)$. The adjustment may be multiplicative, as in our freckles-on-the-Russian-doll problem: $U(\cdot,\, t) \equiv (1+t)\,T(\cdot)$. (The only difference here is that it is now notationally convenient to use $1+t$ instead of t.) Or it may be an additive correction, as in $U(\cdot,\, t) \equiv T(\cdot)+t$. Or t might adjust some particular feature of T, as in the level-error correction for confidence intervals, which we shall discuss shortly. In all cases, the functional $U(\cdot,\, t)$ should be smooth in t. Our aim is to choose t so as to improve on the approximation (1.25).

Ideally, we would like to solve the equation

$$E\{f_{U(F_1,t)}(F_0,\, F_1) \mid F_0\} \;=\; 0 \tag{1.26}$$

for t. If we write $g_t(F,\, G) = f_{U(G,t)}(F,\, G)$, we see that (1.26) is equivalent to

$$E\{g_t(F_0,\, F_1) \mid F_0\} \;=\; 0\,,$$

which is of the same form as the population equation (1.23). Therefore we obtain an approximation by passing to the sample equation,

$$E\{g_t(F_1,\, F_2) \mid F_1\} \;=\; 0\,,$$

or equivalently,

$$E\{f_{U(F_2,t)}(F_1,\, F_2) \mid F_1\} \;=\; 0\,. \tag{1.27}$$

This has solution $\hat{t}_{02} = T_1(F_1)$, say, giving us a new approximate equation of the same form as the first approximation (1.25), and being the result of iterating that earlier approximation,

$$E\{f_{U(F_1,T_1(F_1))}(F_0,\, F_1) \mid F_0\} \;\approx\; 0\,. \tag{1.28}$$

Our hope is that the approximation here is better than that in (1.25), so that in a sense, $U(F_1, T_1(F_1))$ is a better estimate than $T(F_1)$ of the solution t_0 to equation (1.23). Of course, this does not mean that $U(F_1, T_1(F_1))$ is closer to t_0 than $T(F_1)$, only that the left-hand side of (1.28) is closer to zero than the left-hand side of (1.25).

If we revise notation and call $U(F_1, T_1(F_1))$ the "new" $T(F_1)$, we may run through the argument again, obtaining a third approximate solution of (1.23). In principle, these iterations may be repeated as often as desired.

We have given two explicit methods, multiplicative and additive, for modifying our original estimate $\hat{t}_0 = T(F_1)$ of the solution of (1.23) so as to obtain the adjustable form $U(F_1, t)$. Those modifications may be used in a wide range of circumstances. In the special case of confidence intervals, an alternative approach is to modify the nominal coverage probability of the confidence interval. To explain the argument we shall concentrate on the special case of symmetric percentile-method intervals discussed in Example 1.2 of Section 1.3. Corrections for other types of intervals may be introduced in like manner.

An α-level symmetric percentile-method interval for $\theta_0 = \theta(F_0)$ is given by $\big(\theta(F_1) - \hat{t}_0, \theta(F_1) + \hat{t}_0\big)$, where $\hat{t}_0$ is chosen to solve the sample equation

$$P\big\{\theta(F_2) - t \le \theta(F_1) \le \theta(F_2) + t \mid F_1\big\} - \alpha = 0.$$

(In our earlier examples, $\alpha = 0.95$.) This $\hat{t}_0$ is an estimate of the solution $t_0 = T(F_0)$ of the population equation

$$P\big\{\theta(F_1) - t \le \theta(F_0) \le \theta(F_1) + t \mid F_0\big\} - \alpha = 0,$$

that is, of

$$P\big(|\hat{\theta} - \theta_0| \le t \mid F_0\big) = \alpha,$$

where $\hat{\theta} = \theta(F_1)$. Therefore t_0 is just the α-level quantile, x_α, of the distribution of $|\hat{\theta} - \theta_0|$,

$$P\big(|\hat{\theta} - \theta_0| \le x_\alpha \mid F_0\big) = \alpha.$$

Write x_α as $x(F_0)_\alpha$, the quantile when F_0 is the true distribution function. Then $\hat{t}_0 = T(F_1)$ is just $x(F_1)_\alpha$, and we might take $U(\cdot, t)$ to be

$$U(\cdot, t) \equiv x(\cdot)_{\alpha+t}.$$

This is an alternative to multiplicative and additive corrections, which in the present problem are

$$U(\cdot, t) \equiv (1 + t)\, x(\cdot)_\alpha \qquad \text{and} \qquad U(\cdot, t) \equiv x(\cdot)_\alpha + t,$$

respectively. In general, each will give slightly different numerical results, although, as we shall prove shortly, each provides the same order of correction.

Our development of bootstrap iteration has been by analogy with the Russian doll problem. The reader may have noticed that in the case of the Russian dolls the ith iteration involved properties of all dolls from the 1st to the $(i + 1)$th, whereas in the bootstrap problem only the sample distribution function F_1 and the resample distribution function F_2 seem to appear. However, that is a consequence of our notation, which is designed to suppress some of the distracting side-issues of iteration. In actual fact, carrying out i bootstrap iterations usually involves computing $F_1, \ldots, F_{i+1}$, where F_j is defined inductively as the distribution function of a same-size sample drawn from F_{j-1} conditional on F_{j-1}.

Concise definitions of F_j are different in parametric and nonparametric cases. In the former we work within a class $\{F_{(\lambda)}, \lambda \in \Lambda\}$ of distributions that are completely specified up to an unknown vector λ of parameters. The "true" distribution is $F_0 = F_{(\lambda_0)}$, we estimate λ_0 by $\widehat{\lambda} = \lambda[\mathcal{X}]$ where $\mathcal{X} = \mathcal{X}_1$ is an n-sample drawn from F_0, and we take F_1 to be $F_{(\widehat{\lambda})}$. To define F_j, let $\widehat{\lambda}_j = \lambda[\mathcal{X}_j]$ denote the estimate $\widehat{\lambda}$ computed for an n-sample $\mathcal{X}_j$ drawn from F_{j-1}, and put $F_j = F_{(\widehat{\lambda}_j)}$. The nonparametric case is conceptually simpler. There, F_j is the empirical distribution of an n-sample drawn randomly from F_{j-1}, with replacement.

To explain how high-index F_j's enter into computation of bootstrap iterations, we shall discuss calculation of the solution to equation (1.27). That requires calculation of $U(F_2, t)$, defined for example by

$$U(F_2, t) = (1 + t) T(F_2).$$

And for this we must compute $T(F_2)$. Now, $\hat{t}_0 = T(F_1)$ is the solution (in t) of the sample equation

$$E\{f_t(F_1, F_2) \mid F_1\} = 0,$$

and so $T(F_2)$ is the solution (in t) of the resample equation

$$E\{f_t(F_2, F_3) \mid F_2\} = 0.$$

Thus, to find the second bootstrap iterate, the solution of (1.27), we must construct F_1, F_2, and F_3, just as we needed n_1, n_2, and n_3 for the second iterate of the Russian doll principle. Calculation of F_2 "by simulation" typically involves order B sampling operations (B resamples drawn from the original sample), whereas calculation of F_3 "by simulation" involves order B^2 sampling operations (B resamples drawn from each of B resamples) if the same number of operations is used at each level. Thus, i bootstrap iterations could require order B^i computations, and so complexity would increase rapidly with the number of iterations.

In regular cases, expansions of the error in formulae such as (1.25) are usually power series in $n^{-1/2}$ or n^{-1}, often resulting from Edgeworth expansions of the type that we shall discuss in Chapter 2. Each bootstrap iteration reduces the order of magnitude of error by a factor of at least $n^{-1/2}$, as the following proposition shows.

PROPOSITION 1.1. *If, for a smooth functional c,*

$$E\{f_{T(F_1)}(F_0, F_1) \mid F_0\} = c(F_0) n^{-j/2} + O(n^{-(j+1)/2}), \qquad (1.29)$$

and if $(\partial/\partial t)\, E\{f_{U(F_1,t)}(F_0,\, F_1) \mid F_0\}$ evaluated at $t = 0$ is asymptotic to a nonzero constant as $n \to \infty$, then

$$E\{f_{U(F_1,T_1(F_1))}(F_0,\, F_1) \mid F_0\} \;=\; O(n^{-(j+1)/2}).$$

Proposition 1.1 is proved at the end of this section.

In many problems with an element of symmetry, such as two-sided confidence intervals, expansions of error are power series in n^{-1} rather than $n^{-1/2}$, and each bootstrap iteration reduces error by a factor of n^{-1}, not just $n^{-1/2}$. To describe these circumstances, let

$$d(F_0) \;=\; (\partial/\partial t)\, E\{f_{U(F_1,t)}(F_0,\, F_1) \mid F_0\}\big|_{t=0}$$

denote the derivative that was assumed nonzero in Proposition 1.1, and put

$$\Delta \;=\; n^{1/2}\left\{c(F_1)\, d(F_1)^{-1} - c(F_0)\, d(F_0)^{-1}\right\}$$

and

$$e(\Delta) \;=\; (\partial/\partial t)\, E\{f_{U(F_1,t)}(F_0,\, F_1) \mid \Delta\}\big|_{t=0}.$$

The random variable Δ is $O_p(1)$ as $n \to \infty$. Typically it has an asymptotic Normal distribution and $E(\Delta) = O(n^{-1/2})$. Our next proposition presents a little theory for this circumstance.

PROPOSITION 1.2. *If, for a smooth functional* c,

$$E\{f_{T(F_1)}(F_0,\, F_1) \mid F_0\} \;=\; c(F_0)\, n^{-j} + O(n^{-(j+1)}),$$

if $d(F_0)$ *is asymptotic to a nonzero constant, and if*

$$E\{\Delta e(\Delta)\} \;=\; O(n^{-1/2}), \tag{1.30}$$

then

$$E\{f_{U(F_1,T(F_1))}(F_0,\, F_1) \mid F_0\} \;=\; O(n^{-(j+1)}). \tag{1.31}$$

See the end of this section for a proof.

Should it happen that, in contradiction of (1.30),

$$E\{\Delta e(\Delta)\} \;=\; a(F_0) + O(n^{-1/2})$$

where $a(F_0) \neq 0$, then (1.31) fails and instead the right-hand side of (1.31) equals $-a(F_0)\, n^{-j-(1/2)} + O(n^{-(j+1)})$. We show in Examples 1.5 and 1.6 of the next section that (1.31) holds for problems involving bias reduction and two-sided confidence intervals.

In "symmetric" problems where equation (1.30) is satisfied, an expansion of $E\{f_{T(F_1)}(F_0,\, F_1) \mid F_0\}$ usually decreases in powers of n^{-1}. In "asymmetric" problems (for example, one-sided confidence intervals), the expansion of this expectation decreases in powers of $n^{-1/2}$. In both circumstances, each successive bootstrap iteration knocks out the first nonzero term in the expansion, with consequent adjustments to but no changes in the order of later terms. This may be shown by elaboration of the proofs of Propositions 1.1 and 1.2 and will become clear in Section 3.11 where we study the effect of iteration in greater depth.

We close this section with proofs of Propositions 1.1 and 1.2.

Proof of Proposition 1.1.

Since $E\{f_{U(F_1,t)}(F_0, F_1) \mid F_0\}$ has a nonzero first derivative at $t = 0$, and since $U(\cdot, 0) \equiv T(\cdot)$, then in view of (1.29),

$$
\begin{aligned}
E\{f_{U(F_1,t)}(F_0, F_1) \mid F_0\} \\
= c(F_0)\, n^{-j/2} + d(F_0)\, t + O(n^{-(j+1)/2} + t^2)
\end{aligned} \qquad (1.32)
$$

as $n \to \infty$ and $t \to 0$, where $d(F_0) \neq 0$. Therefore the solution of population equation (1.26) is

$$
t = T_1(F_0) = -\{c(F_0)/d(F_0)\}\, n^{-j/2} + O(n^{-(j+1)/2}).
$$

Likewise, the solution of sample equation (1.27) is

$$
t = T_1(F_1) = -\{c(F_1)/d(F_1)\}\, n^{-j/2} + O_p(n^{-(j+1)/2}).
$$

Assuming that the functionals c and d are continuous, we have

$$
c(F_1) = c(F_0) + O_p(n^{-1/2}) \qquad \text{and} \qquad d(F_1) = d(F_0) + O_p(n^{-1/2})
$$

and so

$$
T_1(F_1) = T_1(F_0) + O_p(n^{-(j+1)/2}).
$$

Hence, since $t = T_1(F_0)$ solves (1.26),

$$
\begin{aligned}
E\{f_{U(F_1,T_1(F_1))}(F_0, F_1) \mid F_0\} \\
= E\{f_{U(F_1,T_1(F_0))}(F_0, F_1) \mid F_0\} + O(n^{-(j+1)/2}) \\
= O(n^{-(j+1)/2}),
\end{aligned}
$$

as claimed.   □

Proof of Proposition 1.2.

In these new circumstances, equation (1.32) becomes

$$
E\{f_{U(F_1,t)}(F_0, F_1) \mid F_0\} = c(F_0)\, n^{-j} + d(F_0)\, t + O(n^{-(j+1)} + t^2),
$$

from which we deduce as before that the population equation has solution

$$
t = T_1(F_0) = -\{c(F_0)/d(F_0)\}\, n^{-j} + O(n^{-(j+1)})
$$

and that the sample equation has solution

$$t = T_1(F_1) = -\{c(F_1)/d(F_1)\}\, n^{-j} + O_p(n^{-(j+1)})$$
$$= T_1(F_0) - n^{-(2j+1)/2}\,\Delta + O_p(n^{-(j+1)}).$$

Therefore

$$E\{f_{U(F_1, T_1(F_1))}(F_0, F_1) \mid \Delta\} = E\{f_{U(F_1, T_1(F_0))}(F_0, F_1) \mid \Delta\}$$
$$- n^{(2j+1)/2}\,\Delta e(\Delta) + O(n^{-(j+1)}).$$

Hence, since $t = T_1(F_0)$ solves (1.26),

$$E\{f_{U(F_1, T_1(F_1))}(F_0, F_1) \mid F_0\} = -n^{-(2j+1)/2}\,E\{\Delta\,e(\Delta)\} + O(n^{-(j+1)})$$
$$= O(n^{-(j+1)}),$$

as claimed. $\square$

1.5 Two Examples Revisited

Example 1.5: Bias Reduction

We begin by showing that each bootstrap iteration reduces the order of magnitude of bias by the factor n^{-1}. This follows via Proposition 1.2, once we check condition (1.30). The argument that we shall use to verify (1.30) is valid whenever $f_{U(G,t)}(F, G)$ can be differentiated with respect to t and when the derivative at $t = 0$ is a smooth functional of G. In the case of bias reduction by an additive correction we have

$$f_t(F, G) = \theta(G) - \theta(F) + t \tag{1.33}$$

and $U(G, t) = T(G) + t$. (For equation (1.33), see Example 1.1 in Section 1.3.) Therefore

$$f_{U(G,t)}(F, G) = \theta(G) - \theta(F) + T(G) + t,$$

so that $(\partial/\partial t)\, f_{U(G,t)}(F, G) = 1$, which is assuredly a smooth functional of G. Should the bias correction be made multiplicatively, (1.33) changes to

$$f_t(F, G) = (1 + t)\,\theta(G) - \theta(F),$$

and we might take

$$f_{U(G,t)}(F, G) = (1 + t)\,\{1 + T(G)\}\,\theta(G) - \theta(F),$$

in which case $(\partial/\partial t)\, f_{U(G,t)}(F, G) = \{1 + T(G)\}\,\theta(G)$, again a smooth functional of G.

If

$$a(F, G) \;=\; (\partial/\partial t)\, f_{U(G,t)}(F, G)\big|_{t=0}$$

is smooth in G then

$$a(F_0, F_1) \;=\; a(F_0, F_0) + O_p(n^{-1/2}),$$

so that, since $E(\Delta) = O(n^{-1/2})$,

$$\begin{aligned}
E\{\Delta e(\Delta)\} &= E\{\Delta a(F_0, F_1)\} = E\{\Delta a(F_0, F_1) \mid F_0\} \\
&= E(\Delta)\, a(F_0, F_0) + O(n^{-1/2}) \\
&= O(n^{-1/2}),
\end{aligned}$$

verifying condition (1.30).

To investigate more deeply the effect of bootstrap iteration on bias, recall from (1.33) that, in the case of bias reduction by an additive correction,

$$f_t(F_0, F_1) \;=\; \theta(F_1) - \theta(F_0) + t.$$

Therefore the sample equation,

$$E\{f_t(F_1, F_2) \mid F_1\} \;=\; 0,$$

has solution $t = T(F_1) = \theta(F_1) - E\{\theta(F_2) \mid F_1\}$, and so the once-iterated estimate is

$$\hat{\theta}_1 \;=\; \hat{\theta} + T(F_1) \;=\; \theta(F_1) + T(F_1) \;=\; 2\theta(F_1) - E\{\theta(F_2) \mid F_1\}; \qquad (1.34)$$

see also (1.6). We give below a general formula for the estimate obtained after j iterations.

THEOREM 1.3. *If $\hat{\theta}_j$ denotes the jth iterate of $\hat{\theta}$, and if the adjustment at each iteration is additive, then*

$$\hat{\theta}_j \;=\; \sum_{i=1}^{j+1} \binom{j+1}{i} E\{\theta(F_i) \mid F_1\}, \qquad j \ge 1. \qquad (1.35)$$

Proof.

Write $\theta_j(F_1)$ for $\hat{\theta}_j$. Since the adjustment at each iteration is additive, the population equation at the jth iteration is

$$E\{\theta_{j-1}(F_1) - \theta(F_0) + t \mid F_0\} \;=\; 0.$$

The sample equation (or rather, the resample equation) is therefore

$$E\{\theta_{j-1}(F_2) - \theta(F_1) + t \mid F_1\} \;=\; 0,$$

whose solution is $t = \hat{t}_{0j} = \theta(F_1) - E\{\theta_{j-1}(F_2) \mid F_1\}$. Hence

$$
\begin{aligned}
\theta_j(F_1) &= \theta_{j-1}(F_1) + \hat{t}_{0j} \\
&= \theta(F_1) + \theta_{j-1}(F_1) - E\{\theta_{j-1}(F_2) \mid F_1\}.
\end{aligned}
\tag{1.36}
$$

From this point, our argument is by induction over j. Formula (1.35) holds for $j = 1$, where it is identical to (1.34). Suppose it holds for $j - 1$. Then by (1.36),

$$
\begin{aligned}
\theta_j(F_1) &= \theta(F_1) + \sum_{i=1}^{j} \binom{j}{i} (-1)^{i+1} \left[E\{\theta(F_i) \mid F_1\} - E\{\theta(F_{i+1}) \mid F_1\} \right] \\
&= \sum_{i=1}^{j+1} \left\{ \binom{j}{i} + \binom{j}{i-1} \right\} (-1)^{i+1} E\{\theta(F_i) \mid F_1\} \\
&= \sum_{i=1}^{j+1} \binom{j+1}{i} (-1)^{i+1} E\{\theta(F_i) \mid F_1\},
\end{aligned}
$$

which is formula (1.35). $\square$

Theorem 1.3 has an analogue for multiplicative corrections, where $f_t(F_0, F_1) = (1 + t)\,\theta(F_1) - \theta(F_0)$ and

$$
\theta_j(F_1) = \theta(F_1)\,\theta_{j-1}(F_1) \Big/ E\{\theta_{j-1}(F_2) \mid F_1\}.
$$

Formula (1.35) makes explicitly clear the fact that, generally speaking, carrying out j bootstrap iterations involves computation of $F_1, \dots, F_{j+1}$. This property was discussed at length in Section 1.4. However, there exist circumstances where the estimate $\hat{\theta}_j = \theta_j(F_1)$ may be computed directly, without any simulation at all. To illustrate, let $\mu = \int x\,dF_0(x)$ denote the mean of a scalar (i.e., univariate) population, and put $\theta_0 = \theta(F_0) = \mu^2$, the square of the mean. Let $\bar{X} = n^{-1} \sum X_i$ denote the mean of the basic sample $\mathcal{X} = \{X_1, \dots, X_n\}$. Make all corrections additively, so that $U(\cdot, t) = T(\cdot) + t$, etc. In nonparametric problems at least,

$$
\hat{\theta} = \theta(F_1) = \bar{X}^2
$$

and

$$
E\{\theta(F_1) \mid F_0\} = E\{\mu + n^{-1} \sum (X_i - \mu)\}^2 = \mu^2 + n^{-1}\sigma^2,
\tag{1.37}
$$

where $\sigma^2 = E(X_1 - \mu)^2$ denotes population variance. For the nonparametric bootstrap we obtain, in direct analogy to (1.37),

$$
E\{\theta(F_2) \mid F_1\} = \bar{X}^2 + n^{-1}\hat{\sigma}^2,
\tag{1.38}
$$

where $\hat{\sigma}^2 = \hat{\sigma}^2(F_1) = n^{-1} \sum (X_i - \bar{X})^2$ denotes sample variance. Therefore one bootstrap iteration produces (see (1.34))

$$\hat{\theta}_1 = \theta_1(F_1) = 2\theta(F_1) - E\{\theta(F_2) \mid F_1\}$$
$$= \bar{X}^2 - n^{-1}\hat{\sigma}^2.$$

Noting that $E\{\hat{\sigma}^2(F_1) \mid F_1\} = (1 - n^{-1})\hat{\sigma}^2(F_1)$, we see that a second iteration gives

$$\hat{\theta}_2 = \theta_2(F_1)$$
$$= \theta(F_1) + \theta_1(F_1) - E\{\theta_1(F_2) \mid F_1\}$$
$$= \bar{X}^2 + (\bar{X}^2 - n^{-1}\hat{\sigma}^2) - \{\bar{X}^2 + n^{-1}\hat{\sigma}^2 - n^{-1}(1 - n^{-1})\hat{\sigma}^2\}$$
$$= \bar{X}^2 - n^{-1}(1 + n^{-1})\hat{\sigma}^2.$$

A third iteration gives

$$\hat{\theta}_3 = \theta_3(F_1)$$
$$= \theta(F_1) + \theta_2(F_1) - E\{\theta_2(F_2) \mid F_1\}$$
$$= \bar{X}^2 + \{\bar{X}^2 - n^{-1}(1 + n^{-1})\hat{\sigma}^2\}$$
$$\qquad - \{\bar{X}^2 + n^{-1}\hat{\sigma}^2 - n^{-1}(1 + n^{-1})(1 - n^{-1})\hat{\sigma}^2\}$$
$$= \bar{X}^2 - n^{-1}(1 + n^{-1} + n^{-2})\hat{\sigma}^2.$$

Arguing by induction over j we obtain after j iterations,

$$\hat{\theta}_j = \bar{X}^2 - n^{-1}(1 + n^{-1} + \ldots + n^{-(j-1)})\hat{\sigma}^2$$
$$= \bar{X}^2 - (n-1)^{-1}(1 - n^{-j})\hat{\sigma}^2. \tag{1.39}$$

As $j \to \infty$,

$$\hat{\theta}_j \to \hat{\theta}_\infty = \bar{X}^2 - (n-1)^{-1}\hat{\sigma}^2, \tag{1.40}$$

which is unbiased for θ_0 and on this occasion is the simple jackknife estimator.

More generally, if $\theta_0 = \theta(F_0)$ is a polynomial in population moments with known, bounded coefficients, and if $\hat{\theta} = \theta(F_1)$ is the same polynomial in sample moments, then $\hat{\theta} = \lim_{j \to \infty} \hat{\theta}_j$ exists, is unbiased for θ_0, and has variance of order n^{-1} provided enough population moments are finite. Furthermore, $\hat{\theta}_\infty$ is a polynomial in sample moments. The proof consists of noting that, if μ_j is the jth population moment, then $\mu_{j_1}^{i_1} \ldots \mu_{j_r}^{i_r}$ has a unique unbiased nonparametric estimate expressible as a polynomial in sample moments up to the $(\sum_k i_k j_k)$th, with coefficients equal to inverses of polynomials in n^{-1} and so also to power series in n^{-1}. Repeated iteration does no more than generate higher-order terms in these series.

Fisher's k-statistics provide examples of the use of the infinitely iterated bootstrap and may be constructed as follows. First, define the kth *cumulant* κ_k of a random variable X by

$$\sum_{j\geq1} \tfrac{1}{j!}\kappa_j\, t^j = \log\left\{1+\sum_{j\geq1}\tfrac{1}{j!}E(X^j)\,t^j\right\}$$

$$= \sum_{k\geq1}(-1)^{k+1}\tfrac{1}{k}\left\{\sum_{j\geq1}\tfrac{1}{j!}E(X^j)\,t^j\right\}^k,$$

to be interpreted as an identity in t. (Cumulants will be treated in greater detail in Chapter 2.) Thus, κ_k is a polynomial in population moments. Take this function as the parameter θ_0, draw a random sample of size n from the distribution of X, and take the same function of sample rather than population moments as the estimate $\hat\theta$ of θ_0. This is the *sample cumulant* corresponding to θ_0. Bias-correct using the bootstrap with an additive adjustment, and iterate an infinite number of times. The iterations converge to an unbiased estimate $\hat\theta_\infty$ of θ_0, called a *k-statistic*. These estimates were developed by Fisher (1928) as part of a short-cut route to the calculation of cumulants of functions of sample moments, and may of course be derived in their own right without following the bootstrap route. Their multivariate counterparts are sometimes called *polykays*.

We close this example by studying the parametric case, first with the Normal and then the Exponential distribution. If the population is Normal $N(\mu, \sigma^2)$, and if we use the maximum likelihood estimator $\widehat{\boldsymbol\lambda} = (\bar X, \hat\sigma^2)$ to estimate the parameter vector $\boldsymbol\lambda_0 = (\mu, \sigma^2)$, then the jth iterate continues to be given by formula (1.39), since equation (1.38) (and its analogues for higher-order iterations) hold exactly as before. However, should we take $\widehat{\boldsymbol\lambda} = (\bar X, \tilde\sigma^2)$ where $\tilde\sigma^2 = (n-1)^{-1}\sum(X_i - \bar X)^2$ is the unbiased variance estimate, then equation (1.38) changes to

$$E\{\theta(F_2)\mid F_1\} = \bar X^2 + n^{-1}\tilde\sigma^2,$$

so that

$$\begin{aligned}
\hat\theta_1 = \theta_1(F_1) &= 2\theta(F_1) - E\{\theta(F_2)\mid F_1\}\\
&= \bar X^2 - n^{-1}\tilde\sigma^2\\
&= \bar X^2 - (n-1)^{-1}\hat\sigma^2.
\end{aligned} \tag{1.41}$$

This is exactly the estimate obtained after an infinite number of iterations in the former case — compare formulae (1.40) and (1.41). Since it is already unbiased, further iteration will simply recover the same estimate. That is, $\hat\theta_j = \hat\theta_1$ for $j \geq 1$.

If the population is now Exponential with mean μ, having density $f_\mu(x) = \mu^{-1}e^{-x/\mu}$, then $\sigma^2 = \mu^2$ and (as before) $\hat\theta = \theta(F_1) = \bar X^2$.

Therefore equation (1.37) becomes

$$E\{\theta(F_1) \mid F_0\} = \mu^2(1+n^{-1}). \tag{1.42}$$

Taking our estimate of $\lambda_0 = \mu$ to be the maximum likelihood estimate $\bar{X}$, we obtain in direct analogy to equation (1.42),

$$E\{\theta(F_2) \mid F_1\} = \bar{X}^2(1+n^{-1}).$$

Therefore one bootstrap iteration produces

$$\begin{aligned}
\hat{\theta}_1 = \theta_1(F_1) &= 2\,\theta(F_1) - E\{\theta(F_2) \mid F_1\} \\
&= \bar{X}^2(1-n^{-1}).
\end{aligned}$$

A second iteration gives

$$\begin{aligned}
\hat{\theta}_2 = \theta_2(F_1) &= \theta(F_1) + \theta_1(F_1) - E\{\theta_1(F_2) \mid F_1\} \\
&= \bar{X}^2 + \bar{X}^2(1-n^{-1}) - \bar{X}^2(1+n^{-1})(1-n^{-1}) \\
&= \bar{X}^2(1-n^{-1}+n^{-2}).
\end{aligned}$$

Arguing by induction over j we see that after j iterations,

$$\begin{aligned}
\hat{\theta}_j = \theta_j(F_1) &= \bar{X}^2\left\{1-n^{-1}+n^{-2} - \ldots + (-1)^j\,n^{-j}\right\} \\
&= \bar{X}^2(1+n^{-1})^{-1}\left\{1+(-1)^j\,n^{-(j+1)}\right\}.
\end{aligned}$$

As $j \to \infty$,

$$\hat{\theta}_j \to \hat{\theta}_\infty = \bar{X}^2(1+n^{-1})^{-1},$$

which is unbiased for θ_0.

Convergence of the infinitely-iterated bootstrap is not a universal phenomenon, even in the relatively simple context of bias correction.

Example 1.6: Confidence Interval

It follows from Proposition 1.1, or from the fact that Edgeworth expansions are power series in $n^{-1/2}$ (see Chapter 2), that in regular cases each bootstrap iteration reduces the order of coverage error by a factor of $n^{-1/2}$ or less. For two-sided intervals the reduction in order of error is generally greater than this, being a factor of n^{-1} at each iteration. That follows from the fact that Edgeworth expansions of coverage error of two-sided intervals are power series in n^{-1}, not $n^{-1/2}$ (see Chapter 2), and from Proposition 1.2. We open this example by checking condition (1.30) in Proposition 1.2. For the sake of definiteness we treat the case of symmetric percentile confidence intervals; other cases, such as equal-tailed intervals, are similar. See Section 3.11 for details.

For symmetric intervals we may take the function $f_{T(F_1)}$ to have the form

$$f_{T(F_1)}(F_0, F_1)$$
$$= I\left\{\theta(F_1) - n^{-1/2}T(F_1) \leq \theta(F_0) \leq \theta(F_1) + n^{-1/2}T(F_1)\right\} - \alpha$$
$$= I\left\{n^{1/2}|\theta(F_1) - \theta(F_0)| \leq T(F_1)\right\} - \alpha.$$

Then

$$f_{U(F_1,t)}(F_0, F_1) = I\left\{n^{1/2}|\theta(F_1) - \theta(F_0)| \leq U(F_1, t)\right\} - \alpha$$

and

$$e(\Delta) = (\partial/\partial t) P\left\{n^{1/2}|\theta(F_1) - \theta(F_0)| \leq U(F_1, t) \mid \Delta\right\}\bigg|_{t=0}.$$

(One possible form for $U(\cdot, t)$ is $U(\cdot, t) = T(\cdot) + t$.) Employing a Normal approximation to the joint distribution of $n^{1/2}\{\theta(F_1) - \theta(F_0)\}$ and Δ, we see that there exist bivariate Normal random variables (N_1, N_2) with zero means and such that

$$E\{\Delta e(\Delta)\} = E\left[N_2(\partial/\partial t) P\{|N_1| \leq U(F_0, t) \mid N_2\}\right] + O(n^{-1/2}).$$

The expectation on the right-hand side equals

$$(\partial/\partial t) E\left[N_2 P\{|N_1| \leq U(F_0, t) \mid N_2\}\right]$$
$$= (\partial/\partial t) E\left[N_2 I\{|N_1| \leq U(F_0, t)\}\right]$$
$$= (\partial/\partial t) 0$$
$$= 0.$$

This verifies (1.30).

We now discuss the effect that bootstrap iteration has on the two-sided confidence intervals derived in Example 1.2. The nonparametric case is relatively uninteresting at our present level of detail, the main feature being that each successive iteration reduces the order of coverage error by a factor of n^{-1}. Therefore we confine attention to parametric problems. The parametric intervals developed from pivotal statistics in Example 1.2 were percentile-t intervals and had perfect coverage accuracy. Iteration of these intervals only reproduces the same intervals. Therefore we focus on nonpivotal intervals, derived by the ordinary percentile method in Example 1.2. It turns out that a single bootstrap iteration of any one of these percentile intervals produces a percentile-t interval having perfect coverage accuracy. Further iterations simply reproduce the same interval.

We begin by recalling from Example 1.2 the parametric, percentile confidence interval for $\theta = \mu$, assuming a Normal $N(\mu, \sigma^2)$ population. Let

N denote a Normal $N(0,1)$ random variable. Estimate the parameter $\lambda_0 = (\mu, \sigma^2)$ by the maximum likelihood estimator

$$\hat{\lambda} = (\bar{X}, \hat{\sigma}^2) = \left(\theta(F_1), \sigma^2(F_1) \right),$$

where $\bar{X} = n^{-1} \sum X_i$ and $\hat{\sigma}^2 = n^{-1} \sum (X_i - \bar{X})^2$ are sample mean and sample variance, respectively. The functional f_t is, in the case of a 95% confidence interval,

$$f_t(F_0, F_1) = I\{\theta(F_1) - t \leq \theta(F_0) \leq \theta(F_1) + t\} - 0.95,$$

and the sample equation (1.24) has solution $t = T(F_1) = n^{-1/2} x_{0.95} \, \sigma(F_1)$, where $x_{0.95}$ is given by $P(|N| \leq x_{0.95}) = 0.95$. This gives the percentile interval

$$\left(\bar{X} - n^{-1/2} x_{0.95} \, \hat{\sigma}, \; \bar{X} + n^{-1/2} x_{0.95} \, \hat{\sigma} \right),$$

derived in Example 1.2. For the sake of definiteness we shall make the coverage correction in the form

$$U(F_1, t) = n^{-1/2} \left(x_{0.95} + t \right) \sigma(F_1),$$

although we would draw the same conclusion with other forms of correction. Thus,

$$f_{U(F_1, t)}(F_0, F_1) = I\left\{ n^{1/2} |\theta(F_1) - \theta(F_0)| \, / \, \sigma(F_1) \leq x_{0.95} + t \right\} - 0.95,$$

so that the sample equation (1.27) becomes

$$P\left\{ n^{1/2} |\theta(F_2) - \theta(F_1)| \, / \, \sigma(F_2) \leq x_{0.95} + t \mid F_1 \right\} - 0.95 = 0. \quad (1.43)$$

Observe that

$$W = n^{1/2} \left\{ \theta(F_2) - \theta(F_1) \right\} \big/ \sigma(F_2)$$

$$= \left\{ n^{-1/2} \sum_{i=1}^{n} (X_i^* - \bar{X}) \right\} \left\{ n^{-1} \sum_{i=1}^{n} (X_i^* - \bar{X})^2 \right\}^{-1/2},$$

where conditional on $\mathcal{X}$, $X_1^*, \ldots, X_n^*$ are independent and identically distributed $N(\bar{X}, \hat{\sigma}^2)$ random variables and $\bar{X}^* = n^{-1} \sum X_i^*$. Therefore conditional on $\mathcal{X}$, and also unconditionally, W is distributed as $\{n/(n-1)\}^{1/2} T_{n-1}$ where T_{n-1} has Student's t distribution with $n-1$ degrees of freedom. Therefore the solution $\hat{t}_0$ of equation (1.43) is $\hat{t}_0 = \{n/(n-1)\}^{1/2} w_{0.95} - x_{0.95}$, where $w_\alpha = w_\alpha(n)$ is defined by

$$P(|T_{n-1}| \leq w_\alpha) = \alpha.$$

The resulting bootstrap confidence interval is

$$\left(\theta(F_1) - n^{-1/2} \sigma(F_1) (x_{0.95} + \hat{t}_0), \; \theta(F_1) + n^{-1/2} \sigma(F_1) (x_{0.95} + \hat{t}_0) \right)$$

$$= \left(\bar{X} - (n-1)^{-1/2} w_{0.95} \, \hat{\sigma}, \; \bar{X} + (n-1)^{-1/2} w_{0.95} \, \hat{\sigma} \right).$$

This is identical to the percentile-t (not the percentile) confidence interval derived in Example 1.2 and has perfect coverage accuracy.

Exactly the same conclusion is obtained on iterating the percentile interval constructed in Example 1.2 under the assumption of an Exponential population: a single iteration produces a percentile-t interval with perfect coverage accuracy. This result does require a special property of both examples, namely, that $\hat{\theta} - \theta$ can be made exactly pivotal by rescaling, and is not universal in parametric problems.

1.6 Bibliographical Notes

Before Efron (1979) drew attention to diverse applications of the bootstrap, it existed mainly in disparate, almost anecdotal form, not as a unified, widely applicable technique. Another of Efron's contributions was to marry the bootstrap to modern computational power. He pointed out that many problems which hitherto had not been thought of as avenues for application of the bootstrap, could be tackled via that route if Monte Carlo methods were employed to numerically approximate functionals of the empirical distribution function. The combined effect of these two ideas on statistical practice, statistical theory, and statistical thinking has been profound. The reader is referred to Diaconis and Efron (1983) for a detailed and delightfully readable account of computer intensive methods in statistics.

Others who thought along similar lines include Barnard (1963), Hope (1968), and Marriott (1979), who advocated Monte Carlo methods for hypothesis testing; Hartigan (1969, 1971, 1975), who used resampled subsamples to construct point and interval estimates, and who stressed the connections with Mahalonobis' "interpenetrating samples" and the jackknife of Quenouille (1949, 1956) and Tukey (1958); Simon (1969, Chapters 23-25), who described a variety of Monte Carlo methods; and Maritz and Jarrett (1978), who devised a bootstrap estimate of the variance of the sample median.

Efron's development of the bootstrap, at least insofar as it applies to confidence intervals, diverges from our own almost at the beginning. Efron's account usually stays close to other resampling methods, particularly the jackknife (see, e.g., Gray and Schucany 1972), in its introductory stages. See for example Efron (1979, 1981a, 1982, 1987), Efron and Gong (1983), and Efron and Tibshirani (1986). Parr (1983) makes a careful analytic comparison of bootstrap and jackknife methods for estimating bias and variance, and Beran (1984b) studies jackknife approximations to bootstrap estimates. In comparison, connections with the jackknife do not emerge naturally from our approach. Likewise, the "other percentile method," mentioned briefly in the third paragraph of Example 1.2 in Section 1.3, does not fit at all well into our set-up, but is basic to Efron's development. It may be justified by a Bayesian argument (Efron (1982, p. 81); see also Rubin (1981)). The "other percentile method" is still quite popular among nonstatisticians, and perhaps also with statisticians, who (if those users we know are any guide) seldom employ any corrections such as Efron's (1982, 1987) bias corrections. However, evidence reported by Efron as early as 1980 makes it clear that the uncorrected "other percentile method" can be a poor performer. We doubt that the uncorrected version will survive the test of time, although *with appropriate corrections* it may epitomise future directions for bootstrap research; see Section 3.11. Hall (1988a) terms the "other percentile method" the backwards method because it is analogous to looking up the wrong statistical tables backwards, and called the ordinary percentile method the hybrid method because it can be interpreted as looking up the wrong tables the right way around.

Our development of bootstrap methodology is similar to that of Hall and Martin (1988a). See also Martin (1989). It represents an attempt to find a unified approach that leads canonically to the issues of pivotalness and bootstrap iteration. Those points do not emerge naturally from some earlier accounts. The importance of working with a pivotal statistic when constructing a bootstrap confidence interval, at least in simple situations, is implicit in Hinkley and Wei (1984) and is addressed more explicitly by Babu and Singh (1984), Hall (1986a, 1988a), Hartigan (1986), and Beran (1987), among others. Pivoting does not always mean rescaling; see Ducharme, Jhun, Romano, and Truong (1985) and Fisher and Hall (1989) for discussions of pivotal statistics that involve spherical data. Our discussion of the sample equation, used to define a bootstrap estimate such as a critical point for a confidence interval, has points of contact with the notion of an estimating equation; see Godambe (1960, 1985) and Godambe and Thompson (1984) for the latter.

The iterated bootstrap is introduced and discussed by Efron (1983) in the context of prediction rules, and by Hall (1986a) and Beran (1987) in the case of confidence intervals. A unified account is given by Hall and Martin (1988a). See also Loh (1987), whose method of bootstrap calibration includes bootstrap iteration as a special case.

Our Russian doll analogy for the iterated bootstrap is reminiscent of Mosteller and Tukey's (1977) "misty staircase" of steps in the construction of successively more elaborate, and more accurate, statistical procedures. After calculating the first and second versions of the procedure, "one should have gone on to a tertiary statistic, which indicated the variability or stability of the secondary statistic, then to a quaternary statistic, ... , and so on up and up a staircase, which since the tertiary was a poorer indicator than the secondary and the quaternary was even worse, could only be pictured as becoming mistier and mistier. In practice, workers usually stopped with primary and secondary statistics" (Mosteller and Tukey 1977, p. 2).

Efron (1985) discusses confidence intervals for parametric problems; Athreya (1987), Hall (1988c, 1990f), and Mammen (1991) study properties of the bootstrap when moments are infinite; and Schemper (1987a) describes jackknife and bootstrap methods for estimating variance, skewness, and kurtosis. General accounts of theory for the bootstrap include Bickel and Freedman (1981), Athreya (1983), and Beran (1984a). Contributions to the theory of Edgeworth expansion for the bootstrap, starting with Singh (1981), will be recounted at the end of Chapter 3. Further literature on bootstrap methodology will be cited there and at the end of Chapter 4. Hinkley (1988) surveys bootstrap methods, and DiCiccio and Romano (1988) describe the state of the art for bootstrap confidence intervals. Work on randomization tests, including material in the monographs by Edgington (1987) and Noreen (1989), is directly related to bootstrap methods.

2

Principles of Edgeworth Expansion

2.1 Introduction

In this chapter we define, develop, and discuss Edgeworth expansions as approximations to distributions of estimates $\hat{\theta}$ of unknown quantities θ_0. We call θ_0 a "parameter", for want of a better term. Briefly, if $\hat{\theta}$ is constructed from a sample of size n, and if $n^{1/2}(\hat{\theta} - \theta_0)$ is asymptotically Normally distributed with zero mean and variance σ^2, then in a great many cases of practical interest the distribution function of $n^{1/2}(\hat{\theta} - \theta_0)$ may be expanded as a power series in $n^{-1/2}$,

$$P\{n^{1/2}(\hat{\theta} - \theta_0)/\sigma \leq x\}$$
$$= \Phi(x) + n^{-1/2}p_1(x)\phi(x) + \ldots + n^{-j/2}p_j(x)\phi(x) + \ldots, \quad (2.1)$$

where $\phi(x) = (2\pi)^{-1/2}e^{-x^2/2}$ is the Standard Normal density function and

$$\Phi(x) = \int_{-\infty}^{x} \phi(u)\, du$$

is the Standard Normal distribution function. Formula (2.1) is termed an *Edgeworth expansion*. The functions p_j are polynomials with coefficients depending on cumulants of $\hat{\theta} - \theta_0$. Conceptually, we think of the expansion (2.1) as starting after the Normal approximation $\Phi(x)$, so that $n^{-1/2}p_1(x)\phi(x)$ is the first term rather than the second term.

The most basic result of this type is in the case where $\hat{\theta}$ is a sample mean and θ_0 the population mean. This example evidences all the main features of more general Edgeworth expansions: p_j is a polynomial of degree at most $3j - 1$ and is an odd or even function according to whether j is even or odd, respectively; and the expansion is valid under appropriate moment and smoothness conditions on the sampling distribution. These properties are most easily discussed in the context of means, and so that case forms a convenient introduction. It is given special treatment in Section 2.2, which lays the foundation for more general Edgeworth expansions developed in Section 2.3. For example, we introduce cumulants in Section 2.2. Section

2.3 describes Edgeworth expansions at a formal level, allowing us to give a wide-ranging yet brief account of their applications. Section 2.4 takes up the main issues in more detail and with more conciseness. It describes and analyzes an explicit model under which Edgeworth expansions may be established rigorously.

Equation (2.1) may be inverted, thereby showing that the solution $x = u_\alpha$ of the equation

$$P\{n^{1/2}(\hat{\theta} - \theta_0)/\sigma \leq x\} = \alpha$$

admits an expansion

$$u_\alpha = z_\alpha + n^{-1/2}p_{11}(z_\alpha) + n^{-1}p_{21}(z_\alpha) + \ldots + n^{-j/2}p_{j1}(z_\alpha) + \ldots, \quad (2.2)$$

where the p_{j1}'s are polynomials derivable from the p_j's and z_α is the solution of the equation $\Phi(z_\alpha) = \alpha$. Inverse formulae such as (2.2) are termed *Cornish-Fisher expansions*, and are described in Section 2.5. Section 2.6 treats examples and gives explicit formulae for some of the polynomials p_j in cases of special interest.

In this brief discussion we have focused on the distribution of the non-Studentized statistic, $n^{1/2}(\hat{\theta}-\theta_0)/\sigma$. But in many cases of practical interest the asymptotic variance σ^2 would be unknown and estimated by a function $\hat{\sigma}^2$ of the data. There we would be interested in the distribution function of the Studentized statistic $n^{1/2}(\hat{\theta} - \theta_0)/\hat{\sigma}$. Our treatment will give emphasis to this case, which admits an expansion like (2.1) but with different polynomials p_j. To distinguish the two types of polynomial we shall wherever possible use p_j for polynomials in the non-Studentized case and q_j for polynomials in the Studentized case. This notation will persist throughout the monograph.

The bootstrap will hardly be mentioned during this chapter; we shall allude to it only occasionally, to provide motivation. Thus, the present chapter is quite separate from the previous one. It provides tools for a theoretical analysis of important aspects of the bootstrap principle, to be conducted in the remainder of the monograph. All our future work will call upon ideas from both Chapters 1 and 2.

We should stress that we are interested in Edgeworth expansions primarily for the light they can shed on theory for the bootstrap. Classical statistical theory employs the expansions to provide analytical corrections rather similar to those that the bootstrap gives by numerical means. Such uses of Edgeworth expansions are incidental to our main purpose, although we shall have a little to say about them in Section 3.8.

Finally, it would be remiss of us not to point out that although the Edgeworth expansion theory outlined in this chapter is applicable to a wide range of problems, it is not universal. There do exist important circumstances where our key principles are violated. A case in point is that of Studentized quantiles, as we shall show in Appendix IV.

2.2 Sums of Independent Random Variables

Let X, X_1, X_2, ... be independent and identically distributed random variables with mean $\theta_0 = \mu$ and finite variance σ^2. An estimate of θ_0 is the sample mean,

$$\hat{\theta} = n^{-1} \sum_{i=1}^{n} X_i \,,$$

with variance $n^{-1}\sigma^2$. By the central limit theorem, $S_n = n^{1/2}(\hat{\theta} - \theta_0)/\sigma$ is asymptotically Normally distributed with zero mean and unit variance. Some inferential procedures make direct use of this approximation. For example, if x_α is the solution of the equation

$$P\big(|N| \le x_\alpha\big) = \alpha \,,$$

where N is Normal $N(0,1)$, then $\mathcal{I} = \big(\hat{\theta} - n^{-1/2}\sigma x_\alpha, \ \hat{\theta} + n^{-1/2}\sigma x_\alpha\big)$ is a confidence interval for θ_0 with nominal coverage α. (By "nominal coverage α" we mean that the interval is constructed by an argument that is designed to ensure that at least the asymptotic coverage equals α; that is, $P(\theta_0 \in \mathcal{I}) \to \alpha$.) A great deal of interest centres on the quality of this Normal approximation, for it determines the coverage accuracy of the confidence interval.

Perhaps the simplest way of describing errors in Normal approximations is via characteristic functions. Since S_n is asymptotically Normal $N(0,1)$, the characteristic function χ_n of S_n converges to that of N: as $n \to \infty$,

$$\chi_n(t) = E\{\exp(itS_n)\} \longrightarrow E\{\exp(itN)\} = e^{-t^2/2} \,,$$

$$\text{for} \quad -\infty < t < \infty. \quad (2.3)$$

(Here $i = \sqrt{-1}$.) Now,

$$\chi_n(t) = \{\chi(t/n^{1/2})\}^n \,, \quad (2.4)$$

where χ is the characteristic function of $Y = (X - \mu)/\sigma$. For a general random variable Y with characteristic function χ, the jth cumulant, κ_j, of Y is defined to be the coefficient of $\frac{1}{j!}(it)^j$ in a power series expansion of

$\log \chi(t)$:

$$\chi(t) = \exp\left\{\kappa_1 it + \tfrac{1}{2}\kappa_2(it)^2 + \ldots + \tfrac{1}{j!}\kappa_j(it)^j + \ldots\right\}. \tag{2.5}$$

Equivalently, since

$$\chi(t) = 1 + E(Y)it + \tfrac{1}{2}E(Y^2)(it)^2 + \ldots + \tfrac{1}{j!}E(Y^j)(it)^j + \ldots,$$

cumulants may be defined in terms via the formal identity

$$\sum_{j\geq 1} \tfrac{1}{j!}\kappa_j(it)^j = \log\left\{1 + \sum_{j\geq 1} \tfrac{1}{j!}E(Y^j)(it)^j\right\}$$

$$= \sum_{k\geq 1}(-1)^{k+1}\tfrac{1}{k}\left\{\sum_{j\geq 1}\tfrac{1}{j!}E(Y^j)(it)^j\right\}^k.$$

Equating coefficients of $(it)^j$ we deduce that

$$\left.\begin{aligned}
\kappa_1 &= E(Y), \\
\kappa_2 &= E(Y^2) - (EY)^2 = \mathrm{var}(Y), \\
\kappa_3 &= E(Y^3) - 3\,E(Y^2)E(Y) + 2\,(EY)^3 = E(Y - EY)^3, \\
\kappa_4 &= E(Y^4) - 4\,E(Y^3)E(Y) - 3\,(EY^2)^2 \\
&\qquad\qquad\qquad\qquad + 12\,E(Y^2)(EY)^2 - 6\,(EY)^4 \\
&= E(Y - EY)^4 - 3(\mathrm{var}\,Y)^2,
\end{aligned}\right\} \tag{2.6}$$

and so on, the formula for κ_{10} containing 41 such terms.

Note that κ_j is a homogeneous polynomial in moments, of degree j. Likewise, moments can be expressed as homogeneous polynomials in cumulants, the polynomial for $E(Y^j)$ being of degree j. Cumulants of multivariate distributions may also be defined; see (2.40) and also part (i) of subsection 5.2.3.

We have standardized Y for location and scale, so that $E(Y) = \kappa_1 = 0$ and $\mathrm{var}(Y) = \kappa_2 = 1$. Hence by (2.4) and (2.5), using the series expansion of the exponential function,

$$\begin{aligned}
\chi_n(t) &= \exp\left\{-\tfrac{1}{2}t^2 + n^{-1/2}\tfrac{1}{3!}\kappa_3(it)^3 + \ldots\right. \\
&\qquad\qquad\qquad\left. + n^{-(j-2)/2}\tfrac{1}{j!}\kappa_j(it)^j + \ldots\right\} \\
&= e^{-t^2/2}\left\{1 + n^{-1/2}r_1(it) + n^{-1}r_2(it) + \ldots\right. \\
&\qquad\qquad\qquad\left. + n^{-j/2}r_j(it) + \ldots\right\}, \tag{2.7}
\end{aligned}$$

where r_j is a polynomial with real coefficients, of degree $3j$, depending on $\kappa_3, \ldots, \kappa_{j+2}$ but not on n. This expansion gives a concise account of the rate of convergence in the limit theorem (2.3).

One property of the r_j's is clear on inspection of the argument leading to (2.7) and will prove crucial in future work: r_j is an even polynomial when j is even and is odd when j is odd. It also follows from (2.7) that

$$r_1(u) = \tfrac{1}{6} \kappa_3 u^3 \tag{2.8}$$

and

$$r_2(u) = \tfrac{1}{24} \kappa_4 u^4 + \tfrac{1}{72} \kappa_3^2 u^6 . \tag{2.9}$$

Many of our arguments involving the bootstrap require only these first two polynomials, and so the algebraic details are simpler than might at first appear.

Rewrite (2.7) in the form

$$\chi_n(t) = e^{-t^2/2} + n^{-1/2} r_1(it) e^{-t^2/2} + n^{-1} r_2(it) e^{-t^2/2} + \dots$$
$$+ n^{-j/2} r_j(it) e^{-t^2/2} + \dots . \tag{2.10}$$

Since

$$\chi_n(t) = \int_{-\infty}^{\infty} e^{itx} \, dP(S_n \le x)$$

and

$$e^{-t^2/2} = \int_{-\infty}^{\infty} e^{itx} \, d\Phi(x) , \tag{2.11}$$

where Φ denotes the Standard Normal distribution function, (2.10) is strongly suggestive of the "inverse" expansion

$$P(S_n \le x) = \Phi(x) + n^{-1/2} R_1(x) + n^{-1} R_2(x) + \dots$$
$$+ n^{-j/2} R_j(x) + \dots , \tag{2.12}$$

where $R_j(x)$ denotes a function whose Fourier-Stieltjes transform equals $r_j(it) e^{-t^2/2}$

$$\int_{-\infty}^{\infty} e^{itx} \, dR_j(x) = r_j(it) e^{-t^2/2} .$$

The next step in our argument is to compute R_j.

Repeated integration by parts in formula (2.11) gives

$$e^{-t^2/2} = (-it)^{-1} \int_{-\infty}^{\infty} e^{itx} d\Phi^{(1)}(x)$$

$$= (-it)^{-2} \int_{-\infty}^{\infty} e^{itx} d\Phi^{(2)}(x)$$

$$\vdots$$

$$= (-it)^{-j} \int_{-\infty}^{\infty} e^{itx} d\Phi^{(j)}(x),$$

where $\Phi^{(j)}(x) = (d/dx)^j \Phi(x)$. Therefore

$$\int_{-\infty}^{\infty} e^{itx} d\{(-D)^j \Phi(x)\} = (it)^j e^{-t^2/2}, \qquad (2.13)$$

where D is the differential operator d/dx. Interpret $r_j(-D)$ in the obvious way as a polynomial in D, so that $r_j(-D)$ is itself a differential operator. By (2.13),

$$\int_{-\infty}^{\infty} e^{itx} d\{r_j(-D)\Phi(x)\} = r_j(it)e^{-t^2/2},$$

and so the function that we seek is none other than $r_j(-D)\Phi(x)$,

$$R_j(x) = r_j(-d/dx)\,\Phi(x). \qquad (2.14)$$

For $j \geq 1$,

$$(-D)^j \,\Phi(x) = -He_{j-1}(x)\,\phi(x),$$

where the functions He_j are Hermite polynomials:

$$He_0(x) = 1, \qquad He_1(x) = x, \qquad He_2(x) = x^2 - 1,$$
$$He_3(x) = x(x^2 - 3), \qquad He_4(x) = x^4 - 6x^2 + 3,$$
$$He_5(x) = x(x^4 - 10x^2 + 15), \ldots.$$

See, for example, Magnus, Oberhettinger, and Soni (1966, Section 5.6), Abramowitz and Stegun (1972, Chapter 22), and Petrov (1975, p. 137). They are orthogonal with respect to the weight function ϕ and are normalized so that the coefficient of the highest power in x is unity. Note that He_j is of precise degree j and is even for even j, odd for odd j. We may now deduce from (2.8), (2.9), and (2.14) that

$$R_1(x) = -\tfrac{1}{6}\kappa_3(x^2 - 1)\,\phi(x)$$

and

$$R_2(x) = -x\Big\{\tfrac{1}{24}\kappa_4(x^2 - 3) + \tfrac{1}{72}\kappa_3^2(x^4 - 10x^2 + 15)\Big\}\,\phi(x).$$

For general $j \geq 1$,

$$R_j(x) = p_j(x)\phi(x),$$

where the polynomial p_j is of degree $3j - 1$ and is odd for even j, even for odd j. This follows from the fact that r_j is of degree $3j$ and is even for even j, odd for odd j.

We have just seen that

$$p_1(x) = -\tfrac{1}{6}\kappa_3(x^2 - 1) \tag{2.15}$$

and

$$p_2(x) = -x\left\{\tfrac{1}{24}\kappa_4(x^2 - 3) + \tfrac{1}{72}\kappa_3^2(x^4 - 10x^2 + 15)\right\}. \tag{2.16}$$

Formula (2.12) now assumes the form

$$P(S_n \le x) = \Phi(x) + n^{-1/2}p_1(x)\phi(x) + n^{-1}p_2(x)\phi(x) + \dots$$
$$+ n^{-j/2}p_j(x)\phi(x) + \dots, \tag{2.17}$$

called an *Edgeworth expansion* of the distribution function $P(S_n \le x)$.

Third and fourth cumulants κ_3 and κ_4 are referred to as *skewness* and *kurtosis* respectively. The term of order $n^{-1/2}$ in (2.17) corrects the basic Normal approximation for the main effect of skewness, while the term of order n^{-1} corrects for the main effect of kurtosis and the secondary effect of skewness.

The expansion (2.17) only rarely converges as an infinite series. Indeed, if X has an absolutely continuous distribution then the type of condition that must be imposed to ensure convergence is $E\{\exp(\tfrac{1}{4}Y^2)\} < \infty$ (Cramér 1928). This is a very severe restriction on the tails of the distribution of Y; it fails even if $Y = (X - \mu)/\sigma$ when X is Exponentially distributed. Usually (2.17) is only available as an *asymptotic series*, or an *asymptotic expansion*, meaning that if the series is stopped after a given number of terms then the remainder is of smaller order than the last term that has been included,

$$P(S_n \le x) = \Phi(x) + n^{-1/2}p_1(x)\phi(x) + \dots$$
$$+ n^{-j/2}p_j(x)\phi(x) + o(n^{-j/2}). \tag{2.18}$$

This expansion is valid for fixed j, as $n \to \infty$.

Sufficient regularity conditions for (2.18), with the remainder of the stated order uniformly in all x, are

$$E(|X|^{j+2}) < \infty \quad \text{and} \quad \limsup_{|t| \to \infty} |\chi(t)| < 1.$$

The latter restriction, often called *Cramér's condition*, holds if the distribution of X is nonsingular, or equivalently if that distribution has a nondegenerate absolutely continuous component — in particular, if X has a proper density function. A proof of this fact in the multivariate case will be given at the end of Section 2.4. The reader is referred to bibliographical notes at the end of this chapter for sources of further information.

Thus, our formal inversion of the characteristic function expansion (2.10) is valid if X has sufficiently many finite moments and a smooth distribution. It fails if the distribution of X is unsmooth, for example if X has a lattice distribution. There, extra terms of all orders, $n^{-1/2}, n^{-1}, \ldots$, must be added to (2.17) to take account of the fact that we are approximating an unsmooth step function, $P(S_n \leq x)$, by a smooth function, the right-hand side of (2.17). For example, suppose $E(X) = 0$, $E(X^2) = 1$, $E|X|^3 < \infty$, and X takes only values of the form $a + jb$, where $j = 0, \pm 1, \pm 2, \ldots$ and $b > 0$ is the largest number that allows this property (i.e., b is the maximal span of the lattice). Define $R(x) = [x] - x + \frac{1}{2}$, where $[x]$ denotes the largest integer less than or equal to x. Then the analogue of (2.18) in the case $j = 1$ is

$$P(S_n \leq x) = \Phi(x) + n^{-1/2} p_1(x)\, \phi(x)$$
$$+ n^{-1/2}\, bR\left\{(n^{1/2}x - na)/b\right\} \phi(x) + o(n^{-1/2}),$$

for the same polynomial p_1 as before. (If we strengthen the moment condition on X to $E(X^4) < \infty$ then the remainder $o(n^{-1/2})$ may be replaced by $O(n^{-1})$.) See for example Gnedenko and Kolmogorov (1954, Section 43).

2.3 More General Statistics

Here we treat Edgeworth expansions for statistics more general than the sample mean. Let S_n denote a statistic with a limiting Standard Normal distribution, such as $S_n = n^{1/2}(\hat\theta - \theta_0)/\sigma$, or $S_n = n^{1/2}(\hat\theta - \theta_0)/\hat\sigma$ where $\hat\sigma^2$ is a consistent estimate of the asymptotic variance σ^2 of $n^{1/2}\hat\theta$. Write χ_n for the characteristic function of S_n, and let $\kappa_{j,n}$ be the jth cumulant of S_n. Then

$$\chi_n(t) = E\left\{\exp(itS_n)\right\}$$
$$= \exp\left\{\kappa_{1,n}it + \tfrac{1}{2}\kappa_{2,n}(it)^2 + \ldots + \tfrac{1}{j!}\kappa_{j,n}(it)^j + \ldots\right\}; \qquad (2.19)$$

compare (2.5).

In many cases of practical interest, $\kappa_{j,n}$ is of order $n^{-(j-2)/2}$ and may be expanded as a power series in n^{-1}:

$$\kappa_{j,n} = n^{-(j-2)/2}\left(k_{j,1} + n^{-1}k_{j,2} + n^{-2}k_{j,3} + \ldots\right), \qquad j \geq 1, \quad (2.20)$$

where $k_{1,1} = 0$ and $k_{2,1} = 1$. The latter two relations reflect the fact that S_n is centred and scaled so that $\kappa_{1,n} = E(S_n) \to 0$ and $\kappa_{2,n} = \mathrm{var}(S_n) \to 1$.

When S_n is a normalized sum of independent and identically distributed random variables, formula (2.20) is implicit in the first identity of (2.7), which declares that

$$\kappa_{j,n} = n^{-(j-2)/2}\kappa_j, \qquad j \geq 2, \tag{2.21}$$

where κ_j is the jth cumulant of Y. In its full generality, (2.20) is a deep result with a rather involved proof. Section 2.4 will give a proof of (2.20) under a model where S_n is a smooth function of vector means.

Combining (2.19) and (2.20) we see that

$$\begin{aligned}
\chi_n(t) &= \exp\left[-\tfrac{1}{2}t^2 + n^{-1/2}\{k_{1,2}it + \tfrac{1}{3!}k_{3,1}(it)^3\} \right.\\
&\qquad \left. + n^{-1}\{\tfrac{1}{2!}k_{2,2}(it)^2 + \tfrac{1}{4!}k_{4,1}(it)^4\} + \dots\right]\\
&= e^{-t^2/2}\left(1 + n^{-1/2}\{k_{1,2}it + \tfrac{1}{6}k_{3,1}(it)^3\}\right.\\
&\qquad + n^{-1}\left[\tfrac{1}{2}k_{2,2}(it)^2 + \tfrac{1}{24}k_{4,1}(it)^4 + \tfrac{1}{2}\{k_{1,2}it + \tfrac{1}{6}k_{3,1}(it)^3\}^2\right]\\
&\qquad \left. + O(n^{-3/2})\right).
\end{aligned} \tag{2.22}$$

More generally, (2.19) and (2.20) give

$$\begin{aligned}
\chi_n(t) &= e^{-t^2/2} + n^{-1/2}r_1(it)e^{-t^2/2} + n^{-1}r_2(it)e^{-t^2/2} + \dots\\
&\qquad\qquad + n^{-j/2}r_j(it)e^{-t^2/2} + \dots,
\end{aligned}$$

where r_j is a polynomial of degree no more than $3j$, even for even j and odd for odd j. This expansion is formally identical to (2.10), the latter derived in the special case where S_n was a sum of independent random variables. The expansion may be inverted, as described in Section 2.2, to produce an analogue of the Edgeworth expansion (2.17),

$$\begin{aligned}
P(S_n \leq x) &= \Phi(x) + n^{-1/2}p_1(x)\phi(x) + n^{-1}p_2(x)\phi(x) + \dots\\
&\qquad\qquad + n^{-j/2}p_j(x)\phi(x) + \dots, \tag{2.23}
\end{aligned}$$

where p_j is a polynomial of degree no more than $3j-1$, odd for even j and even for odd j. We call p_1 a skewness correction and p_2 a correction for kurtosis and for the secondary effect of skewness, even though interpretation of skewness and kurtosis may be rather different from the case of sums of independent random variables.

Recall from the previous section that

$$\int_{-\infty}^{\infty} e^{itx}\, d\{-He_{j-1}(x)\phi(x)\} = (it)^j e^{-t^2/2}, \qquad j \geq 1,$$

where He_j is the jth Hermite polynomial. From this fact, and since by (2.22),

$$r_1(it) = k_{1,2}it + \tfrac{1}{6}k_{3,1}(it)^3$$

and

$$r_2(it) = \tfrac{1}{2}(k_{2,2} + k_{1,2}^2)(it)^2 + \tfrac{1}{24}(k_{4,1} + 4k_{1,2}k_{3,1})(it)^4 + \tfrac{1}{72}k_{3,1}^2(it)^6,$$

we have

$$p_1(x) = -\{k_{1,2} + \tfrac{1}{6}k_{3,1}He_2(x)\} = -\{k_{1,2} + \tfrac{1}{6}k_{3,1}(x^2-1)\} \qquad (2.24)$$

and

$$p_2(x) = -\Big\{\tfrac{1}{2}(k_{2,2} + k_{1,2}^2)He_1(x) + \tfrac{1}{24}(k_{4,1} + 4k_{1,2}k_{3,1})He_3(x)$$

$$+ \tfrac{1}{72}k_{3,1}^2 He_5(x)\Big\}$$

$$= -x\Big\{\tfrac{1}{2}(k_{2,2} + k_{1,2}^2) + \tfrac{1}{24}(k_{4,1} + 4k_{1,2}k_{3,1})(x^2-3)$$

$$+ \tfrac{1}{72}k_{3,1}^2(x^4 - 10x^2 + 15)\Big\}. \qquad (2.25)$$

As was the case in Section 2.2, expansions such as (2.23) usually do not converge as infinite series but are available as asymptotic expansions or series, meaning that

$$P(S_n \le x) = \Phi(x) + n^{-1/2}p_1(x)\phi(x) + \ldots + n^{-j/2}p_j(x)\phi(x) + o(n^{-j/2})$$

for $j \ge 1$. In the next section we shall develop a concise model for S_n, with explicit regularity conditions under which this expansion is available uniformly in x.

We motivated our study of Edgeworth expansions in Section 2.2 by pointing out that they may be used to describe coverage accuracy of confidence intervals, among other things. We shall close the present section by illustrating applications. Suppose first that the statistic S_n is $S_n = n^{1/2}(\hat{\theta} - \theta_0)/\sigma$, where σ^2 denotes the asymptotic variance of $n^{1/2}\hat{\theta}$. Then, nominal α-level confidence intervals for θ_0 are

$$\mathcal{I}_1 = (-\infty,\, \hat{\theta} + n^{-1/2}\sigma z_\alpha) \quad \text{and} \quad \mathcal{I}_2 = (\hat{\theta} - n^{-1/2}\sigma x_\alpha,\, \hat{\theta} + n^{-1/2}\sigma x_\alpha),$$

where x_α and z_α are defined by

$$P(|N| \le x_\alpha) = \alpha \quad \text{and} \quad P(N \le z_\alpha) = \alpha,$$

and where the random variable N has the Normal $N(0,1)$ distribution. Remembering that odd/even indexed p_j's are even/odd functions respectively, we see from (2.23) that the actual coverage probabilities of these intervals are

$$
\begin{aligned}
P(\theta_0 \in \mathcal{I}_1) &= P(\theta_0 < \hat{\theta} + n^{-1/2}\sigma z_\alpha) \\
&= P(S_n > -z_\alpha) \\
&= 1 - \left\{ \Phi(-z_\alpha) + n^{-1/2}p_1(-z_\alpha)\phi(-z_\alpha) + O(n^{-1}) \right\} \\
&= \alpha - n^{-1/2}p_1(z_\alpha)\phi(z_\alpha) + O(n^{-1})
\end{aligned}
$$

and

$$
\begin{aligned}
P(\theta_0 \in \mathcal{I}_2) &= P(S_n > -x_\alpha) - P(S_n > x_\alpha) \\
&= P(S_n \leq x_\alpha) - P(S_n \leq -x_\alpha) \\
&= \Phi(x_\alpha) - \Phi(-x_\alpha) \\
&\quad + n^{-1/2}\{p_1(x_\alpha)\phi(x_\alpha) - p_1(-x_\alpha)\phi(-x_\alpha)\} \\
&\quad + n^{-1}\{p_2(x_\alpha)\phi(x_\alpha) - p_2(-x_\alpha)\phi(-x_\alpha)\} \\
&\quad + n^{-3/2}\{p_3(x_\alpha)\phi(x_\alpha) - p_3(-x_\alpha)\phi(-x_\alpha)\} + O(n^{-2}) \\
&= \alpha + 2n^{-1}p_2(x_\alpha)\phi(x_\alpha) + O(n^{-2}).
\end{aligned}
$$

(The function ϕ is of course even.) Thus, one-sided confidence intervals based on the Normal approximation have coverage errors of order $n^{-1/2}$, whereas two-sided intervals have coverage errors of order n^{-1}, without any adjustment or correction.

The same argument shows that

$$
P(\theta_0 \in \mathcal{I}_1) = \alpha + \sum_{j \geq 1} n^{-j/2}(-1)^j p_j(z_\alpha)\,\phi(z_\alpha)
$$

and

$$
P(\theta_0 \in \mathcal{I}_2) = \alpha + 2\sum_{j \geq 1} n^{-j} p_{2j}(x_\alpha)\,\phi(x_\alpha),
$$

where both expansions are to be interpreted as asymptotic series. Thus, the coverage probability of one-sided confidence intervals is a series in $n^{-1/2}$, whereas that for a two-sided interval is a series in n^{-1}. This fact has important ramifications for bootstrap iteration, since it means that each iteration reduces the order of error by a factor $n^{-1/2}$ in the case of one-sided confidence intervals, but by n^{-1} in the case of two-sided intervals. See Example 1.6 in Chapter 1 and Section 3.11 for further details.

These results are consequences of a property that is of such major importance it is worth repeating: odd/even indexed p_j's are even/odd functions, respectively. When this property fails, results and conclusions can be very different from those described here; see Appendix IV.

If σ is unknown then we may estimate it by $\hat{\sigma}$, say. In this case we work with $S_n = n^{1/2}(\hat{\theta} - \theta)/\hat{\sigma}$ instead of $n^{1/2}(\hat{\theta} - \theta)/\sigma$. This new S_n admits the same expansion (2.23), but with different polynomials. We shall denote them by q instead of p; Section 2.6 will give examples of the differences between p_j and q_j. Thus,

$$P\{n^{1/2}(\hat{\theta} - \theta_0)/\hat{\sigma} \leq x\} = \Phi(x) + n^{-1/2}q_1(x)\phi(x) + \dots$$
$$+ n^{-j/2}q_j(x)\phi(x) + \dots,$$

where again odd/even indexed q_j's are even/odd functions, respectively. Examples of one- and two-sided confidence intervals are

$$\mathcal{J}_1 = \left(-\infty, \hat{\theta} + n^{-1/2}\hat{\sigma}z_\alpha\right) \quad \text{and} \quad \mathcal{J}_2 = \left(\hat{\theta} - n^{-1/2}\hat{\sigma}x_\alpha, \hat{\theta} + n^{-1/2}\hat{\sigma}x_\alpha\right),$$

respectively. Both have nominal coverage α. The results on coverage probability stated earlier for $\mathcal{I}_1$ and $\mathcal{I}_2$ continue to hold for $\mathcal{J}_1$ and $\mathcal{J}_2$, except that the p's are now replaced by q's. For example,

$$P(\theta_0 \in \mathcal{J}_1) = \alpha + \sum_{j \geq 1} n^{-j/2}(-1)^j q_j(z_\alpha)\phi(z_\alpha)$$

$$= \alpha - n^{-1/2}q_1(z_\alpha)\phi(z_\alpha) + O(n^{-1})$$

and

$$P(\theta_0 \in \mathcal{J}_2) = \alpha + 2\sum_{j \geq 1} n^{-j}q_{2j}(x_\alpha)\phi(x_\alpha)$$

$$= \alpha + 2n^{-1}q_2(x_\alpha)\phi(x_\alpha) + O(n^{-2}).$$

Edgeworth expansions may be applied to other functionals of the distribution of $\hat{\theta}$. For example, to obtain expansions of moments observe that if $r \geq 1$ is an integer and we define $S_n = n^{1/2}(\hat{\theta} - \theta)/\sigma$, then

$$E\{(\hat{\theta} - \theta)^r\}$$
$$= (n^{-1/2}\sigma)^r E(S_n^r)$$
$$= (n^{-1/2}\sigma)^r \left\{ \int_0^\infty x^r \, dP(S_n \leq x) - \int_0^\infty (-x)^r \, dP(S_n \leq -x) \right\}$$
$$= (n^{-1/2}\sigma)^r r \int_0^\infty x^{r-1} \left\{ P(S_n > x) + (-1)^r P(S_n \leq -x) \right\} dx,$$

on integrating by parts to obtain the last line. When r is even,

$$P(S_n > x) + (-1)^r P(S_n \leq -x) = P(|N| > x) - 2\sum_{j \geq 1} n^{-j}p_{2j}(x)\phi(x),$$

and when r is odd,

$$P(S_n > x) + (-1)^r P(S_n \leq -x) = -2n^{1/2}\sum_{j \geq 1} n^{-j}p_{2j-1}(x)\phi(x).$$

Hence

$$E\{(\hat{\theta} - \theta)^r\} =$$

$$
\begin{cases}
(n^{-1/2}\sigma)^r \left\{ E(N^r) - 2r \sum_{j \geq 1} n^{-j} \int_0^\infty x^{r-1} p_{2j}(x)\phi(x)\,dx \right\} \\
\qquad\qquad\qquad\qquad\qquad\qquad\qquad \text{for even } r, \\
-2r(n^{-1/2}\sigma)^r n^{1/2} \sum_{j \geq 1} n^{-j} \int_0^\infty x^{r-1} p_{2j-1}(x)\phi(x)\,dx \\
\qquad\qquad\qquad\qquad\qquad\qquad\qquad \text{for odd } r.
\end{cases}
\tag{2.26}
$$

These formulae may be combined as

$$E\{(\hat{\theta} - \theta)^r\} = n^{-[(r+1)/2]}(c_1 + c_2 n^{-1} + c_3 n^{-2} + \dots),$$

where $[x]$ denotes the largest integer less than or equal to x and the constants $c_1, c_2, \dots$, depending on r, are determined by polynomials in the Edgeworth expansion for S_n. (The fact that these expansions of moments are series in n^{-1} was used in Section 1.4 to explain the performance of bootstrap iteration in problems of bias reduction. It implies that each iteration reduces the order of error by a factor n^{-1}, not $n^{-1/2}$.)

Formulae for absolute moments may also be obtained in this way. If $r > 0$, not necessarily an integer, then

$$E(|\hat{\theta} - \theta|^r) = (n^{-1/2}\sigma)^r r \left\{ E|N|^r - 2r \sum_{j \geq 1} n^{-j} \int_0^\infty x^{r-1} p_{2j}(x)\phi(x)\,dx \right\}.$$

As in previous expansions, the infinite series here and in (2.26) are to be interpreted as asymptotic series.

A rigorous proof of these results via this route requires a nonuniform (or weighted uniform) estimate, of the form

$$\sup_{-\infty < x < \infty} (1 + |x|)^l \left| P(S_n \leq x) - \left\{ \Phi(x) + \sum_{k=1}^j n^{-k/2} p_k(x)\phi(x) \right\} \right|$$
$$= O(n^{-(j+1)/2})$$

for large l, of the difference between $P(S_n \leq x)$ and its Edgeworth expansion up to the jth term. Such results may be readily derived, but expansions such as (2.26) are often available more generally than are the Edgeworth expansions from which they were obtained. For example, if θ_0 denotes a population mean and $\hat{\theta}$ the corresponding sample mean, and if rth order population moments are finite, then result (2.26) holds without any additional assumptions on the population (e.g., without Cramér's condition). In this case, each series contains only a finite number of nonzero terms. It is for reasons such as this that during most of this monograph we shall work only formally with Edgeworth expansions, without giving absolutely rigorous proofs. (Chapter 5 is an exception.) There is every expectation that many of our results are available under conditions considerably more general than those we would have to impose to obtain rigorous proofs using presently available technology. Sometimes, as in the example of moments of the sample mean, the natural method of proof is not via Edgeworth expansions, but nevertheless that route does serve to unify, explain, and even motivate the results.

2.4 A Model for Valid Edgeworth Expansions

In this section we describe a general model that allows formulae such as the cumulant expansion (2.20) and the Edgeworth expansion (2.23) to be established rigorously. Let $\mathbf{X}$, $\mathbf{X}_1$, $\mathbf{X}_2$, ... be independent and identically distributed random column d-vectors with mean $\boldsymbol{\mu}$, and put $\overline{\mathbf{X}} = n^{-1} \sum \mathbf{X}_i$. Let $A : \mathbb{R}^d \to \mathbb{R}$ be a smooth function satisfying $A(\boldsymbol{\mu}) = 0$. We have in mind a function such as $A(\mathbf{x}) = \{g(\mathbf{x}) - g(\boldsymbol{\mu})\}/h(\boldsymbol{\mu})$, where $\theta_0 = g(\boldsymbol{\mu})$ is the (scalar) parameter estimated by $\hat{\theta} = g(\overline{\mathbf{X}})$ and $h(\boldsymbol{\mu})^2$ is the asymptotic variance of $n^{1/2}\hat{\theta}$; or $A(\mathbf{x}) = \{g(\mathbf{x}) - g(\boldsymbol{\mu})\}/h(\mathbf{x})$, where $h(\overline{\mathbf{X}})$ is an estimate of $h(\boldsymbol{\mu})$. (Thus, we assume h is a known function.)

This "smooth function model" allows us to study problems where θ_0 is a mean, or a variance, or a ratio of means or variances, or a difference of means or variances, or a correlation coefficient, etc. For example, if $\{W_1, \ldots, W_n\}$ were a random sample from a univariate population with mean m and variance β^2, and if we wished to estimate $\theta_0 = m$, then we would take $d = 2$, $\mathbf{X} = (X^{(1)}, X^{(2)})^{\mathrm{T}} = (W, W^2)^{\mathrm{T}}$, $\boldsymbol{\mu} = E(\mathbf{X})$,

$$g(x^{(1)}, x^{(2)}) = x^{(1)}, \qquad h(x^{(1)}, x^{(2)}) = x^{(2)} - (x^{(1)})^2.$$

This would ensure that $g(\boldsymbol{\mu}) = m$, $g(\overline{\mathbf{X}}) = \overline{W}$ (the sample mean), $h(\boldsymbol{\mu}) = \beta^2$, and

$$h(\overline{\mathbf{X}}) = n^{-1} \sum_{i=1}^{n} X_i^{(2)} - \left(n^{-1} \sum_{i=1}^{n} X_i^{(1)}\right)^2 = n^{-1} \sum_{i=1}^{n} (W_i - \overline{W})^2 = \hat{\beta}^2$$

(the sample variance). If instead our target were $\theta_0 = \beta^2$ then we would take $d = 4$, $\mathbf{X} = (W, W^2, W^3, W^4)^{\mathrm{T}}$, $\boldsymbol{\mu} = E(\mathbf{X})$,

$$g(x^{(1)}, \ldots, x^{(4)}) = x^{(2)} - (x^{(1)})^2,$$

$$h(x^{(1)}, \ldots, x^{(4)}) = x^{(4)} - 4x^{(1)}x^{(3)} + 6(x^{(1)})^2 x^{(2)}$$
$$- 3(x^{(1)})^4 - \{x^{(2)} - (x^{(1)})^2\}^2.$$

In this case,

$$g(\boldsymbol{\mu}) \;=\; \beta^2, \qquad\qquad g(\overline{\mathbf{X}}) \;=\; \hat{\beta}^2,$$

$$h(\boldsymbol{\mu}) \;=\; E(W-m)^4 - \beta^4,$$

$$h(\overline{\mathbf{X}}) \;=\; n^{-1}\sum_{i=1}^{n}(W_i-\overline{W})^2 \;=\; \hat{\beta}^4.$$

(Note that $E(W-m)^4 - \beta^4$ equals the asymptotic variance of $n^{1/2}\,\hat{\beta}^2$.)
The cases where θ_0 is a correlation coefficient (a function of five means), or
a variance ratio (a function of four means), among others, may be treated
similarly.

Denote the ith element of a d-vector $\mathbf{v}$ by $v^{(i)}$ or $(\mathbf{v})^{(i)}$, and put
$\mathbf{Z} = n^{1/2}(\overline{\mathbf{X}}-\boldsymbol{\mu})$ and

$$a_{i_1\ldots i_j} \;=\; \left(\partial^j/\partial x^{(i_1)}\ldots\,\partial x^{(i_j)}\right) A(\mathbf{x})\,\big|_{\mathbf{x}=\boldsymbol{\mu}}. \tag{2.27}$$

By Taylor expansion, and since $A(\boldsymbol{\mu})=0$ and $\mathbf{Z}=O_p(1)$,

$$S_n \;=\; n^{1/2}A(\overline{\mathbf{X}}) \;=\; S_{nr} + O_p(n^{-r/2}), \qquad r\ge 1, \tag{2.28}$$

where the O_p notation is defined on p. xii, and

$$S_{nr} \;=\; \sum_{i=1}^{d} a_i Z^{(i)} + n^{-1/2}\tfrac{1}{2}\sum_{i_1=1}^{d}\sum_{i_2=2}^{d} a_{i_1 i_2} Z^{(i_1)} Z^{(i_2)} + \ldots$$

$$+\; n^{-(r-1)/2}\frac{1}{r!}\sum_{i_1=1}^{d}\cdots\sum_{i_r=1}^{d} a_{i_1\ldots i_r} Z^{(i_1)}\ldots Z^{(i_r)}. \tag{2.29}$$

The following theorem declares that the cumulants of S_{nr} admit the expansion (2.20).

THEOREM 2.1. Let $\mathbf{Z} = n^{1/2}(\overline{\mathbf{X}}-\boldsymbol{\mu})$, where $\overline{\mathbf{X}}$ denotes the mean of
n independent random vectors distributed as $\mathbf{X}$, with finite jrth moments
and mean $\boldsymbol{\mu}$. Put

$$U_{nr} \;=\; \sum_{i=1}^{d} b_i Z^{(i)} + n^{-1/2}\sum_{i_1=1}^{d}\sum_{i_2=1}^{d} b_{i_1 i_2} Z^{(i_1)} Z^{(i_2)} + \ldots$$

$$+\; n^{-(r-1)/2}\sum_{i_1=1}^{d}\cdots\sum_{i_r=1}^{d} b_{i_1\ldots i_r} Z^{(i_1)}\ldots Z^{(i_r)},$$

for arbitrary but fixed constants b. Then the jth cumulant of U_{nr} has the
form

$$\kappa_{j,n,r} \;=\; n^{-(j-2)/2}\big(k_{j,1} + n^{-1}k_{j,2} + n^{-2}k_{j,3} + \ldots\big),$$

where the constants $k_{j,l}$ depend only on the b's, on the moments of $\mathbf{X}$ up
to the jrth, and on r, and where the series contains only a finite number

(not depending on n) of nonzero terms. The values of $k_{j,l}$ for $l \leq r + 2 - j$ do not depend on r.

A proof will be given towards the end of this section.

Assume that the first j moments of the $O_p(n^{-r/2})$ term in (2.28) are all of order $n^{-r/2}$, which is guaranteed under moment conditions on $\mathbf{X}$ and mild assumptions about A. Then the jth cumulant $\kappa_{j,n}$ of S_n equals the jth cumulant of S_{nr} plus a remainder of order $n^{-r/2}$.[1] The latter may be made arbitrarily small by choosing r sufficiently large. Arguing thus it can be seen that $\kappa_{j,n}$ admits the expansion (2.20), interpreted as an asymptotic series rather than an infinite series. (See Section 2.2 for a definition of an asymptotic series.)

Result (2.20) was the main requirement for our formal derivation of the Edgeworth expansion (2.23) in the previous section. Theorem 2.2 below gives explicit regularity conditions, assuming the above model for S_n, under which (2.23) is valid as an asymptotic series. Before passing to that theorem we should compute the asymptotic variance of S_n, so that we may rescale to ensure that S_n has unit asymptotic variance. The latter assumption is implicit in (2.23). We should also calculate formulae for the coefficients in the polynomial p_1.

Put

$$\mu_{i_1 \dots i_j} = E\{ (\mathbf{X} - \boldsymbol{\mu})^{(i_1)} \dots (\mathbf{X} - \boldsymbol{\mu})^{(i_j)} \}, \quad j \geq 1. \tag{2.30}$$

Then $\mu_i = 0$ for each i,

$$E(Z^{(i)} Z^{(j)}) = \mu_{ij},$$

$$E(Z^{(i)} Z^{(j)} Z^{(k)}) = n^{-1/2} \mu_{ijk},$$

and

$$E(Z^{(i)} Z^{(j)} Z^{(k)} Z^{(l)}) = \mu_{ij}\mu_{kl} + \mu_{ik}\mu_{jl} + \mu_{il}\mu_{jk} + O(n^{-1}).$$

Take $r = 2$ in (2.28), so that it becomes

$$S_n = \sum_{i=1}^{d} a_i Z^{(i)} + n^{-1/2} \frac{1}{2} \sum_{i=1}^{d} \sum_{j=1}^{d} a_{ij} Z^{(i)} Z^{(j)} + O_p(n^{-1}).$$

---

[1] This is a slight oversimplification, since in principle it may happen that $E(|S_n|^j)$ is not even finite, whereas $E(|S_{nr}|^j)$ is finite. The point is that in a limit theorem for probabilities, as distinct from moments, it is necessary only to note that S_{nr} approximates S_n with an error of $O_p(n^{-r/2})$, not that the moments (or cumulants) agree to this order. Our formal manipulation of cumulants serves only to determine the details of the approximation of S_{nr} to S_n.

Then

$$E(S_n) = n^{-1/2}\tfrac{1}{2}\sum_{i=1}^{d}\sum_{j=1}^{d} a_{ij}\mu_{ij} + O(n^{-1}),$$

$$E(S_n^2) = \sum_{i=1}^{d}\sum_{j=1}^{d} a_i a_j \mu_{ij} + O(n^{-1}),$$

and

$$E(S_n^3) = n^{-1/2}\bigg\{\sum_{i=1}^{d}\sum_{j=1}^{d}\sum_{k=1}^{d} a_i a_j a_k \mu_{ijk}$$

$$+ \tfrac{3}{2}\sum_{i=1}^{d}\sum_{j=1}^{d}\sum_{k=1}^{d}\sum_{l=1}^{d} a_i a_j a_{kl}(\mu_{ij}\mu_{kl} + \mu_{ik}\mu_{jl} + \mu_{il}\mu_{jk})\bigg\}$$

$$+ O(n^{-1}).$$

The first three cumulants of S_n are thus

$$\kappa_{1,n} = E(S_n) = n^{-1/2}A_1 + O(n^{-1}),$$

$$\kappa_{2,n} = E(S_n^2) - (ES_n)^2 = \sigma^2 + O(n^{-1}),$$

and

$$\kappa_{3,n} = E(S_n^3) - 3E(S_n^2)E(S_n) + 2(ES_n)^3 = n^{-1/2}A_2 + O(n^{-1}),$$

where

$$\sigma^2 = \sum_{i=1}^{d}\sum_{j=1}^{d} a_i a_j \mu_{ij}, \tag{2.31}$$

$$A_1 = \tfrac{1}{2}\sum_{i=1}^{d}\sum_{j=1}^{d} a_{ij}\mu_{ij}, \tag{2.32}$$

$$A_2 = \sum_{i=1}^{d}\sum_{j=1}^{d}\sum_{k=1}^{d} a_i a_j a_k \mu_{ijk} + 3\sum_{i=1}^{d}\sum_{j=1}^{d}\sum_{k=1}^{d}\sum_{l=1}^{d} a_i a_j a_{kl}\mu_{ik}\mu_{jl}. \tag{2.33}$$

In particular, the asymptotic variance of S_n is σ^2. If σ does not equal unity then we should redefine $S_n = n^{1/2}A(\bar{\mathbf{X}})/\sigma$. For this definition of S_n, the constants $k_{1,2}$ and $k_{3,1}$ appearing in formula (2.24) for $p_1(x)$ are

$$k_{1,2} = A_1\sigma^{-1} \quad \text{and} \quad k_{3,1} = A_2\sigma^{-3}.$$

This follows from the calculations in the previous paragraph, noting formula (2.20) for $\kappa_{j,n}$.

Given a d-vector $\mathbf{t} = (t^{(1)}, \dots, t^{(d)})^{\mathrm{T}}$, define

$$\|\mathbf{t}\| = \left\{(t^{(1)})^2 + \dots + (t^{(d)})^2\right\}^{1/2}$$

and

$$\chi(\mathbf{t}) = E\left\{ \exp\left(i \sum_{j=1}^{d} t^{(j)} X^{(j)} \right)\right\}.$$

THEOREM 2.2. *Assume that the function A has $j+2$ continuous derivatives in a neighbourhood of $\boldsymbol{\mu} = E(\mathbf{X})$, that $A(\boldsymbol{\mu}) = 0$, that $E(\|\mathbf{X}\|^{j+2}) < \infty$, and that the characteristic function χ of $\mathbf{X}$ satisfies*

$$\limsup_{\|\mathbf{t}\|\to\infty} |\chi(\mathbf{t})| < 1. \tag{2.34}$$

Define $a_{i_1 \dots i_r}$ and $\mu_{i_1 \dots i_r}$ by (2.27) and (2.30), and σ, A_1, and A_2 by (2.31)–(2.33). Suppose $\sigma > 0$. Then for $j \geq 1$,

$$\begin{aligned}
P\{n^{1/2}A(\overline{\mathbf{X}})/\sigma \leq x\} &= \Phi(x) + n^{-1/2}p_1(x)\phi(x) + \dots \\
&\quad + n^{-j/2}p_j(x)\phi(x) + o(n^{-j/2})
\end{aligned} \tag{2.35}$$

uniformly in x, where p_j is a polynomial of degree at most $3j - 1$, odd for even j and even for odd j, with coefficients depending on moments of $\mathbf{X}$ up to order $j + 2$. In particular,

$$p_1(x) = -\left\{A_1\sigma^{-1} + \tfrac{1}{6} A_2\sigma^{-3}(x^2 - 1)\right\}. \tag{2.36}$$

A proof of this result follows straightforwardly from an Edgeworth expansion of the d-variate distribution of $\mathbf{Z} = n^{1/2}(\overline{\mathbf{X}} - \boldsymbol{\mu})$, as follows. It may be proved that under the conditions of Theorem 2.2,

$$P(\mathbf{Z} \in \mathcal{R}) = \sum_{k=0}^{j} n^{-k/2} P_k(\mathcal{R}) + o(n^{-j/2})$$

uniformly in a large class of sets $\mathcal{R} \subseteq \mathbb{R}^d$, where the P_k's are (signed) measures not depending on n and P_0 is the probability measure of a Normal random variable with zero mean and the same variance matrix as $\mathbf{X}$. See Lemma 5.4 for a concise statement of this result. Taking

$$\mathcal{R} = \mathcal{R}(n, x) = \{\mathbf{z} \in \mathbb{R}^d : n^{1/2}A(\boldsymbol{\mu} + n^{-1/2}\mathbf{z}) \leq x\}, \quad -\infty < x < \infty,$$

we obtain immediately an expansion of $P\{n^{1/2}A(\overline{\mathbf{X}})/\sigma \leq x\}$. This may be put into the form (2.35) by Taylor expanding the function $P_k\{\mathcal{R}(n, x)\}$ about $P_k\{\mathcal{R}(\infty, x)\}$, for $0 \leq k \leq j$. Details of this argument are spelled out in part (iii) of subsection 5.2.3. Bhattacharya and Ghosh (1978) give a proof of Theorem 2.2 that follows just this route.

Condition (2.34) is a multivariate form of Cramér's continuity condition. It is satisfied if the distribution of $\mathbf{X}$ is nonsingular (i.e., has a nondegenerate absolutely continuous component), or if $\mathbf{X} = (W, W^2, \ldots, W^d)^{\mathrm{T}}$ where W is a random variable with a nonsingular distribution. See the end of this section for a proof. Cramér's condition implies that any atoms of the distribution of $\overline{\mathbf{X}}$ have exponentially small mass, so that the distributions of statistics such as $n^{1/2} A(\overline{\mathbf{X}})/\sigma$ are virtually continuous. The following result is derived at the end of the section.

THEOREM 2.3. *If the distribution of* $\mathbf{X}$ *satisfies* (2.34) *then for some* $\epsilon > 0$,

$$\sup_{\mathbf{x} \in \mathbb{R}^d} P(\overline{\mathbf{X}} = \mathbf{x}) = O(e^{-\epsilon n})$$

as $n \to \infty$.

The reader will recall that at the beginning of the section we suggested that the asymptotic variance of an estimator $\hat{\theta} = g(\overline{\mathbf{X}})$ of the parameter $\theta_0 = g(\mu)$ might be taken equal to a function τ^2 of the mean μ, and might be estimated by the same function of $\overline{\mathbf{X}}$. We now verify that this is possible. If $A(\mathbf{x}) = g(\mathbf{x}) - g(\mu)$ then by (2.31),

$$\sigma^2 = \sum_{i=1}^{d} \sum_{j=1}^{d} g_{(i)}(\mu) g_{(j)}(\mu) \{ E(X^{(i)} X^{(j)}) - \mu^{(i)} \mu^{(j)} \},$$

where $g_{(i)}(\mathbf{x}) = (\partial/\partial x^{(i)}) g(\mathbf{x})$. Suppose we adjoin to the vector $\mathbf{X}$ all those products $X^{(i)} X^{(j)}$ that are such that $g_{(i)}(\mathbf{x}) g_{(j)}(\mathbf{x})$ is not identically zero and that do not already appear in $\mathbf{X}$, and adjoin analogous terms to the vectors $\mathbf{X}_i$ and $\overline{\mathbf{X}}$. Let μ denote the mean of the new, lengthened $\mathbf{X}$, and put $\overline{\mathbf{X}} = n^{-1} \sum \mathbf{X}_i$. Then σ^2 is assuredly a function of μ, say $h^2(\mu)$, and is estimated $\sqrt{n}$-consistently by $h^2(\overline{\mathbf{X}})$. Values of σ^2 computed for the functions

$$B_1(\mathbf{x}) = \{g(\mathbf{x}) - g(\mu)\} / h(\mu) \quad \text{and} \quad B_2(\mathbf{x}) = \{g(\mathbf{x}) - g(\mu)\} / h(\mathbf{x}),$$

in place of $A(\mathbf{x})$, are both equal to unity. See Section 2.6 for specific examples of this type.

Observe that $g(\mu)$ equals a functional $\theta(F_0)$ of the population distribution function F_0. In all cases of the nonparametric bootstrap applied to estimation of this $\theta(F_0)$, and in many parametric examples, $g(\overline{\mathbf{X}}) = \theta(F_1)$ where F_1 is the sample distribution function, defined in Section 1.2. This means that if $\hat{\theta}$ is defined by $\hat{\theta} = g(\overline{\mathbf{X}})$ then $\hat{\theta} = \theta(F_1)$, so that our model fits into the bootstrap framework developed in Chapter 1. That will prove important in Chapter 3 when we develop Edgeworth expansions in the context of the bootstrap.

The main thrust of work in this section has been to derive Edgeworth expansions for distributions of statistics of the type $A(\overline{\mathbf{X}})$, where the function A does not depend on n. The latter condition may be removed, although then the form of the results can change. Some work in later chapters will be directed towards functions A_n of the type

$$A_n(\mathbf{x}) \;=\; B_0(\mathbf{x}) + n^{-1/2}B_1(\mathbf{x}) + \ldots + n^{-m/2}B_m(\mathbf{x})\,, \qquad (2.37)$$

where $m \geq 0$ is fixed and $B_0, B_1, \ldots, B_m$ satisfy the conditions previously imposed on A. In particular, we ask that B_k not depend on n and that $B_k(\boldsymbol{\mu}) = 0$ for $0 \leq k \leq m$. When A_n has this form, cumulants of the statistic S_n may no longer enjoy the key property (2.20). The reason is that if odd-indexed B_j's are not identically zero, then in a Taylor expansion of S_n of the form (2.28), (2.29) it may no longer be true that the coefficient of $n^{-j/2}$ for odd/even j is always associated with a product of an even/odd number of $Z^{(i)}$'s, respectively. Instead, with $\mathbf{Z} = n^{1/2}(\overline{\mathbf{X}} - \boldsymbol{\mu})$ we have

$$S_n \;=\; \sum_{i=1}^{d} b_{0,i} Z^{(i)} \;+\; n^{-1/2}\left(\tfrac{1}{2}\sum_{i_1=1}^{d}\sum_{i_2=1}^{d} b_{0,i_1 i_2} Z^{(i_1)} Z^{(i_2)} + \sum_{i=1}^{d} b_{1,i} Z^{(i)} \right)$$

$$+\; n^{-1}\left(\tfrac{1}{3!}\sum_{i_1=1}^{d}\sum_{i_2=1}^{d}\sum_{i_3=1}^{d} b_{0,i_1 i_2 i_3} Z^{(i_1)} Z^{(i_2)} Z^{(i_3)} \right.$$

$$\left. +\; \tfrac{1}{2}\sum_{i_1=1}^{d}\sum_{i_2=1}^{d} b_{0,i_1 i_2} Z^{(i_1)} Z^{(i_2)} + \sum_{i=1}^{d} b_{2,i}\, Z^{(i)} \right) + \ldots\,, \qquad (2.38)$$

where

$$b_{j,i_1 \ldots i_k} \;=\; (\partial^k / \partial x^{(i_1)} \ldots \partial x^{(i_k)})\, B_j(\mathbf{x}) \big|_{\mathbf{x}=\boldsymbol{\mu}}\,.$$

This means that the cumulant expansion on the right-hand side of (2.20) now decreases as a power series in $n^{-1/2}$ rather than n^{-1}. However, the leading term *is still of order* $n^{-(j-2)/2}$, as may be seen by reworking the proof of Theorem 2.1 which we shall give shortly.

Thus, when the function $A = A_n$ has the form (2.37), expansion (2.35) in Theorem 2.2 continues to hold if the conditions imposed earlier on A are now required of $B_0, \ldots, B_m$. However, the polynomial p_j in (2.35) can no longer be guaranteed to be odd/even for even/odd j, respectively. Therefore some of the results outlined at the end of Section 2.3 will no longer be valid.

In practice this causes no difficulty, for the following reason. It turns out that we are interested in expansions of

$$P\{n^{1/2}A_n(\overline{\mathbf{X}}) \leq x\}$$

$$= P\left[n^{1/2}\{B_0(\overline{\mathbf{X}}) + n^{-1/2}B_1(\overline{\mathbf{X}}) + \ldots + n^{-m/2}B_m(\overline{\mathbf{X}})\} \leq x \right] \qquad (2.39)$$

when $x = z$, and in the special case where $B_j(\mathbf{x})$ is a polynomial in $\mathbf{x}$ whose coefficients are polynomials in z, all odd/even if j is even/odd, respectively. This means that, in the coefficient of $n^{-l/2}$ in the Taylor expansion (2.39), the sum of the degree in z of any polynomial coefficient in $b_{j,i_1\ldots i_k}$ and the degree k of the product $Z^{(i_1)}\ldots Z^{(i_k)}$ is odd/even for even/odd indices l, respectively, regardless of the value of j. Now, polynomials in z comprising coefficients of $Z^{(i_1)}\ldots Z^{(i_k)}$ in (2.39) become polynomials in z comprising coefficients of x^r (say) in polynomials p_j when expansion (2.35) is developed. The property stated two sentences earlier implies that in $p_l(x)$, the sum of the degree in z of any coefficient of x^r and the value of r is odd/even for even/odd l, respectively. Therefore when we take $x = z$ in an expansion of the probability in (2.39), we obtain an expansion of the form

$$\Phi(z) + n^{-1/2}\pi_1(z)\phi(z) + \ldots + n^{-j/2}\pi_j(z)\phi(z) + \ldots ,$$

in which each π_j is an odd or even polynomial according to whether j is even or odd, respectively. This is the crucial property enjoyed by the p_j's in our earlier study of expansions such as (2.35). Furthermore, in cases of interest π_j is of degree at most $3j - 1$.

Proof of Theorem 2.1.

Put $\mathbf{V} = \bar{\mathbf{X}} - \boldsymbol{\mu}$, so that

$$U = n^{-1/2}U_{nr} = \sum_{k=1}^{r}\sum_{i_1}\cdots\sum_{i_k} b_{i_1\ldots i_k}V^{(i_1)}\ldots V^{(i_k)} .$$

For any k-tuple $(i_1,\ldots,i_k)$ with $1 \leq i_l \leq d$ for $1 \leq l \leq k$, we have

$$E\big(V^{(i_1)}\ldots V^{(i_k)}\big) = c_1 + c_2 n^{-1} + \ldots + c_k n^{-k+1} ,$$

where the constants c_l do not depend on n. Therefore the jth cumulant of U equals

$$C_{j,1} + C_{j,2}n^{-1} + \ldots + C_{j,jr}n^{-jr+1} ,$$

for constants $C_{j,l}$ not depending on n. The jth cumulant of U_{nr} equals $n^{j/2}$ times the jth cumulant of $n^{-1/2}U_{nr}$, and so Theorem 2.1 will follow if we prove that the jth cumulant of $n^{-1/2}U_{nr}$ is of order n^{-j+1}, which amounts to showing that $C_{j,1} = C_{j,2} = \ldots = C_{j,j-1} = 0$. (It is clear, on considering the contributions that different terms in (2.29) make to $\kappa_{j,n,r}$, that $k_{j,l}$ for $l \leq r + 2 - j$ does not depend on r.)

kth order multivariate cumulant $\kappa^{(i_1,\ldots,i_k)}(\mathbf{W})$ of a column d-vector $\mathbf{W}$ is a scalar defined formally by the expansion

$$\sum_{k\geq 1}\frac{1}{k!}\sum_{i_1}\cdots\sum_{i_k}t^{(i_1)}\ldots t^{(i_k)}\kappa^{(i_1,\ldots,i_k)}(\mathbf{W}) = \log\big\{E(e^{\mathbf{t}^{\mathsf{T}}\mathbf{W}})\big\}, \quad (2.40)$$

where $\mathbf{t} = (t^{(1)}, \ldots, t^{(d)})^{\mathrm{T}}$. Of course, the right-hand side here equals

$$\log \left\{ 1 + \sum_{k \geq 1} \frac{1}{k!} \sum_{i_1} \cdots \sum_{i_k} t^{(i_1)} \ldots t^{(i_k)} E(W^{(i_1)} \ldots W^{(i_k)}) \right\}.$$

The identity (2.40) should not be interpreted as requiring that $\mathbf{W}$ have a proper moment generating function; it is intended to give only a formal definition.

An equivalent definition in alternative notation may be given as follows. Suppose the index l appears precisely j_l times in the vector $(i_1, \ldots, i_k)$. Define

$$\kappa(j_1, j_2, \ldots) = \kappa^{(i_1, \ldots, i_k)}(\mathbf{W}), \qquad (2.41)$$

valid since the right-hand side depends on $i_1, \ldots, i_k$ only through $j_1, j_2, \ldots$. Put

$$\mu(j_1, j_2, \ldots) = E(W^{(i_1)} \ldots W^{(i_k)}),$$

$$\mu(0, 0, \ldots) = 1 \quad \text{and} \quad \kappa(0, 0, \ldots) = 1.$$

Then equation (2.40) assumes the form

$$\sum_{j_1, j_2, \ldots} \kappa(j_1! j_2! \ldots) \frac{(t^{(1)})^{j_1} (t^{(2)})^{j_2} \ldots}{j_1! j_2! \ldots}$$

$$= \log \left\{ \sum_{j_1, j_2, \ldots} \mu(j_1, j_2, \ldots) \frac{(t^{(1)})^{j_1} (t^{(2)})^{j_2} \ldots}{j_1! j_2! \ldots} \right\}.$$

List the terms

$$V^{(1)}, \ldots, V^{(d)}, V^{(1)}V^{(1)}, V^{(1)}V^{(2)}, \ldots, V^{(d)}V^{(d)}, V^{(1)}V^{(1)}V^{(1)}, \ldots, (V^{(d)})^r$$

as a vector $\mathbf{W}$, and the terms

$$b_1, \ldots, b_d, b_{11}, b_{12}, \ldots, b_{dd}, b_{111}, \ldots, b_{d\ldots d}$$

as a vector $\mathbf{B}$. Then $U = \sum_i B^{(i)} W^{(i)}$, so that

$$E(U^k) = \sum_{i_1} \cdots \sum_{i_k} B^{(i_1)} \ldots B^{(i_k)} E(W^{(i_1)} \ldots W^{(i_k)}).$$

Let $K_k(U)$ denote the kth cumulant of U. Since cumulants are homogeneous polynomials in moments,

$$K_k(U) = \sum_{i_1} \cdots \sum_{i_k} B^{(i_1)} \ldots B^{(i_k)} \kappa^{(i_1, \ldots, i_k)}(\mathbf{W}).$$

Therefore our proof of Theorem 2.1 will be complete if we show that whenever $k \geq 1$ and $i_1, \ldots, i_k \geq 1$,

$$\kappa^{(i_1, \ldots, i_k)}(\mathbf{W}) = O(n^{-k+1}). \qquad (2.42)$$

Let $(j_1, j_2, \dots)$ be an infinite vector of nonnegative integers such that $\sum_i j_i = k$. Assume that $(i_1, \dots, i_k)$ contains the integer l precisely j_l times, for $l \geq 1$. Define $\kappa(j_1, j_2, \dots)$ as in (2.41). Suppose $W^{(l)}$ is comprised of the product of precisely w_l $V^{(\alpha)}$'s, $\alpha \geq 1$. Put

$$k^\dagger = \sum_{l=1}^{\infty} j_l w_l \geq k.$$

Then $\kappa(j_1, j_2, \dots)$ is a polynomial in moments of the $V^{(\alpha)}$'s, homogeneous and of (precise) degree $k^\dagger$. Each moment of $\mathbf{V}$ of degree b (say) equals a homogeneous polynomial of degree b in cumulants of $\mathbf{V}$. Therefore $\kappa(j_1, j_2, \dots)$ equals a homogeneous polynomial of degree $k^\dagger$ in cumulants of $\mathbf{V}$.

Put

$$\lambda^{(\alpha_1, \dots, \alpha_r)} = \kappa^{(\alpha_1, \dots, \alpha_r)}(\mathbf{V}).$$

A typical term in the formula for $\kappa(j_1, j_2, \dots)$ as a polynomial in cumulants of $\mathbf{V}$, equals a constant (not depending on n) multiple of

$$\prod_{l=1}^{L} \lambda^{(\alpha_{l1}, \dots, \alpha_{lr_l})}, \tag{2.43}$$

where $\sum r_l = k^\dagger$. We may represent this product as a $k \times L$ array, the lth row corresponding to i_l and the lth column to $(\alpha_{l1}, \dots, \alpha_{lr_l})$. For example, suppose $k = 3$, $W^{(i_1)} = V^{(a)}$, $W^{(i_2)} = V^{(b)}$, and $W^{(i_3)} = V^{(c)} V^{(d)}$. Then the sequence $\{(\alpha_{L1}, \dots, \alpha_{1r_1}), \dots, (\alpha_{L1}, \dots, \alpha_{Lr_L})\}$ must have one of the following nine forms:

$$\{(a, b, c, d)\}, \qquad \{(a, b, c), (d)\}, \qquad \{(a, c, d), (b)\},$$
$$\{(a, b), (c, d)\}, \qquad \{(a, c), (b, d)\}, \qquad \{(a, b), (c), (d)\},$$
$$\{(a, c), (b), (d)\}, \qquad \{(c, d), (a), (b)\}, \qquad \{(a), (b), (c), (d)\}.$$

(We have ignored sequences that arise from interchanging a and b, or interchanging c and d, for they are similar in all important respects.) From these sequences we obtain the arrays in Figure 2.1, arranged in the same order as the nine sequences above.

An array is said to be *separable* if the letters within it are contained within two or more rectangular blocks (whose sides are parallel to the axes) such that no two blocks have a part of the same row or column in common. In the examples of Figure 2.1, all arrays except (i), (ii), and (v) are separable. We claim that all separable arrays make zero contribution to $\kappa(j_1, j_2, \dots)$. Accepting this for the time being, we work only with nonseparable arrays. Observe that a nonseparable array may be constructed term by term in such a way that each new entry after the first starts either

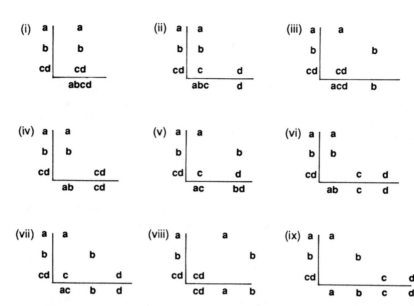

FIGURE 2.1. Arrays constructed from the nine consecutive sequences $\{(a, b, c, d)\}, \dots , \{(a), (b), (c), (d)\}$.

a new row, or a new column, or neither, but not both. Hence if there are R rows, C columns, and L letters in the nonseparable array,

$$L \geq 1 + (R - 1) + (C - 1) = R + C - 1. \qquad (2.44)$$

Since $\mathbf{V}$ is a centred sample mean then the argument leading to (2.21) may be easily modified to show that the cumulant $\lambda^{(\alpha_1, \dots, \alpha_r)}$, corresponding to a vector $(a, b, \dots)$ of r letters, is of order $n^{-(r-1)}$ as $n \to \infty$. Therefore the product at (2.43) is of order $n^{-\nu}$, where

$$\nu = \sum_{\substack{\text{all vectors}(a,b,\dots)\text{ representing} \\ \text{columns in an array}}} (\text{number of letters} - 1)$$

$$= \text{total number of letters} - \text{number of columns}$$

$$= L - C \geq R - 1,$$

using (2.44). But $R = k$, and so the order of each term in $\kappa^{(i_1, \dots, i_k)}(\mathbf{W})$ can be no greater than $n^{-(k-1)}$. The number of terms does not depend on

n, whence

$$\kappa^{(i_1,\dots,i_k)}(\mathbf{W}) = O(n^{-(k-1)}).$$

This proves (2.42).

It remains to prove the claim made two paragraphs earlier, that separable arrays make no contribution to $\kappa(j_1, j_2, \dots) = \kappa^{(i_1,\dots,i_2)}(\mathbf{W})$. This we do by induction over k. The claim is trivially true when $k = 1$, for there are no separable arrays in that case. Given that the claim is true for all orders up to and including $k - 1$, we shall prove it for k.

Put

$$\mu^{(i_1,\dots,i_k)} = E(W^{(i_1)}\dots W^{(i_k)}), \qquad \kappa^{(i_1,\dots,i_k)} = \kappa^{(i_1,\dots,i_k)}(\mathbf{W})$$

and (as before) $\lambda^{(\alpha_1,\dots,\alpha_r)} = \kappa^{(\alpha_1,\dots,\alpha_r)}(\mathbf{V})$. Formula (2.40) gives

$$\mu^{(i_1,\dots,i_k)} = \sum\sum \kappa^{(\beta_{11},\dots,\beta_{1s_1})}\dots\kappa^{(\beta_{a1},\dots,\beta_{as_a})}, \qquad (2.45)$$

where the outer sum is over all $a \geq 1$ and all $s_1,\dots,s_a$ such that $s_1 + \dots + s_a = k$ and the inner sum is over all distinct terms obtained by permuting $(i_1,\dots,i_k)$ to $(\beta_{11},\dots,\beta_{1s_1},\beta_{21},\dots,\beta_{as_a})$. The only term on the right-hand side of (2.45) involving only a single cumulant is $\kappa^{(i_1,\dots,i_k)}$. Therefore we may rewrite (2.45) as

$$\mu^{(i_1,\dots,i_k)} = \kappa^{(i_1,\dots,i_k)} + K,$$

and hence

$$\kappa^{(i_1,\dots,i_k)} = \mu^{(i_1,\dots,i_k)} - K, \qquad (2.46)$$

where K stands for the sum of products of two or more cumulants of the form $\kappa^{(i'_1,\dots,i'_{k'})}$, with $k' \leq k - 1$. By the induction hypothesis, no terms arising from separable arrays make contributions to K. Therefore it suffices to prove that if any term arising from a separable array makes a contribution to $\mu^{(i_1,\dots,i_k)}$ then it cancels from the difference $\mu^{(i_1,\dots,i_k)} - K$.

Consider an array of the type introduced below (2.43), with rows determined by $(i_1,\dots,i_k)$ and columns by vectors $\boldsymbol{\alpha}_l = (\alpha_{l1},\dots,\alpha_{lr_l})$, $1 \leq l \leq L$. Suppose the array is separable into B nonseparable blocks. Without loss of generality, the bth block has its rows determined by

$$i(b) = (i_{t_1+\dots+t_{b-1}+1},\dots,i_{t_1+\dots+t_b})$$

and its columns by vectors $\boldsymbol{\alpha}_{u_1+\dots+u_{b-1}+1},\dots,\boldsymbol{\alpha}_{u_1+\dots+u_b}$, as indicated in Figure 2.2. Of course,

$$\sum_{b=1}^{B} t_b = k \qquad \text{and} \qquad \sum_{b=1}^{B} u_l = L.$$

The entire array represents the product

$$\lambda_0 = \prod_{l=1}^{L} \lambda^{\alpha_l}$$

appearing in an expansion of $\kappa^{(i_1,\dots,i_k)}$. We shall prove in the next paragraph that the coefficient of λ_0 in $\mu^{(i_1,\dots,i_k)}$ equals the product over b, $1 \le b \le B$, of the coefficients of

$$\lambda_b = \lambda^{\alpha_{u_1}+\dots+u_{b-1}+1} \dots \lambda^{\alpha_{u_1}+\dots+u_b}$$

in $\mu^{i(b)}$. By the analogue of (2.46), $\kappa^{i(b)} = \mu^{i(b)} - K(b)$ where $K(b)$ stands for the sum of products of two or more cumulants. Therefore the coefficient of λ_0 in $\mu^{(i_1,\dots,i_k)}$ equals the product over b of the coefficient of λ_b in $\kappa^{i(b)} + K(b)$. If $B \ge 2$ then $\prod_b \kappa^{i(b)}$ is one of the terms comprising K (see (2.45)), and so the contribution from this product cancels from $\mu^{(i_1,\dots,i_k)} - K$ and hence from $\kappa^{(i_1,\dots,i_k)}$ (see (2.46)). And even for $B \ge 1$, any term in $\prod_b \{\kappa^{i(b)} + K(b)\}$ which is not equal to $\prod_b \kappa^{i(b)}$, equals one of the terms in K and so cancels from $\mu^{(i_1,\dots,i_k)} - K = \kappa^{(i_1,\dots,i_k)}$. Thus, all contributions to $\mu^{(i_1,\dots,i_k)}$ coming from separable arrays, cancel from the difference $\mu^{(i_1,\dots,i_k)} - K$. This is the required result.

Finally, observe that to obtain the term in λ_0 in an expansion of $\mu^{(i_1,\dots,i_k)}$, we may assume that the sets of variables that comprise individual components in the expectations that form the product λ_0 are stochastically independent. This ensures that we focus attention on the product, and since that is the only term in which we are interested, the effect on other terms is immaterial. Thus, the following construction is motivated. Let $\mathcal{I}_b$ denote the set of indices s appearing in one or another of the vectors α_r that determine the columns of the bth block. Put $\mathcal{S}_b = \{V^{(s)}, \ s \in \mathcal{I}_b\}$. Now replace the elements of $\mathcal{S}_b$ by new random variables that are such that the elements of $\mathcal{S}_b$ have the same joint distribution as before, but are stochastically independent of elements of all sets $\mathcal{S}_{b'}$ with $b' \ne b$. Let $\mathcal{S}_b^\dagger$, $1 \le b \le B$, denote the resulting sets. Consider an expansion of $\mu^{(i_1,\dots,i_k)}$ as a polynomial in cumulants of $\mathbf{V}$. If we replace each element of $\cup \mathcal{S}_b$ by its counterpart in $\cup \mathcal{S}_b^\dagger$, then the term in λ is easily recognizable. Furthermore, under the hypothesis of independence,

$$\mu^{(i_1,\dots,i_k)} = \prod_{b=1}^{B} \mu^{i(b)} .$$

When the moments on the right-hand side are expanded as polynomials in cumulants of $\mathbf{V}$, each group of indices β in λ^β can arise from *only one* of the B blocks, by virtue of the definition of the right-hand side as a product. Therefore the coefficient of $\lambda_0 = \prod_b \lambda^{\alpha_b}$ in $\mu^{(i_1,\dots,i_k)}$ equals the product over $1 \le b \le B$ of the coefficients of λ_b in $\mu^{i(b)}$.

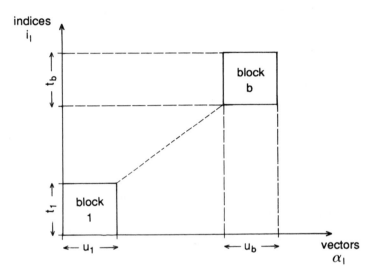

FIGURE 2.2. Blocks comprising a separable array.

Derivation of Cramér's Condition

Here we prove that Cramér's condition, (2.34), holds if the distribution of the d-vector $\mathbf{X}$ has a nondegenerate, absolutely continuous component, and also in other circumstances.

Suppose the d-variate distribution function F of $\mathbf{X}$ may be written as $F = \delta G + (1 - \delta)H$, where $\delta > 0$, G and H are d-variate distribution functions, and G is absolutely continuous with density g. Then if $\mathbf{x}$ and $\mathbf{t}$ are both column vectors,

$$|\chi(\mathbf{t})| \leq \delta \left| \int_{\mathbb{R}^d} e^{i\mathbf{t}^T\mathbf{x}} g(\mathbf{x})\, d\mathbf{x} \right| + 1 - \delta \,,$$

and so to establish (2.34) it suffices to show that

$$\lim_{\|\mathbf{t}\| \to \infty} \left| \int_{\mathbb{R}^d} e^{i\mathbf{t}^T\mathbf{x}} g(\mathbf{x})\, d\mathbf{x} \right| \;=\; 0\,. \tag{2.47}$$

(Throughout this proof, $i = \sqrt{-1}$.)

Given $\epsilon > 0$, let

$$g_1(\mathbf{x}) = \sum_{j=1}^{m} c_j\, I(\mathbf{x} \in \mathcal{S}_j)$$

denote a simple function, supported on bounded rectangular prisms $\mathcal{S}_j$, such that

$$\int_{\mathbb{R}^d} |g(\mathbf{x}) - g_1(\mathbf{x})|\, d\mathbf{x} \le \epsilon.$$

It suffices to prove (2.47) with g_1 replacing g, and so it suffices to show that for each bounded rectangular prism $\mathcal{S}$,

$$\lim_{\|\mathbf{t}\|\to\infty} \left| \int_{\mathcal{S}} e^{i\mathbf{t}^T\mathbf{x}}\, d\mathbf{x} \right| = 0.$$

The last-written integral may be worked out exactly and shown to converge to zero at rate $O(\|\mathbf{t}\|^{-1})$ as $\|\mathbf{t}\| \to \infty$.

In the examples of Section 2.6 we shall take $\mathbf{X} = (W, W^2, \dots, W^d)^T$, where W is a (scalar) random variable and $d \ge 1$. This circumstance is not covered by the previous proof, but we claim nevertheless that Cramér's condition holds provided the distribution of W has a nondegenerate absolutely continuous component. (Similarly it may be proved that if $\mathbf{X}$ can be expressed as a nondegenerate multivariate polynomial in a vector $\mathbf{Y}$ whose distribution has a nondegenerate absolutely continuous component, then $\mathbf{X}$ satisfies Cramér's condition.) To verify the claim, suppose the univariate distribution function F of W may again be written as $F = \delta G + (1 - \delta)H$, where $\delta > 0$, G and H are univariate distribution functions, and G is absolutely continuous with density g. Then

$$|\chi(\mathbf{t})| \le \delta \left| \int_{-\infty}^{\infty} \exp\left(i \sum_{j=1}^{d} t^{(j)} x^j \right) g(x)\, dx \right| + 1 - \delta,$$

and so it suffices to prove that

$$\lim_{\|\mathbf{t}\|\to\infty} \left| \int_{-\infty}^{\infty} \exp\left(i \sum_{j=1}^{d} t^{(j)} x^j \right) g(x)\, dx \right| = 0.$$

By approximating g by a simple function g_1 supported on bounded intervals, much as in the previous proof, we see that we need only show that for each bounded interval $\mathcal{S}$,

$$\lim_{\|\mathbf{t}\|\to\infty} \left| \int_{\mathcal{S}} \exp\left(i \sum_{j=1}^{d} t^{(j)} x^j \right) dx \right| = 0. \tag{2.48}$$

Take $\mathbf{t} = \mathbf{t}(u)$, where u is an index diverging to infinity. Using a subsequence argument, we see that we may assume without loss of generality that for some $1 \leq j \leq d$, and with

$$r(j, k, u) = t^{(k)}(u)/t^{(j)}(u),$$

the limit

$$r(j, k) = \lim_{u \to \infty} r(j, k, u)$$

exists for each $1 \leq k \leq d$ and does not exceed unity. For definiteness, let $\mathcal{S} = (0, 1)$ (an open interval), and take the real part of the integral in (2.48). The function

$$a(x) = \sum_{k=1}^{d} r(j, k) x^k$$

is strictly monotone on each of d intervals $I_l = (b_{l-1}, b_l)$, $1 \leq l \leq d$, where $0 = b_0 \leq b_1 \leq \ldots \leq b_d = 1$. The same applies to the function

$$a(x, u) = \sum_{k=1}^{d} r(j, k, u) x^k,$$

the intervals now being $I_l(u) = (b_{l-1}(u), b_l(u))$ where $b_l(u) \to b_l$ as $u \to \infty$. The integral whose value we wish to prove converges to zero is

$$\int_0^1 \cos\left\{ t^{(j)} \sum_{k=1}^{d} r(j, k, u) x^k \right\} dx = \sum_{l=1}^{d} \int_{I_l(u)} \cos\left\{ t^{(j)} a(x, u) \right\} dx.$$

Change variable from x to $y = a(x, u)$ in each of the d integrals on the right-hand side, thereby showing that each of those integrals converges to zero as $t^{(j)} \to \infty$. This proves that the real part of the integral in (2.48) converges to zero, and similarly it may be shown that the imaginary part converges to zero. This completes the proof.

Proof of Theorem 2.3.

Since

$$\sup_{\mathbf{x} \in \mathbb{R}^d} P(\bar{\mathbf{X}} = \mathbf{x}) \leq \max_{1 \leq j \leq d} \sup_{x \in \mathbb{R}} P(\bar{X}^{(j)} = x)$$

and

$$\max_{1 \leq j \leq d} \limsup_{|t| \to \infty} E\left\{ \exp(it\, X^{(j)}) \right\} \leq \limsup_{\|t\| \to \infty} E\left\{ \exp(i t^{\mathrm{T}} \mathbf{X}) \right\},$$

it suffices to prove the theorem in the case $d = 1$. In that circumstance, define

$$C_1 = \sup_{|t| \geq 1} |\chi(t)| < 1.$$

and $\epsilon = -\frac{1}{2} \log C_1$, and put $\eta = e^{-\epsilon n}$. Let the random variable N be independent of $\bar{X}$ and have the Standard Normal distribution, and write f for the density of $n\bar{X} + \eta^2 N$. By Fourier inversion,

$$\sup_x f(x) = (2\pi)^{-1} \sup_x \left| \int_{-\infty}^{\infty} \chi(t)^n \exp\left(-\tfrac{1}{2}\eta^4 t^2\right) dt \right|$$

$$\leq \pi^{-1} \left\{ 1 + C_1^n \int_0^{\infty} \exp\left(-\tfrac{1}{2}\eta^4 t^2\right) dt \right\}$$

$$= \pi^{-1} \left(1 + \int_0^{\infty} e^{-t^2/2} dt \right) = C_2,$$

say. Therefore

$$\sup_x P(\bar{X} = x) \leq \sup_x P(|n\bar{X} + \eta^2 N + x| \leq \eta) + P(|N| > \eta^{-1})$$

$$\leq \sup_x \int_{x-\eta}^{x+\eta} f(y)\, dy + \eta$$

$$= (2C_2 + 1)\eta = O(e^{-\epsilon n}),$$

as had to be shown.   □

2.5 Cornish-Fisher Expansions

Let S_n denote a statistic whose distribution admits the Edgeworth expansion (2.23), usually interpreted as an asymptotic series. Write $w_\alpha = w_\alpha(n)$ for the α-level quantile of S_n, determined by

$$w_\alpha = \inf\{x : P(S_n \leq x) \geq \alpha\},$$

and let z_α be the α-level Standard Normal quantile, given by $\Phi(z_\alpha) = \alpha$. Implicit in (2.23) is an expansion of w_α in terms of z_α and an expansion of z_α in terms of w_α,

$$w_\alpha = z_\alpha + n^{-1/2}p_{11}(z_\alpha) + n^{-1}p_{21}(z_\alpha) + \ldots + n^{-j/2}p_{j1}(z_\alpha) + \ldots$$

and

$$z_\alpha = w_\alpha + n^{-1/2}p_{12}(w_\alpha) + n^{-1}p_{22}(w_\alpha) + \ldots + n^{-j/2}p_{j2}(w_\alpha) + \ldots,$$

where the functions p_{j1} and p_{j2} are polynomials. These expansions are to be interpreted as asymptotic series, and in that sense are available uniformly in $\epsilon < \alpha < 1 - \epsilon$ for any $0 < \epsilon < \frac{1}{2}$. Both are called *Cornish-Fisher expansions*.

The polynomials p_{j1} and p_{j2} are of degree at most $j + 1$, odd for even j and even for odd j, and depend on cumulants only up to order $j + 2$.

They are completely determined by the p_j's in (2.23), being defined by the formal relations

$$\Phi\left\{z_\alpha + \sum_{j \geq 1} n^{-j/2} p_{j1}(z_\alpha)\right\}$$

$$+ \sum_{i \geq 1} n^{-i/2} p_i \left\{z_\alpha + \sum_{j \geq 1} n^{-j/2} p_{j1}(z_\alpha)\right\}$$

$$\times \phi\left\{z_\alpha + \sum_{j \geq 1} n^{-j/2} p_{j1}(z_\alpha)\right\} = \alpha, \quad 0 < \alpha < 1, \quad (2.49)$$

and

$$\sum_{j \geq 1} n^{-j/2} p_{j1}(x) + \sum_{i \geq 1} n^{-i/2} p_{i2}\left\{x + \sum_{j \geq 1} n^{-j/2} p_{j1}(x)\right\} = 0.$$

In particular, it follows from (2.49) that p_{j1} is determined by $p_1, \ldots, p_j$. To derive formulae for p_{11} and p_{21}, note that

$$\alpha = \Phi(z_\alpha) + \left\{n^{-1/2} p_{11}(z_\alpha) + n^{-1} p_{21}(z_\alpha)\right\} \phi(z_\alpha)$$

$$- \tfrac{1}{2}\left\{n^{-1/2} p_{11}(z_\alpha)\right\}^2 z_\alpha \phi(z_\alpha)$$

$$+ n^{-1/2}\left[p_1(z_\alpha)\phi(z_\alpha) + n^{-1/2} p_{11}(z_\alpha)\{p_1'(z_\alpha) - z_\alpha p_1(z_\alpha)\}\phi(z_\alpha)\right]$$

$$+ n^{-1} p_2(z_\alpha)\phi(z_\alpha) + O(n^{-3/2})$$

$$= \alpha + n^{-1/2}\{p_{11}(z_\alpha) + p_1(z_\alpha)\}\phi(z_\alpha)$$

$$+ n^{-1}\left[p_{21}(z_\alpha) - \tfrac{1}{2} z_\alpha p_{11}(z_\alpha)^2\right.$$

$$\left. + p_{11}(z_\alpha)\{p_1'(z_\alpha) - z_\alpha p_1(z_\alpha)\} + p_2(z_\alpha)\right]\phi(z_\alpha)$$

$$+ O(n^{-3/2}).$$

From this we may conclude that

$$p_{11}(x) = -p_1(x)$$

and

$$p_{21}(x) = p_1(x) p_1'(x) - \tfrac{1}{2} x p_1(x)^2 - p_2(x). \quad (2.50)$$

Formulae for the other polynomials p_{i1}, and for the p_{i2}'s, may be derived similarly; however, they will not be needed in our work.

Cornish-Fisher expansions under explicit regularity conditions may be deduced from results such as Theorem 2.2. For example, the following theorem is easily proved.

THEOREM 2.4. *Assume the conditions of Theorem 2.2 on the function A and the distribution of* $\mathbf{X}$, *and define*

$$w_\alpha = \inf\left\{x : P[n^{1/2}\, A(\overline{\mathbf{X}}) \leq x] \geq \alpha\right\}.$$

Let $p_1, \ldots, p_j$ *denote the polynomials appearing in* (2.35), *and define* $p_{11}, \ldots, p_{j1}$ *by* (2.49). *Then*

$$w_\alpha = z_\alpha + n^{-1/2} p_{11}(z_\alpha) + \cdots + n^{-j/2} p_{j1}(z_\alpha) + o(n^{-j/2})$$

uniformly in $\epsilon < \alpha < 1 - \epsilon$ *for each* $\epsilon > 0$.

Expansions such as this usually cannot be extended to the range $0 < \alpha < 1$ since the right-hand side of the expansion is generally not a monotone function of α. For example, in the case of a univariate mean in which $A(\bar{X}) = \bar{X} - \mu$, we have $p_{11}(x) = \frac{1}{6}\gamma(x^2 - 3)$ where γ denotes skewness; see Example 2.1. Therefore

$$w_\alpha = z_\alpha + n^{-1/2} \tfrac{1}{6}\gamma(z_\alpha^2 - 3) + o(n^{-1/2}).$$

If $\gamma > 0$ then as $\alpha \to 0$, $w_\alpha \to -\infty$ but

$$z_\alpha + n^{-1/2}\tfrac{1}{6}\gamma(z_\alpha^2 - 1) \longrightarrow +\infty\,;$$

and if $\gamma < 0$ then as $\alpha \to 1$, $w_\alpha \to +\infty$ but

$$z_\alpha + n^{-1/2}\tfrac{1}{6}\gamma(z_\alpha^2 - 1) \longrightarrow -\infty.$$

This makes it clear that the Cornish-Fisher expansion is not available uniformly in $0 < \alpha < 1$. However, under sufficiently stringent moment conditions it generally can be established uniformly in $n^{-c} \leq \alpha \leq 1 - n^{-c}$, for given $c > 0$.

2.6 Two Examples

In both examples, $W, W_1, W_2, \ldots$ denote independent and identically distributed random variables having mean m and variance $\beta^2 > 0$. Let $\overline{W} = n^{-1}\sum W_i$ denote the mean and $\hat{\beta}^2 = n^{-1}\sum(W_i - \overline{W})^2$ the variance of the sample $\{W_1, \ldots, W_n\}$.

Example 2.1: Estimating a Mean

In Section 2.2 we treated the estimation of $\theta_0 = m$ by $\hat{\theta} = \overline{W}$. The Edgeworth expansion (2.18) is a special case of expansion (2.35) from Theorem

2.2, with $A(x) = (x - m)/\beta$ for scalar x,

$$P\{n^{1/2}(\overline{W} - m)/\beta \leq x\} = \Phi(x) + n^{-1/2}p_1(x)\phi(x) + \ldots$$
$$+ n^{-j/2}p_j(x)\phi(x) + \ldots \quad (2.51)$$

The expansion is valid as an asymptotic series to j terms if $E(|W|^{j+2}) < \infty$ and

$$\limsup_{|t| \to \infty} |E \exp(itW)| < \infty,$$

the latter condition holding if the distribution of W is nonsingular.

Third and fourth cumulants of $(W - m)/\beta$, i.e., κ_3 and κ_4, are

$$\gamma = E(W - m)^3\beta^{-3} \quad \text{and} \quad \kappa = E(W - m)^4\beta^{-4} - 3,$$

respectively. Formulae for the polynomials p_1 and p_2 are given by (2.15) and (2.16),

$$p_1(x) = -\tfrac{1}{6}\gamma(x^2 - 1), \quad (2.52)$$

$$p_2(x) = -x\left\{\tfrac{1}{24}\kappa(x^2 - 3) + \tfrac{1}{72}\gamma^2(x^4 - 10x^2 + 15)\right\}. \quad (2.53)$$

As explained in Section 2.3, the above expansion of the distribution function of $n^{1/2}(\overline{W} - m)/\beta$ may be used to approximate coverage probabilities of confidence intervals for m, constructed when the population variance β^2 is known. However, when β is unknown, as is almost always the case in practice, we need instead an Edgeworth expansion of the distribution function of $n^{1/2}(\overline{W} - m)/\hat{\beta}$. Define

$$\mathbf{X} = (W, W^2)^{\mathrm{T}}, \qquad \boldsymbol{\mu} = E(\mathbf{X}) = (m, \beta^2 + m^2),$$

$$\mathbf{X}_i = (W_i, W_i^2)^{\mathrm{T}}, \qquad \overline{\mathbf{X}} = n^{-1}\sum\mathbf{X}_i, \qquad \mathbf{x} = (x^{(1)}, x^{(2)})^{\mathrm{T}},$$

and

$$A(\mathbf{x}) = A(x^{(1)}, x^{(2)}) = (x^{(1)} - m)\{x^{(2)} - (x^{(1)})^2\}^{-1/2}.$$

This is of the form $A(\mathbf{x}) = \{g(\mathbf{x}) - g(\boldsymbol{\mu})\}/h(\mathbf{x})$ discussed in Section 2.4, with $g(\mathbf{x}) = x^{(1)}$ and $h(\mathbf{x})^2 = x^{(2)} - (x^{(1)})^2$. Now, $A(\overline{\mathbf{X}}) = (\overline{W} - m)/\hat{\beta}$. According to Theorem 2.2,

$$P\{n^{1/2}(\overline{W} - m)/\hat{\beta} \leq x\} = \Phi(x) + n^{-1/2}q_1(x)\phi(x) + \ldots$$
$$+ n^{-j/2}q_j(x)\phi(x) + \ldots,$$

the above expansion being valid as an asymptotic series to j terms if $E(W^{2j+2}) < \infty$ and

$$\limsup_{|t^{(1)}|+|t^{(2)}| \to \infty} |E \exp(it^{(1)}W + it^{(2)}W^2)| < 1. \quad (2.54)$$

The latter condition holds if W has an absolutely continuous distribution, and the moment condition on W may be weakened to $E(|W|^{j+2}) < \infty$; see the end of Section 2.4 for a proof of the former result and notes at the end of this chapter for details of the latter.

We denote the polynomials here by q, to distinguish them from the p's appearing in (2.51). To derive q_1, assume without loss of generality that $E(W) = 0$ and $E(W^2) = 1$. Then in the notation of (2.31)–(2.33),

$$a_1 = 1, \quad a_2 = a_{11} = a_{22} = 0, \quad a_{12} = a_{21} = -\tfrac{1}{2},$$

$$\mu_{11} = 1, \quad \text{and} \quad \mu_{12} = \mu_{21} = \mu_{111} = \gamma.$$

Therefore by (2.31)–(2.33), $\sigma^2 = 1$ (as expected), $A_1 = -\tfrac{1}{2}\gamma$, and $A_2 = -2\gamma$, whence by (2.36),

$$q_1(x) = \tfrac{1}{6}\gamma(2x^2 + 1). \tag{2.55}$$

(Compare with the formula for p_1 at (2.52).)

The second polynomial q_2 could be derived in a similar way, by first developing the analogues of formulae (2.31)–(2.33). However, it is often simpler to obtain cumulant expansions such as (2.20) directly, making use of special properties of the statistic under investigation. To illustrate this approach, which involves a form of the *delta method*, which we shall describe in more detail in Section 2.7, observe that we may again assume without loss of generality that $E(W) = 0$ and $E(W^2) = 1$. Then, with $S_n = n^{1/2} A(\overline{\mathbf{X}})$,

$$S_n = n^{1/2}\overline{W}/\hat{\beta} = n^{1/2}\overline{W}\left\{1 + n^{-1}\sum_{i=1}^{n}(W_i^2 - 1) - \overline{W}^2\right\}^{-1/2}$$

$$= n^{1/2}\overline{W}\left[1 - \tfrac{1}{2}n^{-1}\sum_{i=1}^{n}(W_i^2 - 1) + \tfrac{1}{2}\overline{W}^2\right.$$

$$\left. + \tfrac{3}{8}\left\{n^{-1}\sum_{i=1}^{n}(W_i^2 - 1)\right\}^2 + O_p(n^{-3/2})\right]. \tag{2.56}$$

Hence[2]

$$E(S_n) = E\left[n^{1/2}\overline{W}\left\{1 - \tfrac{1}{2}n^{-1}\sum_{i=1}^{n}(W_i^2 - 1)\right\}\right] + O(n^{-1})$$

$$= -\tfrac{1}{2}n^{-1/2}\gamma + O(n^{-1}),$$

---

[2]To be strictly rigorous, we should be taking cumulants of a Taylor series approximant of S_n (see (2.29)) rather than of S_n itself.

$$E(S_n^2) = E\left(n\overline{W}^2\left[1 - n^{-1}\sum_{i=1}^n (W_i^2 - 1) + \tfrac{1}{2}\overline{W}^2\right.\right.$$

$$\left.\left. + \tfrac{3}{8}\left\{n^{-1}\sum_{i=1}^n (W_i^2 - 1)\right\}^2\right]\right) + O(n^{-3/2})$$

$$= 1 - n^{-1}(\kappa + 2) + 3n^{-1} + n^{-1}(2\gamma^2 + \kappa + 2) + O(n^{-3/2})$$

$$= 1 + n^{-1}(2\gamma^2 + 3) + O(n^{-3/2}),$$

$$E(S_n^3) = E\left[n^{3/2}\overline{W}^3\left\{1 - \tfrac{3}{2}n^{-1}\sum_{i=1}^n (W_i^2 - 1)\right\}\right] + O(n^{-1})$$

$$= -\tfrac{7}{2}n^{-1/2}\gamma + O(n^{-1}),$$

and

$$E(S_n^4) = E\left(n^2\overline{W}^4\left[1 - 2n^{-1}\sum_{i=1}^n (W_i^2 - 1) + 2\overline{W}^2\right.\right.$$

$$\left.\left. + 3\left\{n^{-1}\sum_{i=1}^n (W_i^2 - 1)\right\}^2\right]\right) + O(n^{-3/2})$$

$$= (3 + n^{-1}\kappa) - 2n^{-1}\{4\gamma^2 + 6(\kappa + 2)\} + 30n^{-1}$$

$$+ 3n^{-1}\{12\gamma^2 + 3(\kappa + 2)\} + O(n^{-3/2})$$

$$= 3 + n^{-1}(28\gamma^2 - 12\kappa + 24) + O(n^{-3/2}).$$

In fact, we know from Theorem 2.1 that the remainders in $E(S_n)$, $E(S_n^2)$, $E(S_n^3)$, and $E(S_n^4)$ are actually $O(n^{-3/2})$, $O(n^{-2})$, $O(n^{-3/2})$, and $O(n^{-2})$, respectively, each a factor of $n^{-1/2}$ smaller than those given above. Therefore, by (2.6),

$$\kappa_{1,n} = E(S_n) = -\tfrac{1}{2}n^{-1/2}\gamma + O(n^{-3/2}),$$

$$\kappa_{2,n} = E(S_n^2) - (ES_n)^2 = 1 + \tfrac{1}{4}n^{-1}(7\gamma^2 + 12) + O(n^{-2}),$$

$$\kappa_{3,n} = E(S_n^3) - 3E(S_n^2)E(S_n) + 2(ES_n)^3 = -2n^{-1/2}\gamma + O(n^{-3/2}),$$

and

$$\kappa_{4,n} = E(S_n^4) - 4E(S_n^3)E(S_n) - 3(ES_n^2)^2 + 12E(S_n^2)(ES_n)^2 - 6(ES_n)^4$$

$$= n^{-1}(12\gamma^2 - 2\kappa + 6) + O(n^{-2}).$$

Hence, in the notation of (2.20),

$$k_{1,2} = -\tfrac{1}{2}\gamma, \quad k_{2,2} = \tfrac{1}{4}(7\gamma^2 + 12), \quad k_{3,1} = -2\gamma, \quad k_{4,1} = 12\gamma^2 - 2\kappa + 6.$$

It now follows from (2.25) that

$$q_2(x) = x\left\{\tfrac{1}{12}\kappa(x^2 - 3) - \tfrac{1}{18}\gamma^2(x^4 + 2x^2 - 3) - \tfrac{1}{4}(x^2 + 3)\right\}. \quad (2.57)$$

Compare this with formula (2.53) for the polynomial p_2 in the non-Studentized case.

In the argument above we took the sample variance to be our estimate of population variance, regardless of context. While this is appropriate for the nonparametric bootstrap, and sometimes for the parametric bootstrap (e.g., estimation of the mean of a normal population), it is not universally applicable. For example, consider estimating the mean, m, of an Exponential population. The sample mean $\hat{m} = \overline{W}$ is the maximum likelihood estimate of both the population mean and the population standard deviation. Therefore the appropriate pivotal statistic is $n^{1/2}(\overline{W} - m)/\overline{W}$, not $n^{1/2}(\overline{W} - m)/\hat{\beta}$. Noting that "$n^{1/2}(\overline{W} - m)/\overline{W} \leq x$" is equivalent to

$$\text{``} n^{1/2}(\overline{W} - m)/m \ \leq \ x\,(1 + n^{-1/2}\,\beta^{-1}\,x)^{-1} \text{''} \,,$$

we may deduce that

$$P\{n^{1/2}(\overline{W} - m)/\overline{W} \leq x\} \ = \ \Phi(x) + \sum_{j\geq 1} n^{-j/2}\, q_j(x)\, \phi(x) \,,$$

where

$$q_1(x) \ = \ \tfrac{1}{3}\,(2x^2 + 1)\,, \qquad q_2(x) \ = \ -\tfrac{1}{36}\,(8x^4 - 11x^2 + 3)\,. \qquad (2.58)$$

Example 2.2: Estimating a Variance

Now we take the unknown parameter to be $\theta_0 = \beta^2$, the population variance. Let its estimate be $\hat{\theta} = \hat{\beta}^2$, the sample variance. The asymptotic variance of $\hat{\theta}$ is $n^{-1}\tau^2$, where

$$\tau^2 \ = \ E(W - m)^4 \ - \ \beta^4 \,.$$

Initially, define the 2-vectors $\mathbf{X}$, $\boldsymbol{\mu}$, $\mathbf{X}_i$, and $\overline{\mathbf{X}}$ exactly as in Example 2.1 (for example, $\mathbf{X} = (W, W^2)^{\mathrm{T}}$ and $\boldsymbol{\mu} = E(\mathbf{X})$), and put

$$A(\mathbf{x}) \ = \ \{x^{(2)} - (x^{(1)})^2 - \beta^2\}/\tau \,,$$

where $\mathbf{x} = (x^{(1)}, x^{(2)})^{\mathrm{T}}$. Then $A(\overline{\mathbf{X}}) = (\hat{\beta}^2 - \beta^2)/\tau$. Applying Theorem 2.2 to this function A we see that

$$P\{n^{1/2}(\hat{\beta}^2 - \beta^2)/\tau \leq x\} \ = \ \Phi(x) + n^{-1/2}p_1(x)\phi(x) + \cdots$$
$$+ \ n^{-j/2}p_j(x)\phi(x) + \cdots \,,$$

the above expansion being valid as an asymptotic series to j terms if $E(W^{2j+2}) < \infty$ and Cramér's condition (2.54) holds. In determining p_1 we may assume without loss of generality that $E(W) = 0$ and $E(W^2) = 1$. Then,

$$a_2 = \tau^{-1}\,, \quad a_1 = a_{22} = a_{12} = a_{21} = 0\,, \quad a_{11} = -2\tau^{-1}\,, \quad \mu_{11} = 1\,, \quad \mu_{22} = \tau^4\,,$$

$$\mu_{12} = \mu_{21} = \gamma = E(W - m)^3\beta^{-3}\,, \qquad \text{and} \qquad \mu_{222} = \lambda = E(w^2 - 1)^3\,.$$

Therefore by (2.31)–(2.33), $\sigma^2 = 1$, $A_1 = -\tau^{-1}$, and $A_2 = \tau^{-3}(\lambda - 6\gamma^2)$. Hence by (2.36),

$$p_1(x) = \tau^{-1} + \tau^{-3}\left(\gamma^2 - \tfrac{1}{6}\lambda\right)(x^2 - 1). \tag{2.59}$$

Care should be taken when interpreting τ and λ in this formula. Remember that in the last few lines we supposed $m = 0$ and $\beta^2 = 1$. If these assumptions are not valid then τ^2 and λ in (2.59) should be replaced by

$$(\tau^\dagger)^2 = E(W - m)^4 \beta^{-4} - 1 \quad \text{and} \quad \lambda^\dagger = E\{(W - m)^2\beta^{-2} - 1\}^3,$$

respectively.

In almost all cases of practical interest, τ^2 will be unknown and will have to be estimated. A natural estimate is

$$\hat{\tau}^2 = n^{-1}\sum_{i=1}^{n}(W_i - \overline{W})^4 - \hat{\beta}^4$$

$$= n^{-1}\sum_{i=1}^{n} W_i^4 - 4\overline{W}n^{-1}\sum_{i=1}^{n} W_i^3 + 6\overline{W}^2 n^{-1}\sum_{i=1}^{n} W_i^2 - 3\overline{W}^4 - \hat{\beta}^4.$$

Redefine

$$\mathbf{X} = (W, W^2, W^3, W^4)^{\mathrm{T}}, \qquad \boldsymbol{\mu} = E(\mathbf{X}), \qquad \mathbf{X}_i = (X_i, X_i^2, X_i^3, X_i^4)^{\mathrm{T}},$$

$$\overline{\mathbf{X}} = n^{-1}\sum_i \mathbf{X}_i, \qquad \text{and} \qquad \mathbf{x} = (x^{(1)}, x^{(2)}, x^{(3)}, x^{(4)})^{\mathrm{T}},$$

so that we are now working with 4-vectors rather than 2-vectors. Put

$$A(\mathbf{x}) = \{x^{(2)} - (x^{(1)})^2 - \beta^2\}\Big[x^{(4)} - 4x^{(1)}x^{(3)} + 6(x^{(1)})^2 x^{(2)}$$

$$- 3(x^{(1)})^4 - \{x^{(2)} - (x^{(1)})^2\}^2\Big]^{-1/2}$$

$$= \{x^{(2)} - (x^{(1)})^2 - \beta^2\}\Big\{x^{(4)} - (x^{(2)})^2 - 4x^{(1)}x^{(3)}$$

$$+ 8(x^{(1)})^2 x^{(2)} - 4(x^{(1)})^4\Big\}^{-1/2}.$$

This is of the form $A(\mathbf{x}) = \{g(\mathbf{x}) - g(\boldsymbol{\mu})\}/h(\mathbf{x})$ discussed in Section 2.4, with

$$g(\mathbf{x}) = x^{(2)} - (x^{(1)})^2,$$

$$h(\mathbf{x}) = x^{(4)} - 4x^{(1)}x^{(3)} + 6(x^{(1)})^2 x^{(2)} - 3(x^{(1)})^4 - \{x^{(2)} - (x^{(1)})^2\}^2.$$

In this notation, $A(\overline{\mathbf{X}}) = (\hat{\beta}^2 - \beta^2)/\hat{\tau}$. By Theorem 2.2,

$$P\{n^{1/2}(\hat{\beta}^2 - \beta^2)/\hat{\tau} \leq x\} = \Phi(x) + n^{-1/2}q_1(x)\phi(x) + \dots$$

$$+ n^{-j/2}q_j(x)\phi(x) + \dots,$$

where we have denoted the polynomials by q to distinguish them from their counterparts p in the non-Studentized case. The expansion is valid as an asymptotic series to j terms if $E(W^{4j+2}) < \infty$ and Cramér's condition holds for the 4-vector $\mathbf{X}$. The latter constraint holds if W has a nonsingular distribution; see the proof at the end of Section 2.4.

The polynomial q_1 may be determined from formula (2.36) and found to be

$$q_1(x) \; = \; -\left\{ C_1 + \tfrac{1}{6} C_2(x^2 - 1) \right\},$$

where, with $\nu_j = E\{(W - m)/\beta\}^j$,

$$C_1 \; = \; \tfrac{1}{2}(\nu_4 - 1)^{-3/2}\left(4\nu_3^2 + \nu_4 - \nu_6 \right),$$

and

$$C_2 \; = \; 2(\nu_4 - 1)^{-3/2}\left(3\nu_3^2 + 3\nu_4 - \nu_6 - 2 \right).$$

In the same notation the polynomial p_1 given at (2.59) is

$$p_1(x) \; = \; -\left\{ B_1 + \tfrac{1}{6} B_2(x^2 - 1) \right\},$$

where

$$B_1 \; = \; -(\nu_4 - 1)^{-1/2},$$

and

$$B_2 \; = \; (\nu_4 - 1)^{-3/2}\left(\nu_6 - 3\nu_4 - 6\nu_3^2 + 2 \right).$$

Therefore p_1 and q_1 are usually quite different.

2.7 The Delta Method

The delta method is a frequently-used device for calculating Edgeworth expansions. In a general form it amounts to the following rule: if S_n, T_n are two asymptotically Normal statistics that satisfy

$$S_n \; = \; T_n + O_p(n^{-j/2}) \qquad (2.60)$$

for an integer $j \geq 1$, then Edgeworth expansions of the distributions of S_n and T_n generally disagree only in terms of order $n^{-j/2}$ (or smaller), i.e.,

$$P(S_n \leq x) \; = \; P(T_n \leq x) + O(n^{-j/2}). \qquad (2.61)$$

Often S_n is a linear approximation to T_n, and some authors restrict use of the term "delta method" to this simple special case. However, we shall have cause to use more general forms of the rule.

In many cases of practical interest, validity of the delta method can be established rigorously as follows. Assume that the $O_p(n^{-j/2})$ term in (2.60) can be broken up into two parts: a relatively simple component

Δ_1 of order $n^{-j/2}$, and a remainder Δ_2 of order $n^{-(j+1)/2}$. Often Δ_1 is a relatively uncomplicated linear approximation to the $O_p(n^{-j/2})$ term, and then it is straightforward to prove (under suitable regularity conditions, usually moment and smoothness conditions) that Edgeworth expansions of T_n and $T_n + \Delta_1$ differ in terms of order $n^{-j/2}$, i.e.,

$$P(T_n \leq x) = P(T_n + \Delta_1 \leq x) + O(n^{-j/2}).$$

By making use of the fact that Δ_2 is an order of magnitude smaller than $n^{-j/2}$, it is usually a simple matter, despite the complexity of Δ_2, to show that

$$P(|\Delta_2| > n^{-j/2}) = O(n^{-j/2}).$$

Sometimes Markov's inequality suffices to establish this result. Therefore

$$
\begin{aligned}
P(S_n \leq x) &= P(T_n + \Delta_1 + \Delta_2 \leq x) \\
&\leq P(T_n + \Delta_1 \leq x + n^{-j/2}) + P(|\Delta_2| > n^{-j/2}) \\
&\leq P(T_n \leq x + n^{-j/2}) + O(n^{-j/2}) \\
&= P(T_n \leq x) + O(n^{-j/2}),
\end{aligned}
$$

the last identity following from an Edgeworth expansion of the distribution of T_n. Similarly,

$$P(S_n \leq x) \geq P(T_n \leq x) + O(n^{-j/2}),$$

and these two formulae give (2.61).

We used the delta method in Section 2.6 to derive early terms in an Edgeworth expansion of the distribution of the Studentized mean. In particular, formula (2.56) is equivalent to (2.60), with $j = 3$, $S_n = n^{1/2} \overline{W}/\hat{\beta}$, and

$$T_n = n^{1/2} \overline{W} \left[1 - \tfrac{1}{2} n^{-1} \sum_{i=1}^{n} (W_i^2 - 1) + \tfrac{1}{2} \overline{W}^2 + \tfrac{3}{8} \left\{ n^{-1} \sum_{i=1}^{n} (W_i^2 - 1) \right\}^2 \right].$$

2.8 Edgeworth Expansions for Probability Densities

If the distribution function of an absolutely continuous random variable admits an Edgeworth expansion then we might expect that an expansion of the density function would be obtainable by differentiating the distribution expansion. For example, if the distribution function of S_n allows the formula

$$
\begin{aligned}
F_n(x) &= P(S_n \leq x) \\
&= \Phi(x) + n^{-1/2} p_1(x) \phi(x) + \ldots + n^{-j/2} p_j(x) \phi(x) + \ldots,
\end{aligned}
$$

then it is natural to ask whether $f_n(x) = F_n'(x)$ might be expressed as

$$f_n(x) = \phi(x) + n^{-1/2} r_1(x) \phi(x) + \ldots + n^{-j/2} r_j(x) \phi(x) + \ldots , \quad (2.62)$$

where $r_j(x) = \phi(x)^{-1} (d/dx) \{p_j(x) \phi(x)\}$. If p_j is a polynomial of degree $3j - 1$, of opposite parity to j, then r_j is a polynomial of degree $3j$, of the same parity as j. Of course, (2.62) should be interpreted as an asymptotic series, not as a convergent infinite series. That is to say, the remainder term in (2.62) is of smaller order than the last included term.

We call (2.62) an Edgeworth expansion of the density of S_n. It is usually available under mild additional assumptions on the smoothness of the underlying sampling distribution. For example, we have the following analogue of Theorem 2.2. Let $\mathbf{X}, \mathbf{X}_1, \mathbf{X}_2, \ldots$ be independent and identically distributed random variables with finite variance and mean μ; let $A = \mathbb{R}^d \rightarrow \mathbb{R}$ denote a smooth function satisfying $A(\mu) = 0$; define σ^2 as in (2.31); put $\overline{\mathbf{X}} = n^{-1} \sum_{i \le n} \mathbf{X}_i$ and $S_n = n^{1/2} A(\overline{\mathbf{X}})/\sigma$; and write f_n for the density of S_n.

THEOREM 2.5. *Assume that the function A has $j+2$ continuous derivatives in a neighbourhood of μ and satisfies*

$$\sup_{\substack{-\infty < x < \infty \\ n \ge 1}} \int_{\mathcal{S}(n,x)} (1 + \|\mathbf{y}\|)^{-(j+2)} \, d\mathbf{y} \; < \; \infty , \quad (2.63)$$

where $\mathcal{S}(n, x) = \{\mathbf{y} \in \mathbb{R}^d : n^{1/2} A(\mu + n^{-1/2}\mathbf{y})/\sigma = x\}$. Assume that $E(\|\mathbf{X}\|^{j+2}) < \infty$ and that the characteristic function χ of $\mathbf{X}$ satisfies

$$\int_{\mathbb{R}^d} |\chi(\mathbf{t})|^c \, d\mathbf{t} \; < \; \infty \quad (2.64)$$

for some $c \ge 1$. Then

$$f_n(x) = \phi(x) + n^{-1/2} r_1(x) \phi(x) + \ldots$$
$$+ n^{-j/2} r_j(x) \phi(x) + o(n^{-j/2}) \quad (2.65)$$

uniformly in x, where $r_j(x) = \phi(x)^{-1} (d/dx) \{p_j(x) \phi(x)\}$ and p_j is as in the Edgeworth expansion (2.35).

Condition (2.64) replaces Cramér's condition (2.34) in Theorem 2.2. It may be shown that (2.64) is equivalent to the hypothesis that for some $n \ge 1$, $\overline{\mathbf{X}}$ has a bounded density. The latter condition implies that for some n, $\overline{\mathbf{X}}$ is absolutely continuous, and this leads immediately to (2.34). Therefore (2.64) is stronger than Cramér's condition. The assumption (2.63) may usually be checked very easily, noting that A is smooth and satisfies $A(\mu) = 0$.

A proof of (2.65) follows directly from an Edgeworth expansion of the density g_n of $n^{1/2}(\overline{\mathbf{X}} - \boldsymbol{\mu})$, as follows. It may be shown that under the conditions of Theorem 2.5, the d-variate density g_n of $\mathbf{Z}$ satisfies

$$\sup_{\mathbf{y} \in \mathbb{R}^d} (1 + \|\mathbf{y}\|)^{j+2} \left| g_n(\mathbf{y}) - \sum_{k=0}^{j} n^{-k/2} \pi_k(\mathbf{y}) \phi_{\mathbf{0},\boldsymbol{\Sigma}}(\mathbf{y}) \right| = o(n^{-j/2}), \quad (2.66)$$

where π_k is a d-variate polynomial of degree $3j$ with the same parity as j, $\pi_0 \equiv 1$, $\boldsymbol{\Sigma} = \text{var}(\mathbf{X})$, and $\phi_{\mathbf{0},\boldsymbol{\Sigma}}$ denotes the d-variate $N(\mathbf{0}, \boldsymbol{\Sigma})$ density. See Bhattacharya and Rao (1976, p. 192). The density f_n of $n^{1/2} A(\overline{\mathbf{X}})$ is given by

$$\begin{aligned} f_n(x) &= \int_{\mathcal{S}(n,x)} g_n(\mathbf{y}) \, d\mathbf{y} \\ &= \sum_{k=0}^{j} n^{-k/2} \int_{\mathcal{S}(n,x)} \pi_k(\mathbf{y}) \phi_{\mathbf{0},\boldsymbol{\Sigma}}(\mathbf{y}) \, d\mathbf{y} + R_n(x), \quad (2.67) \end{aligned}$$

where, in view of (2.63) and (2.66),

$$\sup_{n,x} |R_n(x)| = o(n^{-j/2}).$$

By Taylor-expanding the integral in (2.67) we may show that

$$\left| \sum_{k=0}^{j} n^{-k/2} \int_{\mathcal{S}(n,x)} \pi_k(\mathbf{y}) \phi_{\mathbf{0},\boldsymbol{\Sigma}}(\mathbf{y}) \, d\mathbf{y} - \sum_{k=0}^{j} n^{-k/2} r_j(x) \phi(x) \right| = o(n^{-j/2}),$$

from which follows (2.65).

Edgeworth expansions of the densities of multivariate statistics may be derived by similar arguments. They will be considered briefly in Section 4.2.

2.9 Bibliographical Notes

Chebyshev (1890) and Edgeworth (1896, 1905) conceived the idea of expanding a distribution function in what is now known as an Edgeworth expansion. They discussed formal construction of the expansion in the case of sums of independent random variables. Cramér (1928; 1946, Chapter VII) developed a detailed and rigorous theory for that circumstance. Formal expansions in other cases were the subject of much statistical work for many years (e.g., Geary 1936; Gayen 1949; Wallace 1958) before they were made rigorous, largely by work of Chibishov (1972, 1973a, 1973b), Sargan (1975, 1976), Bhattacharya and Ghosh (1978), and contemporaries. Hsu (1945) was a pioneer in this area. Wallace (1958) and Bickel (1974) have surveyed the subject of Edgeworth expansions and their applications. Barndorff-Nielsen and Cox (1989, Chapter 4) give an excellent account of the formal theory of univariate Edgeworth expansions. Cox and Reid (1987) discuss the relationship between stochastic expansions and deterministic expansions.

Edgeworth expansions may be viewed as rearrangements of Gram-Charlier or Bruns-Charlier expansions (Edgeworth 1907; Cramér 1970, p. 86ff). However, Gram-Charlier series do not enjoy the relatively good convergence properties of their Edgeworth counterparts (Cramér 1946, p. 221ff), and are seldom used today. The relationship between Edgeworth expansions and Gram-Charlier expansions is described in detail by Johnson and Kotz (1970, p. 17ff).

Formal determination of terms in an Edgeworth expansion is a relatively straightforward if tedious matter, once the basic cumulant expansion (2.20) has been recognized. That formula, and results such as Theorem 2.1, may be traced back to Fisher (1928) although without a comprehensive proof. James (1955, 1958), Leonov and Shiryaev (1959), and James and Mayne (1962) provide more thorough arguments. See also Kendall and Stuart (1977, Chapters 12 and 13) and McCullagh (1987, Chapter 3). McCullagh (1987, pp. 80, 81) recounts the history of Theorem 2.1 and related results. Barndorff-Nielsen and Cox (1979) discuss the topics of Edgeworth and saddle-point approximation. Draper and Tierney (1973) give formulae for high-order terms in Edgeworth and Cornish-Fisher expansions.

Petrov (1975, Chapter VI) provides a comprehensive account of Edgeworth expansions for distributions of sums of independent random variables. Petrov's Chapter VI, Section 1 is particularly useful as an introduction to the structure of polynomials appearing in expansions. The reader is referred to Magnus, Oberhettinger, and Soni (1966, Section 5.6) and Abramowitz and Stegun (1972, Chapter 22) for details of Hermite polynomials.

A difficulty with work on Edgeworth expansion is that algebra can become extremely tedious and notation unwieldy. Some of the former difficulties are alleviated by Withers (1983, 1984), who gives general formulae for early terms in models which include that treated in Section 2.4. Tensor notation is sometimes used to reduce notational complexity. It was introduced to this context by Kaplan (1952), and is described in detail by Speed (1983) and McCullagh (1984; 1987, Chapter 3). It is employed in modified form during our proof of Theorem 2.1. We have given a classical proof based on combinatorial analysis, rather than a proof where tensor notation and lattice theory support much of the argument, since we shall have no further use for tensor and lattice methods and since the former proof "from first principles" better illustrates the properties on which Theorem 2.1 is based.

Cramér's continuity condition is central to much of the theory, even in the relatively simple case of sums of independent random variables. However, one-term expansions for sums with remainder $o(n^{-1/2})$ are valid under the weaker smoothness assumption of nonlatticeness. That this is so was first proved by Esseen (1945) in his pathbreaking work on convergence rates. For lattice-valued random variables, additional terms of all orders must be added to take account of errors in approximating to discrete distributions by smooth distributions. Esseen (1945) contains a slip concerning the form of the first correction term, but Gnedenko and Kolmogorov (1954, Section 43) are reliable there. Bhattacharya and Rao (1976, Chapter 5) give a comprehensive account of Edgeworth expansions for sums of independent lattice-valued random variables and random vectors.

The case of sums of vector-valued random variables is treated in detail by Bhattacharya and Rao (1976); see also Chambers (1967), Sargan (1976), Phillips (1977), and Barndorff-Nielsen and Cox (1989, Chapter 6). Multivariate expansions form the basis for Bhattacharya and Ghosh's (1978) rigorous treatment of Edgeworth expansions for general statistics. This approach yields a unified treatment but does not always provide results under minimal moment conditions. Our Theorem 2.2 is a case in point. When applied to the Studentized mean it gives an expansion up to and including terms of order $n^{-j/2}$, assuming $(2j + 2)$th moments on the underlying distribution. A more refined argument, making extensive use of special properties of the Studentized mean, shows that $(j + 2)$th moments suffice (Chibishov 1984, Hall 1987c, Bhattacharya and Ghosh 1989).

The inverse Edgeworth expansions described in Section 2.5 were first studied by Cornish and Fisher (1937) and Fisher and Cornish (1960). See also Finney (1963). Kendall and Stuart (1977, Section 16.21), Johnson and Kotz (1970, p. 33ff), and Barndorff-Nielsen and Cox (1989, Section 4.4) give very accessible accounts.

Bhattacharya and Rao (1976, p. 189ff) provide a detailed account of Edgeworth expansions for sums of independent random variables. Their Theorem 19.1 points out that condition (2.64) in our Theorem 2.5 is equivalent to each of the following conditions:

for some $n \geq 1$, $\overline{\mathbf{X}}$ has a bounded density;

$f_n(x) \to \phi(x)$ uniformly in x as $n \to \infty$.

3

An Edgeworth View of the Bootstrap

3.1 Introduction

Our purpose in this chapter is to bring together salient parts of the technology described in Chapters 1 and 2, with the aim of explaining properties of bootstrap methods for estimating distributions and constructing confidence intervals. We shall emphasize the role played by pivotal methods, introduced in Section 1.3 (Example 1.2).

Recall that a statistic is (asymptotically) pivotal if its limiting distribution does not depend on unknown quantities. In several respects the bootstrap does a better job of estimating the distribution of a pivotal statistic than it does for a nonpivotal statistic. The advantages of pivoting can be explained very easily by means of Edgeworth expansion, as follows. If a pivotal statistic T is asymptotically Normally distributed, then in regular cases we may expand its distribution function as

$$G(x) = P(T \leq x) = \Phi(x) + n^{-1/2} q(x) \phi(x) + O(n^{-1}), \qquad (3.1)$$

where q is an even quadratic polynomial and Φ, ϕ are the Standard Normal distribution, density functions respectively. See Sections 2.3 and 2.4. As an example we might take $T = n^{1/2}(\hat{\theta} - \theta)/\hat{\sigma}$, where $\hat{\theta}$ is an estimate of the unknown parameter θ_0, and $\hat{\sigma}^2$ is an estimate of the asymptotic variance σ^2 of $n^{1/2}\hat{\theta}$. The bootstrap estimate of G admits an analogous expansion,

$$\widehat{G}(x) = P(T^* \leq x \mid \mathcal{X}) = \Phi(x) + n^{-1/2}\hat{q}(x)\phi(x) + O_p(n^{-1}), \quad (3.2)$$

where T^* is the bootstrap version of T, computed from a resample $\mathcal{X}^*$ instead of the sample $\mathcal{X}$, and the polynomial $\hat{q}$ is obtained from q on replacing unknowns, such as skewness, by bootstrap estimates. (The notation "$O_p(n^{-1})$" denotes a random variable that is of order n^{-1} "in probability"; see pp. xii and 88 for definitions.) The distribution of T^* conditional on $\mathcal{X}$ is called the bootstrap distribution of T^*. Section 3.3 will develop expansions such as (3.2). The estimates in the coefficients of $\hat{q}$ are typically distant $O_p(n^{-1/2})$ from their respective values in q, and so $\hat{q} - q = O_p(n^{-1/2})$.

Therefore, subtracting (3.1) and (3.2), we conclude that

$$P(T^* \leq x \mid \mathcal{X}) - P(T \leq x) = O_p(n^{-1}).$$

That is, the bootstrap approximation to G is in error by only n^{-1}. This is a substantial improvement on the Normal approximation, $G \simeq \Phi$, which by (3.1) is in error by $n^{-1/2}$.

On the other hand, were we to use the bootstrap to approximate the distribution of a nonpivotal statistic U, such as $U = n^{1/2}(\hat{\theta} - \theta_0)$, we would typically commit an error of size $n^{-1/2}$ rather than n^{-1}. To appreciate why, observe that the analogues of (3.1) and (3.2) in this case are

$$
\begin{aligned}
H(x) &= P(U \leq x) \\
&= \Phi(x/\sigma) + n^{-1/2}p(x/\sigma)\,\phi(x/\sigma) + O(n^{-1})\,, \\
\widehat{H}(x) &= P(U^* \leq x \mid \mathcal{X}) \\
&= \Phi(x/\hat{\sigma}) + n^{-1/2}\hat{p}(x/\hat{\sigma})\phi(x/\hat{\sigma}) + O_p(n^{-1})
\end{aligned}
\tag{3.3}
$$

respectively, where p is a polynomial, $\hat{p}$ is obtained from p on replacing unknowns by their bootstrap estimates, σ^2 equals the asymptotic variance of U, $\hat{\sigma}^2$ is the bootstrap estimate of σ^2, and U^* is the bootstrap version of U. Again, $\hat{p} - p = O_p(n^{-1/2})$, and also $\hat{\sigma} - \sigma = O_p(n^{-1/2})$, whence

$$\widehat{H}(x) - H(x) = \Phi(x/\hat{\sigma}) - \Phi(x/\sigma) + O_p(n^{-1}). \tag{3.4}$$

Now, the difference between $\hat{\sigma}$ and σ is usually of precise order $n^{-1/2}$. Indeed, $n^{1/2}(\hat{\sigma} - \sigma)$ typically has a limiting Normal $N(0, \zeta^2)$ distribution, for some $\zeta > 0$. Thus, $\Phi(x/\hat{\sigma}) - \Phi(x/\sigma)$ is generally of size $n^{-1/2}$, not n^{-1}. Hence by (3.4), the bootstrap approximation to H is in error by terms of size $n^{-1/2}$, not n^{-1}. This relatively poor performance is due to the presence of σ in the limiting distribution function $\Phi(x/\sigma)$, i.e., to the fact that U is not pivotal.

In the present chapter we elaborate on and extend this simple argument. Following the introduction of notation in Section 3.2, we describe Edgeworth expansions and Cornish-Fisher expansions for bootstrap distributions (Section 3.3), properties of quantiles and critical points of bootstrap distributions (Section 3.4), coverage accuracy of bootstrap confidence intervals (Section 3.5), symmetric confidence intervals and short confidence intervals (Sections 3.6 and 3.7), explicit corrections for skewness and for higher-order effects (Section 3.8), the use of skewness-reducing transformations (Section 3.9), bias correction of bootstrap confidence intervals (Section 3.10), and bootstrap iteration (Section 3.11). Finally, Section 3.12 treats bootstrap methods for hypothesis testing.

The present chapter unites the two main streams in this monograph, the bootstrap and Edgeworth expansion. The same theme, of using Edgeworth expansion methods to elucidate properties of the bootstrap, will also be present in Chapter 4. In this respect the present contribution is a primer, a foretaste of things to come in the areas of regression, curve estimation, and hypothesis testing.

3.2 Different Types of Confidence Interval

Let F_0 denote the population distribution function, $\theta(\cdot)$ a functional of distribution functions, and $\theta_0 = \theta(F_0)$ a true parameter value, such as the rth power of a population mean $\theta_0 = \{\int x \, dF_0(x)\}^r$. Write F_1 for the distribution function "of a sample" $\mathcal{X}$ drawn from the population. Interpretations of F_1 differ for parametric and nonparametric cases; see Section 1.2. In a nonparametric setting F_1 is just the empirical distribution function of $\mathcal{X}$; in a parametric context where $F_0 = F_{(\lambda_0)}$ and λ_0 is a vector of unknown parameters, we estimate λ_0 by $\widehat{\lambda}$ (e.g., by using maximum likelihood) and take $F_1 = F_{(\widehat{\lambda})}$. The bootstrap estimate of θ_0 is $\hat{\theta} = \theta(F_1)$. Define F_2 to be the distribution function of a resample $\mathcal{X}^*$ drawn from the "population" with distribution function F_1. Again, F_2 is different in parametric and nonparametric settings. The reader is referred to Section 1.2 for details.

A theoretical α-level percentile confidence interval for θ_0,

$$I_1 = (-\infty, \hat{\theta} + t_0),$$

is obtained by solving the population equation (1.1) for $t = t_0$, using the function

$$f_t(F_0, F_1) = I\{\theta(F_0) \le \theta(F_1) + t\} - \alpha.$$

Thus, t_0 is defined by

$$P(\theta \le \hat{\theta} + t_0) = \alpha.$$

The bootstrap version of this interval is $\widehat{I}_1 = (-\infty, \hat{\theta} + \hat{t}_0)$, where $t = \hat{t}_0$ is the solution of the sample equation (1.5). Equivalently, $\hat{t}_0$ is given by

$$P\{\theta(F_1) \le \theta(F_2) + \hat{t}_0 \mid F_1\} = \alpha.$$

We call $\widehat{I}_1$ a bootstrap percentile confidence interval, or simply a percentile interval.

To construct a percentile-t confidence interval for θ_0, define $\sigma^2(F_0)$ to be the asymptotic variance of $n^{1/2} \hat{\theta}$, and put $\hat{\sigma}^2 = \sigma^2(F_1)$. A theoretical α-level percentile-t confidence interval is $J_1 = (-\infty, \hat{\theta} + t_0 \hat{\sigma})$, where on the present occasion t_0 is given by

$$P(\theta_0 \le \hat{\theta} + t_0 \hat{\sigma}) = \alpha.$$

This is equivalent to solving the population equation (1.1) with

$$f_t(F_0, F_1) = I\{\theta(F_0) \leq \theta(F_1) + t\sigma(F_1)\} - \alpha.$$

The bootstrap interval is obtained by solving the corresponding sample equation, and is $\widehat{J}_1 = (-\infty, \ \hat{\theta} + \hat{t}_0 \hat{\sigma})$, where $\hat{t}_0$ is now defined by

$$P\{\hat{\theta} \leq \theta(F_2) + \hat{t}_0 \, \sigma(F_2) \mid F_1\} = \alpha.$$

To simplify notation in future sections we shall often denote $\theta(F_2)$ and $\sigma(F_2)$ by $\hat{\theta}^*$ and $\hat{\sigma}^*$, respectively.

Exposition will be clearer if we represent t_0 and $\hat{t}_0$ in terms of quantiles. Thus, we define u_α, v_α, $\hat{u}_\alpha$, and $\hat{v}_\alpha$ by

$$P\big[n^{1/2}\{\theta(F_1) - \theta(F_0)\}/\sigma(F_0) \leq u_\alpha \mid F_0\big] = \alpha, \qquad (3.5)$$

$$P\big[n^{1/2}\{\theta(F_1) - \theta(F_0)\}/\sigma(F_1) \leq v_\alpha \mid F_0\big] = \alpha, \qquad (3.6)$$

$$P\big[n^{1/2}\{\theta(F_2) - \theta(F_1)\}/\sigma(F_1) \leq \hat{u}_\alpha \mid F_1\big] = \alpha, \qquad (3.7)$$

and

$$P\big[n^{1/2}\{\theta(F_2) - \theta(F_1)\}/\sigma(F_2) \leq \hat{v}_\alpha \mid F_1\big] = \alpha. \qquad (3.8)$$

Write $\sigma = \sigma(F_0)$ and $\hat{\sigma} = \sigma(F_1)$. Then definitions of $I_1, J_1, \widehat{I}_1$, and $\widehat{J}_1$ equivalent to those given earlier are

$$I_1 = (-\infty, \ \hat{\theta} - n^{-1/2}\sigma\, u_{1-\alpha}), \quad J_1 = (-\infty, \ \hat{\theta} - n^{-1/2}\hat{\sigma} v_{1-\alpha}), \qquad (3.9)$$

$$\widehat{I}_1 = (-\infty, \ \hat{\theta} - n^{-1/2}\hat{\sigma}\hat{u}_{1-\alpha}), \quad \widehat{J}_1 = (-\infty, \ \hat{\theta} - n^{-1/2}\hat{\sigma}\hat{v}_{1-\alpha}). \qquad (3.10)$$

All are confidence intervals for θ_0, with coverage probabilities approximately equal to α. Additional confidence intervals $\widehat{I}_{11}$ and $\widehat{I}_{12}$ will be introduced in Section 3.4. The intervals $\widehat{I}_1, \widehat{I}_{11}$, and $\widehat{I}_{12}$ are of the percentile type, whereas $\widehat{J}_1$ is a percentile-t interval.

In the nonparametric case the statistic $\theta(F_2)$, conditional on F_1, has a discrete distribution. This means that equations (3.7) and (3.8) will usually not have exact solutions, although as we point out in Section 1.3 and Appendix I, the errors due to discreteness are exponentially small functions of n. The reader concerned by the problem of discreteness might like to define $\hat{u}_\alpha$ and $\hat{v}_\alpha$ by

$$\hat{u}_\alpha = \inf\Big\{u : P\big[n^{1/2}\{\theta(F_2) - \theta(F_1)\}/\sigma(F_1) \leq u \mid F_1\big] \geq \alpha\Big\}$$

and

$$\hat{v}_\alpha = \inf\Big\{v : P\big[n^{1/2}\{\theta(F_2) - \theta(F_1)\}/\sigma(F_2) \leq v \mid F_1\big] \geq \alpha\Big\}.$$

Two-sided, equal-tailed confidence intervals are constructed by forming the intersection of two one-sided intervals. Two-sided analogues of I_1 and

J_1 are

$$I_2 = \left(\hat{\theta} - n^{-1/2} \sigma\, u_{(1+\alpha)/2} \,,\; \hat{\theta} - n^{-1/2} \sigma\, u_{(1-\alpha)/2} \right)$$

and

$$J_2 = \left(\hat{\theta} - n^{-1/2} \hat{\sigma} v_{(1+\alpha)/2} \,,\; \hat{\theta} - n^{-1/2} \hat{\sigma} v_{(1-\alpha)/2} \right)$$

respectively, with bootstrap versions

$$\widehat{I}_2 = \left(\hat{\theta} - n^{-1/2} \hat{\sigma} \hat{u}_{(1+\alpha)/2} \,,\; \hat{\theta} - n^{-1/2} \hat{\sigma} \hat{u}_{(1-\alpha)/2} \right)$$

and

$$\widehat{J}_2 = \left(\hat{\theta} - n^{-1/2} \hat{\sigma} \hat{v}_{(1+\alpha)/2} \,,\; \hat{\theta} - n^{-1/2} \hat{\sigma} \hat{v}_{(1-\alpha)/2} \right).$$

The intervals I_2 and J_2 have equal probability in each tail; for example,

$$P\left(\theta_0 \le \hat{\theta} - n^{-1/2} \sigma\, u_{(1+\alpha)/2} \right) = P\left(\theta_0 > \hat{\theta} - n^{-1/2} \sigma\, u_{(1-\alpha)/2} \right) = \tfrac{1}{2}(1-\alpha).$$

Intervals $\widehat{I}_2$ and $\widehat{J}_2$ have approximately the same level of probability in each tail.

Two-sided symmetric confidence intervals are best introduced separately, and will be studied in Section 3.6. The reader will by now have guessed that our notation for confidence intervals conveys pertinent information about interval type: intervals denoted by I are constructed by the percentile method; intervals denoted by J are percentile-t; and the first index in a numerical subscript (1 or 2) on I or J indicates the number of sides to the interval. Unusual intervals, such as symmetric intervals or short intervals, will be designated by alphabetical rather than numerical indices (J_{SY} for symmetric, J_{SH} for short). A circumflex on I or J indicates the bootstrap version of the interval, with u_α or v_α replaced by $\hat{u}_\alpha$ or $\hat{v}_\alpha$, respectively. The symbols $\mathcal{I}_1$ and $\mathcal{I}_2$ denote generic one- and two-sided intervals, respectively. When there can be no ambiguity we abbreviate $\mathcal{I}_1$ and $\mathcal{I}_2$ to $\mathcal{I}$.

All the intervals defined above have at least asymptotic coverage α, in the sense that if $\mathcal{I}$ is any one of the intervals,

$$P(\theta_0 \in \mathcal{I}) \longrightarrow \alpha$$

as $n \to \infty$. Outlines of proofs are given in Section 3.5. We call α the *nominal coverage* of the confidence interval $\mathcal{I}$. The *coverage error* of $\mathcal{I}$ is the difference between true coverage and nominal coverage,

$$\text{coverage error} = P(\theta_0 \in \mathcal{I}) - \alpha.$$

We close this section by reminding the reader of some standard notation. Let $\{\delta_n,\ n \ge 1\}$ be a sequence of positive numbers. A random variable

Y_n is said to be of order δ_n in probability, written $Y_n = O_p(\delta_n)$, if the sequence $\{Y_n/\delta_n, \ n \geq 1\}$ is tight, that is, if

$$\lim_{\lambda \to \infty} \limsup_{n \to \infty} P(|Y_n/\delta_n| > \lambda) = 0.$$

Tightness is usually indicated by writing $Y_n/\delta_n = O_p(1)$. We say that $Y_n = O(\delta_n)$ almost surely if $|Y_n/\delta_n|$ is bounded with probability one; that $Y_n = o_p(\delta_n)$ if $Y_n/\delta_n \to O$ in probability; and that $Y_n = o(\delta_n)$ almost surely if $Y_n/\delta_n \to 0$ almost surely. Should Y_n be a function of x, any one of these statements is said to apply to $Y_n(x)$ uniformly in x if it applies to $\sup_x |Y_n(x)|$. For example, "$Y_n(x) = O_p(\delta_n)$ uniformly in x" means that

$$\sup_x |Y_n(x)|/\delta_n = O_p(1).$$

3.3 Edgeworth and Cornish-Fisher Expansions

Let

$$P\{n^{1/2}(\hat{\theta} - \theta_0)/\sigma \leq x\} = \Phi(x) + n^{-1/2}p_1(x)\phi(x) + \cdots$$
$$+ n^{-j/2}p_j(x)\phi(x) + \cdots \quad (3.11)$$

and

$$P\{n^{1/2}(\hat{\theta} - \theta_0)/\hat{\sigma} \leq x\} = \Phi(x) + n^{-1/2}q_1(x)\phi(x) + \cdots$$
$$+ n^{-j/2}q_j(x)\phi(x) + \cdots \quad (3.12)$$

denote Edgeworth expansions for non-Studentized and Studentized statistics, respectively, discussed in Chapter 2. Here, p_j and q_j are polynomials of degree at most $3j - 1$ and are odd or even functions according to whether j is even or odd. Inverting these formulae to obtain Cornish-Fisher expansions, as described in Section 2.5, we see that the quantiles u_α and v_α defined in (3.5) and (3.6) admit the expressions

$$u_\alpha = z_\alpha + n^{-1/2}p_{11}(z_\alpha) + n^{-1}p_{21}(z_\alpha) + \cdots$$
$$+ n^{-j/2}p_{j1}(z_\alpha) + \cdots \quad (3.13)$$

and

$$v_\alpha = z_\alpha + n^{-1/2}q_{11}(z_\alpha) + n^{-1}q_{21}(z_\alpha) + \cdots$$
$$+ n^{-j/2}q_{j1}(z_\alpha) + \cdots, \quad (3.14)$$

where z_α is the solution of $\Phi(z_\alpha) = \alpha$ and p_{j1} and q_{j1} are polynomials defined in terms of the p_j's and q_j's by equations such as (2.49). Both p_{j1}

and q_{j1} are of degree at most $j + 1$ and are odd functions for even j and even functions for odd j. We proved in Section 2.5 that

$$p_{11}(x) = -p_1(x), \quad p_{21}(x) = p_1(x)p_1'(x) - \tfrac{1}{2}xp_1(x)^2 - p_2(x), \quad (3.15)$$

$$q_{11}(x) = -q_1(x), \quad q_{21}(x) = q_1(x)q_1'(x) - \tfrac{1}{2}xq_1(x)^2 - q_2(x); \quad (3.16)$$

see (2.50).

As shown in Sections 2.3 and 2.4, coefficients of polynomials p_j and q_j (and hence also of p_{j1} and q_{j1}) depend on cumulants of $(\hat{\theta} - \theta_0)/\sigma$ and $(\hat{\theta} - \theta_0)/\hat{\sigma}$, respectively. Under the "smooth function model" described in Section 2.4, where $\hat{\theta}$ and $\hat{\sigma}$ are smooth functions of the mean of a d-variate sample $\mathcal{X} = \{\mathbf{X}_1, \ldots, \mathbf{X}_n\}$, those coefficients are polynomials in moments of the vector $\mathbf{X}$. When we pass from the "population-based" expansions (3.11) and (3.12) to their "sample-based" versions, the role of the population distribution function F_0 is replaced by that of the sample distribution function F_1, and so population moments in coefficients of the p_j's and q_j's change to sample moments. Let $\hat{p}_j$ and $\hat{q}_j$ denote those polynomials obtained from p_j and q_j on replacing population moments by sample moments, and put $\hat{\theta}^* = \theta(F_2)$ and $\hat{\sigma}^* = \sigma(F_2)$. The sample versions of expansions (3.11) and (3.12) are then

$$P\{n^{1/2}(\hat{\theta}^* - \hat{\theta})/\hat{\sigma} \leq x \,|\, \mathcal{X}\}$$
$$= \Phi(x) + n^{-1/2}\hat{p}_1(x)\phi(x) + \ldots + n^{-j/2}\hat{p}_j(x)\phi(x) + \ldots \quad (3.17)$$

and

$$P\{n^{1/2}(\hat{\theta}^* - \hat{\theta})/\hat{\sigma}^* \leq x \,|\, \mathcal{X}\}$$
$$= \Phi(x) + n^{-1/2}\hat{q}_1(x)\phi(x) + \ldots + n^{-j/2}\hat{q}_j(x)\phi(x) + \ldots, \quad (3.18)$$

respectively. Formally, these expansions are obtained from (3.11) and (3.12) on replacing the pair (F_0, F_1) by (F_1, F_2). Rigorous proofs of (3.17) and (3.18), under explicit regularity conditions, are given in Section 5.2.

Both expansions are intended to be interpreted as asymptotic series. For example, under appropriate moment and smoothness assumptions,

$$P\{n^{1/2}(\hat{\theta}^* - \hat{\theta})/\hat{\sigma} \leq x \,|\, \mathcal{X}\}$$
$$= \Phi(x) + n^{-1/2}\hat{p}_1(x)\phi(x) + \ldots$$
$$+ n^{-j/2}\hat{p}_j(x)\phi(x) + o_p(n^{-j/2}). \quad (3.19)$$

Furthermore, the remainder term $o_p(n^{-j/2})$ is of that size uniformly in x. In fact, under appropriate regularity conditions it equals $o(n^{-j/2})$ almost surely, uniformly in x:

$$n^{j/2} \sup_{-\infty < x < \infty} \left| P\{n^{1/2}(\hat{\theta}^* - \hat{\theta})/\hat{\sigma} \le x \,|\, \mathcal{X}\} \right.$$
$$\left. - \{\Phi(x) + n^{-1/2}\hat{p}_1(x)\phi(x) + \ldots + n^{-j/2}\hat{p}_j(x)\phi(x)\} \right| \longrightarrow 0$$

almost surely as $n \to \infty$. Results of this type will be given detailed proofs in Chapter 5.

If we invert the Edgeworth expansions (3.17) and (3.18) we obtain sample versions of the Cornish-Fisher expansions (3.9) and (3.10),

$$\hat{u}_\alpha = z_\alpha + n^{-1/2}\hat{p}_{11}(z_\alpha) + n^{-1}\hat{p}_{21}(z_\alpha) + \ldots$$
$$+ n^{-j/2}\hat{p}_{j1}(z_\alpha) + \ldots \quad (3.20)$$

and

$$\hat{v}_\alpha = z_\alpha + n^{-1/2}\hat{q}_{11}(z_\alpha) + n^{-1}\hat{q}_{21}(z_\alpha) + \ldots$$
$$+ n^{-j/2}\hat{q}_{j1}(z_\alpha) + \ldots . \quad (3.21)$$

Here, $\hat{p}_{j1}$ and $\hat{q}_{j1}$ differ from p_{j1} and q_{j1} only in that F_0 is replaced by F_1; that is, population moments are replaced by sample moments. In particular, $\hat{p}_{11}$, $\hat{p}_{21}$, $\hat{q}_{11}$, and $\hat{q}_{21}$ are given by formulae (3.15) and (3.16), provided that hats are placed over each p and q. Of course, Cornish-Fisher expansions are to be interpreted as asymptotic series and apply uniformly in values of α bounded away from zero and one. For example,

$$n^{j/2} \sup_{\epsilon < \alpha < 1-\epsilon} \left| \hat{u}_\alpha - \{z_\alpha + n^{-1/2}\hat{p}_{11}(z_\alpha) + \ldots + n^{-j/2}\hat{p}_{j1}(z_\alpha)\} \right| \longrightarrow 0 \quad (3.22)$$

almost surely (and hence also in probability) as $n \to \infty$, for each $0 < \epsilon < \frac{1}{2}$.

Rigorous proofs of results such as (3.20)–(3.22) are given in Section 5.2.

A key assumption underlying these results is smoothness of the sampling distribution. For example, under the "smooth function model" introduced in Section 2.4, the sampling distribution would typically be required to satisfy Cramér's condition, implying that the mass of the heaviest atom of the distribution of either $(\hat{\theta} - \theta_0)/\sigma$ or $(\hat{\theta} - \theta_0)/\hat{\sigma}$ is an exponentially small function of n. See Theorem 2.3. Should the sampling distribution be of the lattice type then the results described above fail to hold, since the bootstrap does not accurately approximate the "continuity correction" terms in an Edgeworth expansion.

To appreciate why, we consider the case of approximating the distribution of a sample mean. Let $\mathcal{X} = \{X_1, \ldots, X_n\}$ denote a sample of independent and identically distributed random variables from a population with mean zero, unit variance, and finite third moment, and assume that X takes only values of the form $a + jb$, for $j = 0, \pm 1, \pm 2, \ldots$, where $b > 0$ denotes the maximal span of the lattice. Define $R(x) = [x] - x + \frac{1}{2}$, and write $\bar{X}$ for the sample mean. Then, as we noted at the end of Section 2.2,

$$P(n^{1/2}\,\bar{X} \leq x)$$
$$= \Phi(x) + n^{-1/2}\,p_1(x)\,\phi(x)$$
$$\quad\quad + n^{-1/2}\,bR\{(n^{1/2}x - na)/b\}\,\phi(x) + o(n^{-1/2}).$$

To construct the bootstrap approximation, let $\mathcal{X}^* = \{X_1^*, \ldots, X_n^*\}$ denote a resample drawn randomly, with replacement, from $\mathcal{X}$. Write $\bar{X}^*$ for the resample mean, and let $\hat{\sigma}^2$ denote the sample variance. Since $(\sum X_i^* - n\bar{X})/\hat{\sigma}$ takes only values of the form $jb/\hat{\sigma}$ for $j = 0, \pm 1, \pm 2, \ldots$, then the bootstrap version of the above Edgeworth expansion involves replacing (a, b) by $(0, b/\hat{\sigma})$, as well as changing p_1 to $\hat{p}_1$. Thus,

$$P\{n^{1/2}(\bar{X}^* - \bar{X})/\hat{\sigma} \leq x \mid \mathcal{X}\}$$
$$= \Phi(x) + n^{-1/2}\,\hat{p}_1(x)\,\phi(x)$$
$$\quad\quad + n^{-1/2}\,\hat{\sigma}^{-1}\,bR(n^{1/2}\,\hat{\sigma}x/b)\,\phi(x) + o_p(n^{-1/2}).$$

Subtracting the two expansions and noting that $\hat{p}_1 - p_1 = O_p(n^{-1/2})$ and $\hat{\sigma} - 1 = O_p(n^{-1/2})$, we see that

$$P(n^{1/2}\,\bar{X} \leq x) - P\{n^{1/2}(\bar{X}^* - \bar{X})/\hat{\sigma} \leq x \mid \mathcal{X}\}$$
$$= n^{-1/2}b\left[R\{(n^{1/2}x - na)/b\} - R(n^{1/2}\,\hat{\sigma}x/b)\right]\phi(x) + o_p(n^{-1/2}).$$

The right-hand side here does not equal $o_p(n^{-1/2})$. Indeed, for any nonempty open interval $\mathcal{I} \subseteq \mathbb{R}$,

$$\limsup_{n\to\infty}\,\sup_{x\in\mathcal{I}}\,\left|R\{(n^{1/2}x - na)/b\} - R(n^{1/2}\hat{\sigma}x/b)\right|\phi(x) \longrightarrow \sup_{x\in\mathcal{I}}\,\phi(x)$$

with probability one. Therefore the bootstrap approximation to the distribution of $\bar{X}$ is in error by terms of size $n^{-1/2}$, and in this sense does not improve on the Normal approximation. By way of contrast, in the case where the sampling distribution is smooth we have by (3.11) and (3.19) that

$$P(n^{1/2}\,\bar{X} \leq x) - P\{n^{1/2}(\bar{X}^* - \bar{X})/\hat{\sigma} \leq x \mid \mathcal{X}\} = O_p(n^{-1}),$$

so that the bootstrap approximation is in error by n^{-1} rather than $n^{-1/2}$. In subsequent work we always assume that the sampling distribution is smooth.

3.4 Properties of Bootstrap Quantiles and Critical Points

In this section we discuss the accuracy of bootstrap estimates of quantiles of the distributions of S and T and the effect that this has on critical points of bootstrap confidence intervals.

The α-level quantiles of the distributions of S and T are u_α and v_α, respectively, with bootstrap estimates $\hat{u}_\alpha$ and $\hat{v}_\alpha$. Subtracting expansions (3.13) and (3.14) from (3.20) and (3.21) we deduce that

$$\hat{u}_\alpha - u_\alpha = n^{-1/2}\{\hat{p}_{11}(z_\alpha) - p_{11}(z_\alpha)\}$$
$$+ n^{-1}\{\hat{p}_{21}(z_\alpha) - p_{21}(z_\alpha)\} + \dots \quad (3.23)$$

and

$$\hat{v}_\alpha - v_\alpha = n^{-1/2}\{\hat{q}_{11}(z_\alpha) - q_{11}(z_\alpha)\}$$
$$+ n^{-1}\{\hat{q}_{21}(z_\alpha) - q_{21}(z_\alpha)\} + \dots . \quad (3.24)$$

The polynomial $\hat{p}_{j1}$ is obtained from p_{j1} on replacing population moments by sample moments, and the latter are distant $O_p(n^{-1/2})$ from their population counterparts. Therefore $\hat{p}_{j1}$ is distant $O_p(n^{-1/2})$ from p_{j1}. Thus, by (3.23),

$$\hat{u}_\alpha - u_\alpha = O_p\left(n^{-1/2} \cdot n^{-1/2} + n^{-1}\right) = O_p(n^{-1}),$$

and similarly $\hat{v}_\alpha - v_\alpha = O_p(n^{-1})$.

This establishes one of the important properties of bootstrap, or sample, critical points: the bootstrap estimates of u_α and v_α are in error by only order n^{-1}. In comparison, the traditional Normal approximation argues that u_α and v_α are both close to z_α and is in error by $n^{-1/2}$; for example,

$$z_\alpha - u_\alpha = z_\alpha - \{z_\alpha + n^{-1/2}p_{11}(z_\alpha) + \dots\} = -n^{-1/2}p_{11}(z_\alpha) + O(n^{-1}).$$

Approximation by Student's t distribution hardly improves on the Normal approximation, since the α-level quantile t_α of Student's t distribution with $n - \nu$ degrees of freedom (for any fixed ν) is distant order n^{-1}, not order $n^{-1/2}$, away from z_α. Thus, the bootstrap has definite advantages over traditional methods employed to approximate critical points.

This property of the bootstrap will only benefit us if we use bootstrap critical points in the right way. To appreciate the importance of this remark, go back to equations (3.9) and (3.10) defining the confidence intervals I_1, J_1, $\hat{I}_1$, and $\hat{J}_1$. Since $\hat{v}_{1-\alpha} = v_{1-\alpha} + O_p(n^{-1})$, the upper endpoint of the interval $\hat{J}_1 = (-\infty, \hat{\theta} - n^{-1/2}\hat{\sigma}\hat{v}_{1-\alpha})$ differs from the upper endpoint of $J_1 = (-\infty, \hat{\theta} - n^{-1/2}\hat{\sigma}v_{1-\alpha})$ by only $O_p(n^{-3/2})$.

We say that $\widehat{J}_1$ is second-order correct for J_1 and that $\hat{\theta} - n^{-1/2}\hat{\sigma}\hat{v}_{1-\alpha}$ is second-order correct for $\hat{\theta} - n^{-1/2}\hat{\sigma}v_{1-\alpha}$ since the latter two quantities are in agreement up to and including terms of order $(n^{-1/2})^2 = n^{-1}$. In contrast, $\widehat{I}_1 = (-\infty, \ \hat{\theta} - n^{-1/2}\hat{\sigma}\hat{u}_{1-\alpha})$ is generally only first-order correct for $I_1 = (-\infty, \ \hat{\theta} - n^{-1/2}\sigma\, u_{1-\alpha})$ since the upper endpoints agree only in terms of order $n^{-1/2}$, not n^{-1},

$$
\begin{aligned}
(\hat{\theta} - n^{-1/2}\hat{\sigma}\hat{u}_{1-\alpha}) \ &- \ (\hat{\theta} - n^{-1/2}\sigma\, u_{1-\alpha}) \\
&= \ n^{-1/2}(\sigma\, u_{1-\alpha} - \hat{\sigma}\hat{u}_{1-\alpha}) \\
&= \ n^{-1/2}u_{1-\alpha}(\sigma - \hat{\sigma}) + O_p(n^{-3/2}),
\end{aligned}
$$

and (usually) $n^{1/2}(\hat{\sigma} - \sigma)$ is asymptotically Normally distributed with zero mean and nonzero variance. Likewise, $\widehat{I}_1$ is usually only first-order correct for J_1 since terms of order $n^{-1/2}$ in Cornish-Fisher expansions of $\hat{u}_{1-\alpha}$ and $v_{1-\alpha}$ generally do not agree,

$$
\begin{aligned}
(\hat{\theta} - n^{-1/2}\hat{\sigma}\hat{u}_{1-\alpha}) \ &- \ (\hat{\theta} - n^{-1/2}\hat{\sigma}\, v_{1-\alpha}) \\
&= \ n^{-1/2}\hat{\sigma}(v_{1-\alpha} - \hat{u}_{1-\alpha}) \\
&= \ n^{-1}\hat{\sigma}\{p_1(z_\alpha) - q_1(z_\alpha)\} + O_p(n^{-3/2}).
\end{aligned}
$$

However, there do exist circumstances where p_1 and q_1 are identical, in which case it follows from the formula above that $\widehat{I}_1$ is first-order correct for J_1. Estimation of slope in regression provides an example and will be discussed in Section 4.2.

As we shall show in the next section, these "correctness" properties of critical points have important consequences for coverage accuracy. The second-order correct confidence interval $\widehat{J}_1$ has coverage error of order n^{-1}, whereas the first-order correct interval $\widehat{I}_1$ has coverage error of size $n^{-1/2}$.

As noted in Section 3.2, the intervals $\widehat{I}_1$, $\widehat{J}_1$ represent bootstrap versions of I_1, J_1 respectively. Recall from Example 1.2 of Section 1.3 that percentile intervals such as $\widehat{I}_1$ are based on the nonpivotal statistic $\hat{\theta} - \theta_0$ and that it is this nonpivotalness that causes the asymptotic standard deviation σ to appear in the definition of I_1. That is why $\widehat{I}_1$ is not second-order correct for I_1, and so our problems may be traced back to the issue of pivotalness raised in Section 1.3. The percentile-t interval $\widehat{J}_1$ is based on the (asymptotically) pivotal statistic $(\hat{\theta} - \theta_0)/\hat{\sigma}$, hence its virtuous properties. However, the reader should note the caveats about percentile-t in subsection 3.10.1.

If the asymptotic variance σ^2 should be known then we may use it to standardize, and construct confidence intervals based on $(\hat{\theta} - \theta_0)/\sigma$ (which is asymptotically pivotal), instead of on either $\hat{\theta} - \theta_0$ or $(\hat{\theta} - \theta_0)/\hat{\sigma}$. Application of the principle enunciated in Section 1.2 now produces the interval

$$\widehat{I}_{11} = (-\infty, \ \hat{\theta} - n^{-1/2}\sigma\,\hat{u}_{1-\alpha}),$$

which is second-order correct for I_1 and has coverage error of order n^{-1}.

We should warn the reader not to read too much into the notion of "correctness order" for confidence intervals. While it is true that second-order correctness is a desirable property, and that intervals that fail to exhibit it do not correct even for the elementary skewness errors in Normal approximations, it does not follow that we should seek third- or fourth-order correct confidence intervals. Indeed, such intervals are usually unattainable as the next paragraph will show. Techniques such as bootstrap iteration, which reduce the order of coverage error, do not accomplish this goal by achieving high orders of correction but rather by adjusting the error of size $n^{-3/2}$ inherent to almost all sample-based critical points.

Recall from (3.14) that

$$v_{1-\alpha} = z_{1-\alpha} + n^{-1/2}q_{11}(z_{1-\alpha}) + n^{-1}q_{21}(z_{1-\alpha}) + \dots .$$

Coefficients of the polynomial q_{11} are usually unknown quantities. In view of results such as the Cramér-Rao lower bound (e.g., Cox and Hinkley 1974, p. 254ff), the coefficients cannot be estimated with an accuracy better than order $n^{-1/2}$. This means that $v_{1-\alpha}$ cannot be estimated with an accuracy better than order n^{-1}, and that the upper endpoint of the confidence interval $J_1 = (-\infty, \ \hat{\theta} - n^{-1/2}\hat{\sigma}v_{1-\alpha})$ cannot be estimated with an accuracy better than order $n^{-3/2}$. Therefore, except in unusual circumstances, any practical confidence interval $\widehat{J}_1$ that tries to emulate J_1 will have an endpoint differing in a term of order $n^{-3/2}$ from that of J_1, and so will not be third-order correct. Exceptional circumstances are those where we have enough parametric information about the population to know the coefficients of q_{11}. For example, in the case of estimating a mean, q_{11} vanishes if the underlying population is symmetric; see Example 2.1 of Section 2.6. If we know that the population is symmetric, we may construct confidence intervals that are better than second-order correct. For example, we may resample in a way that ensures that the bootstrap distribution is symmetric, by sampling with replacement from the collection $\{\pm(X_1 - \bar{X}), \dots, \pm(X_n - \bar{X})\}$ rather than $\{X_1 - \bar{X}, \dots, X_n - \bar{X}\}$. But in most problems, both parametric and nonparametric, second-order correctness is the best we can hope to achieve.

In Section 1.3 we mentioned the "other percentile confidence interval," defined there by

$$\widehat{I}_{12} = (-\infty, \ \widehat{H}^{-1}(\alpha))$$

where $\widehat{H}(x) = P(\widehat{\theta}^* \leq x \,|\, \mathcal{X})$. In our present notation this interval is just

$$\widehat{I}_{12} = (-\infty, \ \widehat{\theta} + n^{-1/2}\widehat{\sigma}\widehat{u}_\alpha)$$

since the equation

$$\alpha = \widehat{H}(x) = P\{n^{1/2}(\widehat{\theta}^* - \widehat{\theta})/\widehat{\sigma} \leq n^{1/2}(x - \widehat{\theta})/\widehat{\sigma} \,|\, \mathcal{X}\}$$

is solved when $n^{1/2}(x - \widehat{\theta})/\widehat{\sigma} = \widehat{u}_\alpha$. (Here we have ignored rounding errors due to possible discreteness of the distribution of $\widehat{\theta}^*$. They are exponentially small as functions of n, as shown in Appendix I.) It is clear from this representation that the confidence interval $\widehat{I}_{12}$ suffers two drawbacks. Firstly, it employs a quantile $\widehat{u}$ instead of $\widehat{v}$, even though σ is unknown. In a classical, Normal-theory based context this would be tantamount to looking up Normal tables when we should be consulting Student's t tables. Secondly, the interval $\widehat{I}_{12}$ uses $\widehat{u}_\alpha$ instead of $-\widehat{u}_{1-\alpha}$ or $-\widehat{v}_{1-\alpha}$; this amounts to looking up the wrong tail in the wrong tables of a distribution. We do find the "other percentile method" to be unattractive, unless it is corrected or adjusted in an appropriate way.

All our remarks about critical points carry over to two-sided confidence intervals, such as the equal-tailed intervals defined in Section 3.2. For example, the two-sided equal-tailed percentile-t confidence interval $\widehat{J}_2$ is second-order correct for J_2 since respective endpoints differ only in terms of order $n^{-3/2}$. The percentile interval $\widehat{I}_2$ is only first-order correct for I_2. If the asymptotic variance σ^2 is known then application of the bootstrap resampling principle to $(\widehat{\theta} - \theta_0)/\sigma$ produces a two-sided analogue of interval $\widehat{I}_{11}$,

$$\widehat{I}_{21} = \left(\widehat{\theta} - n^{-1/2}\sigma\,\widehat{u}_{(1+\alpha)/2}, \ \ \widehat{\theta} - n^{-1/2}\sigma\,\widehat{u}_{(1-\alpha)/2} \right),$$

which is second-order correct for I_2.   However, the "other percentile interval,"

$$\widehat{I}_{22} = \left(\widehat{H}^{-1}\{\tfrac{1}{2}(1-\alpha)\}, \ \ \widehat{H}^{-1}\{\tfrac{1}{2}(1+\alpha)\} \right)$$

$$= \left(\widehat{\theta} + n^{-1/2}\widehat{\sigma}\widehat{u}_{(1-\alpha)/2}, \ \ \widehat{\theta} + n^{-1/2}\widehat{\sigma}\widehat{u}_{(1+\alpha)/2} \right),$$

is the two-sided, equal-tailed analogue of $\widehat{I}_{12}$ and is more difficult to justify. It is only first-order correct for I_2.

3.5 Coverage Error of Confidence Intervals

In this section we show how to compute asymptotic formulae for coverage and coverage error of a variety of different confidence intervals, including bootstrap confidence intervals. Subsection 3.5.1 will formulate a general version of the confidence interval problem, subsection 3.5.2 will develop methods for solving that general problem, and subsection 3.5.3 will state the general solution. The latter result will be applied to a variety of one-sided and two-sided confidence intervals in subsections 3.5.4 and 3.5.5, respectively. Finally, subsection 3.5.6 will present an example that illustrates the main features of our argument.

3.5.1 The Problem

A general one-sided confidence interval for θ_0 may be expressed as $\mathcal{I}_1 = (-\infty, \hat{\theta} + \hat{t})$, where $\hat{t}$ is determined from the data. In most circumstances, if $\mathcal{I}_1$ has nominal coverage α then $\hat{t}$ admits the representation

$$\hat{t} = n^{-1/2}\hat{\sigma}(z_\alpha + \hat{c}_\alpha), \qquad (3.25)$$

where $\hat{c}_\alpha$ is a random variable and converges to zero as $n \to \infty$. For example, this would typically be the case if $T = n^{1/2}(\hat{\theta} - \theta_0)/\hat{\sigma}$ had an asymptotic Standard Normal distribution. However, should the value σ^2 of asymptotic variance be known, we would most likely use an interval in which $\hat{t}$ had the form

$$\hat{t} = n^{-1/2}\sigma(z_\alpha + \hat{c}_\alpha).$$

Intervals $\widehat{I}_1$, $\widehat{I}_{12}$, J_1, and $\widehat{J}_1$ are of the former type, with $\hat{c}_\alpha$ in (3.25) assuming the respective values $-\hat{u}_{1-\alpha} - z_\alpha$, $\hat{u}_\alpha - z_\alpha$, $-v_{1-\alpha} - z_\alpha$, $-\hat{v}_{1-\alpha} - z_\alpha$. So also are the "Normal approximation interval" $(-\infty, \hat{\theta} + n^{-1/2}\hat{\sigma}z_\alpha)$ and "Student's t approximation interval" $(-\infty, \hat{\theta} + n^{-1/2}\hat{\sigma}t_\alpha)$, where t_α is the α-level quantile of Student's t distribution with $n - 1$ degrees of freedom. Intervals I_1 and $\widehat{I}_{11}$ are of the latter type. The main purpose of the correction term $\hat{c}_\alpha$ is to adjust for skewness. To a lesser extent it corrects for higher-order departures from Normality.

Suppose that $\hat{t}$ is of the form (3.25). Then coverage probability equals

$$\alpha_{1,n} = P(\theta_0 \in \mathcal{I}_1) = P\{\theta_0 \leq \hat{\theta} + n^{-1/2}\hat{\sigma}(z_\alpha + \hat{c}_\alpha)\}$$
$$= 1 - P\{n^{1/2}(\hat{\theta} - \theta_0)\hat{\sigma}^{-1} + \hat{c}_\alpha < -z_\alpha\}. \qquad (3.26)$$

We wish to develop an expansion of this probability. For that purpose it is necessary to have an Edgeworth expansion of the distribution function of

$$n^{1/2}(\hat{\theta} - \theta_0)\hat{\sigma}^{-1} + \hat{c}_\alpha \,,$$

or at least a good approximation to it. In some circumstances, $\hat{c}_\alpha$ is easy to work with directly; for example $\hat{c}_\alpha = 0$ in the case of the Normal approximation interval. But for bootstrap intervals, $\hat{c}_\alpha$ is defined only implicitly as the solution of an equation, and that makes it rather difficult to handle. So we first approximate it by a Cornish-Fisher expansion, as discussed in the next subsection.

Strictly speaking, the coverage probability in (3.26) is that of the closed interval $(-\infty,\ \hat{\theta} + \hat{t}\,]$, not the open interval $(-\infty,\ \hat{\theta} + \hat{t}\,)$. In ignoring the difference between these probabilities,

$$\delta = P\{\theta_0 \in (-\infty,\ \hat{\theta} + \hat{t}\,]\,\} - P\{\theta_0 \in (-\infty,\ \hat{\theta} + \hat{t}\,)\}\,,$$

we are in effect assuming that $\hat{\theta} + \hat{t}$ has a continuous distribution. While this is very often the case in practice, there do exist circumstances where the error is nonzero but nevertheless negligibly small. Assuming Cramér's condition and the smooth function model, which are imposed in our rigorous account of coverage probability (see Section 5.3), it follows that the heaviest atom of the distribution of $\hat{\theta} + \hat{t}$ has mass only $O(e^{-\epsilon n})$, for some $\epsilon > 0$. See Theorem 2.3. In this event we have $\delta = O(e^{-\epsilon n})$, and so although δ may not be zero it can be neglected relative to the remainder terms in our expansions of coverage probability, which are all $O(n^{-c})$ for some $c > 0$. We shall therefore ignore, throughout this monograph, the effect of openness or closedness of confidence intervals on coverage probability, taking it to be either zero or negligible.

3.5.2 The Method of Solution

We begin with several examples, to illustrate the range of situations that have to be treated. Let $\hat{p}_{j1}$ and $\hat{q}_{j1}$, $j \geq 1$, be as in (3.20) and (3.21). When $\mathcal{I}_1 = (-\infty,\ \hat{\theta} + \hat{t}) = (-\infty,\ \hat{\theta} + n^{-1/2}\hat{\sigma}(z_\alpha + \hat{c}_\alpha))$ is the percentile interval $\widehat{I}_1$, we have

$$\begin{aligned}
\hat{c}_\alpha &= -\hat{u}_{1-\alpha} - z_\alpha = z_{1-\alpha} - \hat{u}_{1-\alpha} \\
&= -n^{-1/2}\hat{p}_{11}(z_{1-\alpha}) - n^{-1}\hat{p}_{21}(z_{1-\alpha}) - \cdots \qquad [\,\text{use } (3.20)\,] \\
&= -n^{-1/2}\hat{p}_{11}(z_\alpha) + n^{-1}\hat{p}_{21}(z_\alpha) + O_p(n^{-3/2})\,,
\end{aligned}$$

since $z_{1-\alpha} = -z_\alpha$ and $\hat{p}_{j1}$ is an odd/even function for even/odd j. Similarly, when $\mathcal{I}_1$ is the percentile-t interval $\widehat{J}_1$,

$$
\begin{aligned}
\hat{c}_\alpha &= -\hat{v}_{1-\alpha} - z_\alpha \\
&= -n^{-1/2}\hat{q}_{11}(z_\alpha) + n^{-1}\hat{q}_{21}(z_\alpha) + O_p(n^{-3/2}),
\end{aligned}
$$

and when $\mathcal{I}_1 = \widehat{I}_{12}$,

$$
\hat{c}_\alpha = n^{-1/2}\hat{p}_{11}(z_\alpha) + n^{-1}\hat{p}_{21}(z_\alpha) + O_p(n^{-3/2}).
$$

In general, write

$$
\hat{c}_\alpha = n^{-1/2}\hat{s}_1(z_\alpha) + n^{-1}\hat{s}_2(z_\alpha) + O_p(n^{-3/2}), \tag{3.27}
$$

where s_1 and s_2 are polynomials with coefficients equal to polynomials in population moments and $\hat{s}_j$ is obtained from s_j on replacing population moments by sample moments. Then $\hat{s}_j = s_j + O_p(n^{-1/2})$ and

$$
\begin{aligned}
P\{n^{1/2}(\hat{\theta} - \theta_0)\hat{\sigma}^{-1} + \hat{c}_\alpha \le x\} \\
= P\Big[n^{1/2}(\hat{\theta} - \theta_0)\hat{\sigma}^{-1} + n^{-1/2}\{\hat{s}_1 - s_1(z_\alpha)\} \\
\le x - \sum_{j=1}^{2} n^{-j/2}s_j(z_\alpha)\Big] + O(n^{-3/2}). \tag{3.28}
\end{aligned}
$$

Here we have used the delta method; see Section 2.7.

Therefore, to evaluate the coverage probability $\alpha_{1,n}$ at (3.26) up to a remainder of order $n^{-3/2}$, we need only derive an Edgeworth expansion of the distribution function of

$$
S_n = n^{1/2}(\hat{\theta} - \theta_0)\hat{\sigma}^{-1} + n^{-1}\Delta_n, \tag{3.29}
$$

where $\Delta_n = n^{1/2}\{\hat{s}_1(z_\alpha) - s_1(z_\alpha)\}$. That is usually simpler than finding an Edgeworth expansion for $n^{1/2}(\hat{\theta} - \theta_0)\hat{\sigma}^{-1} + \hat{c}_\alpha$.

The method of solution of this problem was outlined in Chapter 2 following equation (2.37). (In that notation, and assuming the smooth function model, we have $S_n = n^{1/2} A_n(\overline{\mathbf{X}})$ where $A_n(\mathbf{x}) = B_0(\mathbf{x}) + n^{-1} B_2(\mathbf{x})$ for functions B_0, B_2.) We require the first four cumulants of S_n. Let $K_k(U)$ denote the kth cumulant of a random variable U, and put

$T_n = n^{1/2}(\hat\theta - \theta_0)/\hat\sigma$. Our claim is that[1]

$$K_{2j+1}(S_n) = K_{2j+1}(T_n) + O(n^{-3/2}), \qquad j \geq 0, \qquad (3.30)$$

$$K_2(S_n) = K_2(T_n) + 2n^{-1}E(T_n\Delta_n) + O(n^{-2}), \qquad (3.31)$$

and

$$K_4(S_n) = K_4(T_n) + O(n^{-2}). \qquad (3.32)$$

To appreciate why, note that $\Delta_n = O_p(1)$ and so by (3.29) moments and cumulants of S_n differ from those of T_n only in terms of order n^{-1} or smaller. Furthermore, odd-order moments and cumulants of S_n admit expansions in odd powers of $n^{-1/2}$. This follows from the fact that in the version of expansion (2.38) for S_n, only products of an odd/even number of $Z^{(i)}$'s appear in coefficients of $n^{-j/2}$ for even/odd j, respectively. Therefore odd-order moments and cumulants of T_n must differ from those of S_n only in terms of order $n^{-3/2}$. This proves (3.30) and also that

$$E(S_n^{2j+1}) = E(T_n^{2j+1}) + O(n^{-3/2}) = O(n^{-1/2}), \quad j \geq 0. \quad (3.33)$$

From the latter result follows (3.31), since by (2.6),

$$\begin{aligned}
K_2(S_n) &= E(S_n^2) - (ES_n)^2 \\
&= E(T_n + n^{-1}\Delta_n)^2 - \{E(T_n) + O(n^{-3/2})\}^2 \\
&= E(T_n^2) - (ET_n)^2 + 2n^{-1}E(T_n\Delta_n) + O(n^{-2}) \\
&= K_2(T_n) + 2n^{-1}E(T_n\Delta_n) + O(n^{-2}).
\end{aligned}$$

To prove (3.32), observe that by (2.6), (3.33), and the fact that $S_n = T_n + O_p(n^{-1})$,

$$\begin{aligned}
K_4(S_n) &= E(S_n^4) - 4E(S_n)E(S_n^3) \\
&\quad - 3(ES_n^2)^2 + 12E(S_n^2)(ES_n)^2 - 6(ES_n)^4 \\
&= \left\{E(T_n^4) + 4n^{-1}E(T_n^3\Delta_n)\right\} - 4E(T_n)E(T_n^3) \\
&\quad - 3\left\{E(T_n^2) + 2n^{-1}E(T_n\Delta_n)\right\}^2 + 12E(T_n^2)(ET_n)^2 \\
&\quad - 6(ET_n)^4 + O(n^{-2}) \\
&= K_4(T_n) + 4n^{-1}\left\{E(T_n^3\Delta_n) - 3E(T_n^2)E(T_n\Delta_n)\right\} \\
&\quad + O(n^{-2}). \qquad (3.34)
\end{aligned}$$

---

[1]Strictly speaking, we should be taking cumulants of Taylor series approximations $S_{n,5}$ and $T_{n,5}$ to S_n and T_n, respectively, defined by the expansion (2.29) and whose distributions agree with those of S_n and T_n up to terms of order n^{-2}. However, it is tedious to acknowledge these caveats on all occasions. See also footnote 1 to Chapter 2.

In many problems, such as those admitting the models introduced in Section 2.4,

$$E(T_n^3 \Delta_n) = 3 E(T_n^2) E(T_n \Delta_n) + O(n^{-1}).$$

To see why, note that

$$T_n = n^{-1/2} \sum_{i=1}^{n} U_i + O_p(n^{-1/2}) \quad \text{and} \quad \Delta_n = n^{-1/2} \sum_{i=1}^{n} V_i + O_p(n^{-1/2}),$$

where the pairs (U_i, V_i), $1 \le i \le n$, are independent and identically distributed with zero mean. Therefore

$$E(T_n^2) = E(U_1^2) + O(n^{-1}),$$
$$E(T_n \Delta_n) = E(U_1 V_1) + O(n^{-1})$$

and

$$E(T_n^3 \Delta_n) = n^{-2} E \left\{ \left(\sum_{i=1}^{n} U_i \right)^3 \left(\sum_{i=1}^{n} V_i \right) \right\} + O(n^{-1})$$

$$= 3 E(U_1^2) E(U_1 V_1) + O(n^{-1})$$

$$= 3 E(T_n^2) E(T_n \Delta_n) + O(n^{-1}).$$

The desired result (3.32) follows from this identity and (3.34).

Formulae (3.30)–(3.32) show that of the first four cumulants of S_n, only the second differs from its counterpart for T_n in terms of order n^{-1} or larger. Let a_α denote the real number such that[2]

$$E(T_n \Delta_n) = E \left[n^{1/2} (\hat{\theta} - \theta_0) \hat{\sigma}^{-1} \cdot n^{1/2} \{ \hat{s}_1(z_\alpha) - s_1(z_\alpha) \} \right]$$

$$= a_\alpha + O(n^{-1}). \tag{3.35}$$

If s_1 is an even polynomial of degree 2, which would typically be the case, then $a_\alpha = \pi(z_\alpha)$, where π is an even polynomial of degree 2 with coefficients not depending on α. Using the Edgeworth expansion (2.23), and applying formulae (2.24) and (2.25) expressing the polynomials p_1 and p_2 from (2.23) in terms of the first four cumulants, we deduce from (3.30)–(3.32) that

$$P(S_n \le x) = P(T_n \le x) - n^{-1} a_\alpha x \phi(x) + O(n^{-3/2}). \tag{3.36}$$

(Among the coefficients $k_{i,j}$ appearing in (2.24) and (2.25), only $k_{2,2}$ has a different formula in the case of T_n than it does for S_n, and by (3.31) and (3.35), $k_{2,2}(S_n) = k_{2,2}(T_n) + 2a_\alpha$.)

It is now a simple matter to obtain expansions of coverage probability for our general one-sided confidence interval $\mathcal{I}_1$. Taking $x = -z_\alpha$ in (3.28),

---

[2]The random variables in these expectations should, strictly speaking, be replaced by Taylor series approximations; see footnote 1.

and noting (3.26) and (3.36), we conclude that if $\hat{c}_\alpha$ is given by (3.27) then the confidence interval

$$\mathcal{I}_1 = \left(-\infty,\ \hat{\theta} + n^{-1/2}\hat{\sigma}\left(z_\alpha + \hat{c}_\alpha\right)\right) \tag{3.37}$$

has coverage probability

$$\alpha_{1,n} = P(\theta_0 \in \mathcal{I}_1)$$

$$= P\left\{n^{1/2}(\hat{\theta} - \theta_0)/\hat{\sigma} > -z_\alpha - \sum_{j=1}^{2} n^{-j/2} s_j(z_\alpha)\right\}$$

$$- n^{-1} a_\alpha z_\alpha \phi(z_\alpha) + O(n^{-3/2}).$$

Assuming the usual Edgeworth expansion for a Studentized statistic, i.e.,

$$P\{n^{1/2}(\hat{\theta} - \theta_0)/\hat{\sigma} \le x\}$$

$$= \Phi(x) + n^{-1/2} q_1(x)\phi(x) + n^{-1} q_2(x)\phi(x) + O(n^{-3/2}) \tag{3.38}$$

(see equation (3.12)), putting $x = -z_\alpha - \sum_{j=1,2} n^{-j/2} s_j(z_\alpha)$, and Taylor-expanding, we finally obtain

$$\alpha_{1,n} = \alpha + n^{-1/2}\{s_1(z_\alpha) - q_1(z_\alpha)\}\phi(z_\alpha)$$

$$+ n^{-1}\left[q_2(z_\alpha) + s_2(z_\alpha) - \tfrac{1}{2} z_\alpha s_1(z_\alpha)^2\right.$$

$$\left. + s_1(z_\alpha)\{z_\alpha q_1(z_\alpha) - q_1'(z_\alpha)\} - a_\alpha z_\alpha\right]\phi(z_\alpha)$$

$$+ O(n^{-3/2}). \tag{3.39}$$

(Remember that q_j is an odd/even function for even/odd j, respectively.)

The function s_1 is generally an even polynomial and s_2 an odd polynomial. This follows from the Cornish-Fisher expansion (3.27) of $\hat{c}_\alpha$; note the examples preceding (3.27), and also the properties of polynomials in Cornish-Fisher expansions, enunciated in Section 2.5. Since s_1 is an even function, then by the definition of a_α in (3.35), a_α is an even polynomial in z_α. Therefore the coefficient of $n^{-1}\phi(z_\alpha)$ in (3.39) is an odd polynomial in z_α. The coefficient of $n^{-1/2}\phi(z_\alpha)$ is clearly an even polynomial.

3.5.3 The Solution

There is no difficulty developing the expansion (3.39) to an arbitrary number of terms, obtaining a series in powers of $n^{-1/2}$ where the coefficient of $n^{-j/2}\phi(z_\alpha)$ equals an odd or even polynomial depending on whether j is even or odd. The following proposition summarizes that result.

PROPOSITION 3.1. *Consider the confidence interval*

$$\mathcal{I}_1 = \mathcal{I}_1(\alpha) = \left(-\infty,\ \hat{\theta} + n^{-1/2}\hat{\sigma}(z_\alpha + \hat{c}_\alpha)\right),$$

where

$$\hat{c}_\alpha = n^{-1/2}\hat{s}_1(z_\alpha) + n^{-1}\hat{s}_2(z_\alpha) + \dots,$$

where the $\hat{s}_j$'s are obtained from polynomials s_j on replacing population moments by sample moments and odd/even indexed s_j's are even/odd polynomials, respectively. Suppose

$$P\{n^{1/2}(\hat{\theta}-\theta_0)/\hat{\sigma} \le x\} = \Phi(x) + n^{-1/2}q_1(x)\phi(x) + n^{-1}q_2(x)\phi(x) + \dots,$$

where odd/even indexed polynomials q_j are even/odd functions, respectively. Then

$$\begin{aligned}
P\{\theta_0 \in \mathcal{I}_1(\alpha)\} \\
= \alpha + n^{-1/2}r_1(z_\alpha)\phi(z_\alpha) + n^{-1}r_2(z_\alpha)\phi(z_\alpha) + \dots,\quad (3.40)
\end{aligned}$$

where odd/even indexed polynomials r_j are even/odd functions, respectively. In particular,

$$r_1 = s_1 - q_1$$

and

$$\begin{aligned}
r_2(z_\alpha) = q_2(z_\alpha) + s_2(z_\alpha) - \tfrac{1}{2}z_\alpha s_1(z_\alpha)^2 \\
+ s_1(z_\alpha)\{z_\alpha q_1(z_\alpha) - q_1'(z_\alpha)\} - a_\alpha z_\alpha,
\end{aligned}$$

where a_α is defined in (3.35).

The coverage expansion (3.40) should of course be interpreted as an asymptotic series. It is often not available uniformly in α, but does hold uniformly in $\epsilon < \alpha < 1-\epsilon$ for any $0 < \epsilon < \frac{1}{2}$. However, if $\hat{c}_\alpha$ is monotone increasing in α then (3.40) will typically be available uniformly in $0 < \alpha < 1$. Further details are given in Section 5.3, where a version of Proposition 3.1 is derived under explicit regularity conditions.

3.5.4 Coverage Properties of One-Sided Intervals

An immediate consequence of (3.40) is that *a necessary and sufficient condition for our confidence interval $\mathcal{I}_1$, defined at (3.37), to have coverage error of order n^{-1} for all values of α, is that it be second-order correct relative to the interval J_1 introduced in Section 3.2.* To appreciate why, go back to the definition (3.27) of $\hat{c}_\alpha$, which implies that

$$\mathcal{I}_1 = \left(-\infty, \ \hat{\theta} + n^{-1/2}\hat{\sigma}(z_\alpha + \hat{c}_\alpha)\right)$$

$$= \left(-\infty, \ \hat{\theta} + n^{-1/2}\hat{\sigma}\{z_\alpha + n^{-1/2}s_1(z_\alpha)\} + O_p(n^{-3/2})\right).$$

Since $q_{11} = -q_1$ (see equation (3.16)) then by (3.9) and (3.14),

$$J_1 = \left(-\infty, \ \hat{\theta} - n^{-1/2}\hat{\sigma}v_{1-\alpha}\right)$$

$$= \left(-\infty, \ \hat{\theta} - n^{-1/2}\hat{\sigma}\{z_{1-\alpha} + n^{-1/2}q_{11}(z_{1-\alpha})\} + O_p(n^{-3/2})\right)$$

$$= \left(-\infty, \ \hat{\theta} + n^{-1/2}\hat{\sigma}\{z_\alpha + n^{-1/2}q_1(z_\alpha)\} + O_p(n^{-3/2})\right).$$

The upper endpoint of this interval agrees with that of $\mathcal{I}_1$ in terms of order n^{-1}, for all α, if and only if $s_1 = q_1$, that is, if and only if the term of order $n^{-1/2}$ vanishes from (3.40). Therefore the second-order correct interval $\widehat{J}_1$ has coverage error of order n^{-1}, but the intervals $\widehat{I}_1$ and $\widehat{I}_{12}$, which are only first-order correct, have coverage error of size $n^{-1/2}$ except in special circumstances.

So far we have worked only with confidence intervals of the form $(-\infty, \hat{\theta} + \hat{t})$, where $\hat{t} = n^{-1/2}\hat{\sigma}(z_\alpha + \hat{c}_\alpha)$ and $\hat{c}_\alpha$ is given by (3.27). Should the value σ^2 of asymptotic variance be known then we would most likely construct confidence intervals using $\hat{t} = n^{-1/2}\sigma(z_\alpha + \hat{c}_\alpha)$, again for $\hat{c}_\alpha$ given by (3.27). This case may be treated by reworking arguments above. We should change the symbol q to p at each appearance, because we are now working with the Edgeworth expansion (3.11) rather than (3.12). With this alteration, formula (3.39) for coverage probability continues to apply,

$$P\left\{\theta_0 \in \left(-\infty, \ \hat{\theta} + n^{-1/2}\sigma(z_\alpha + \hat{c}_\alpha)\right)\right\}$$

$$= \alpha + n^{-1/2}\{s_1(z_\alpha) - p_1(z_\alpha)\}\phi(z_\alpha)$$

$$+ n^{-1}\left[p_2(z_\alpha) + s_2(z_\alpha) - \tfrac{1}{2}z_\alpha s_1(z_\alpha)^2\right.$$

$$\left. + s_1(z_\alpha)\{z_\alpha p_1(z_\alpha) - p_1'(z_\alpha)\} - a_\alpha z_\alpha\right]\phi(z_\alpha)$$

$$+ O(n^{-3/2}).$$

(Our definition of a_α at (3.35) is unaffected if $\hat{\sigma}^{-1}$ is replaced by σ^{-1}, since $\hat{\sigma}^{-1} = \sigma^{-1} + O_p(n^{-1/2})$.) Likewise, the analogue of Proposition 3.1 is valid; it is necessary only to replace $\hat{\sigma}$ by σ in the definition of $\mathcal{I}_1$ and q_j by p_j at all appearances of the former. Therefore our conclusions in the case where σ is known are similar to those when σ is unknown: a necessary and sufficient condition for the confidence interval $(-\infty, \hat{\theta} + n^{-1/2}\sigma(z_\alpha + \hat{c}_\alpha))$

to have coverage error of order n^{-1} for all values of α is that it be second-order correct relative to I_1. The interval $\widehat{I}_{11}$, defined in Section 3.4, is second-order correct for I_1.

Similarly it may be proved that if $\mathcal{I}_1$ is jth order correct relative to a one-sided confidence interval $\mathcal{I}_1'$, meaning that the upper endpoints agree in terms of size $n^{-j/2}$ or larger, then $\mathcal{I}_1$ and $\mathcal{I}_1'$ have the same coverage probability up to but not necessarily including terms of order $n^{-j/2}$. The converse of this result is false for $j \geq 3$. Indeed, there are many important examples of confidence intervals whose coverage errors differ by $O(n^{-3/2})$ but which are not third-order correct relative to one another; see Sections 3.8 to 3.11. There are relatively few instances where third- or higher-order correctness is encountered in practical problems, although we shall meet with examples in Sections 3.6 and 4.3.

3.5.5 Coverage Properties of Two-Sided Intervals

Coverage properties of two-sided confidence intervals are rather different from those in the one-sided case. For two-sided intervals, parity properties of polynomials in expansions such as (3.40) cause terms of order $n^{-1/2}$ to cancel completely from expansions of coverage error. Therefore coverage error is always of order n^{-1} or smaller, even for the most basic Normal approximation method. In the case of symmetric two-sided intervals constructed using the percentile-t bootstrap, coverage error is of order n^{-2}. The remainder of the present section will treat two-sided equal-tailed intervals. The reader is referred to Section 3.6 for two-sided symmetric intervals and to Section 3.7 for two-sided short intervals.

We begin by recalling our definition of the general one-sided interval $\mathcal{I}_1 = \mathcal{I}_1(\alpha)$ whose nominal coverage is α:

$$\mathcal{I}_1(\alpha)$$
$$= \left(-\infty, \ \hat{\theta} + n^{-1/2}\hat{\sigma}(z_\alpha + \hat{c}_\alpha) \right)$$
$$= \left(-\infty, \ \hat{\theta} + n^{-1/2}\hat{\sigma}\{z_\alpha + n^{-1/2}\hat{s}_1(z_\alpha) + n^{-1}\hat{s}_2(z_\alpha)\} + O_p(n^{-2}) \right).$$

The equal-tailed interval based on this scheme and having nominal coverage α is

$$\mathcal{I}_2(\alpha)$$
$$= \mathcal{I}_1\left(\tfrac{1}{2}(1+\alpha)\right) \setminus \mathcal{I}_1\left(\tfrac{1}{2}(1-\alpha)\right) \tag{3.41}$$
$$= \left(\hat{\theta} + n^{-1/2}\hat{\sigma}\left(z_{(1-\alpha)/2} + \hat{c}_{(1-\alpha)/2}\right), \ \hat{\theta} + n^{-1/2}\hat{\sigma}\left(z_{(1+\alpha)/2} + \hat{c}_{(1+\alpha)/2}\right) \right).$$

(Here $\mathcal{I}\backslash\mathcal{J}$ denotes the intersection of set $\mathcal{I}$ with the complement of set $\mathcal{J}$.) Apply Proposition 3.1 with $z = z_{(1+\alpha)/2} = -z_{(1-\alpha)/2}$, noting particularly that r_j is an odd or even function accordingly as j is even or odd, to obtain an expansion of the coverage probability of $\mathcal{I}_2(\alpha)$:

$$
\begin{aligned}
\alpha_{2,n} &= P\{\theta_0 \in \mathcal{I}_2(\alpha)\} \\
&= \Phi(z) + n^{-1/2}r_1(z)\phi(z) + n^{-1}r_2(z)\phi(z) + \cdots \\
&\quad - \Big\{\Phi(-z) + n^{-1/2}r_1(-z)\phi(-z) + n^{-1}r_2(-z)\phi(-z) + \cdots\Big\} \\
&= \alpha + 2n^{-1}r_2(z)\phi(z) + 2n^{-2}r_4(z)\phi(z) + \cdots \\
&= \alpha + 2n^{-1}\Big[q_2(z) + s_2(z) - \tfrac{1}{2}zs_1(z)^2 \\
&\quad + s_1(z)\{zq_1(z) - q_1'(z)\} - a_{(1+\alpha)/2}z\Big]\phi(z) \\
&\quad + O(n^{-2}).
\end{aligned}
\tag{3.42}
$$

The property of second-order correctness, which as we have seen is equivalent to $s_1 = q_1$, has relatively little effect on the coverage probability in (3.42). This contrasts with the case of one-sided confidence intervals.

For percentile confidence intervals,

$$
s_1 = -p_{11} = p_1 \tag{3.43}
$$

and

$$
s_2(x) = p_{21}(x) = p_1(x)p_1'(x) - \tfrac{1}{2}xp_1(x)^2 - p_2(x), \tag{3.44}
$$

while for percentile-t intervals,

$$
s_1 = -q_{11} = q_1 \tag{3.45}
$$

and

$$
s_2(x) = q_{21}(x) = q_1(x)q_1'(x) - \tfrac{1}{2}xq_1(x)^2 - q_2(x). \tag{3.46}
$$

These formulae follow from (3.15), (3.16), and the examples preceding (3.27). There is no significant simplification of (3.42) when (3.43) and (3.44) are used to express s_1 and s_2. However, in the percentile-t case we see from (3.42), (3.45), and (3.46) that

$$
\alpha_{2,n} = \alpha - 2n^{-1}a_{(1+\alpha)/2}\,z_{(1+\alpha)/2}\,\phi(z_{(1+\alpha)/2}) + O(n^{-2}),
$$

which represents a substantial simplification.

When the asymptotic variance σ^2 is known, our formula for equal-tailed, two-sided, α-level confidence intervals should be changed from that in (3.41) to

$$
\Big(\hat{\theta} + n^{-1/2}\sigma\big(z_{(1-\alpha)/2} + \hat{c}_{(1-\alpha)/2}\big),\ \hat{\theta} + n^{-1/2}\sigma\big(z_{(1+\alpha)/2} + \hat{c}_{(1+\alpha)/2}\big)\Big),
$$

for a suitable random function $\hat{c}_\alpha$. If $\hat{c}_\alpha$ is given by (3.27) then the coverage probability of this interval is given by (3.42), except that q should be changed to p at each appearance in that formula. The value of $a_{(1+\alpha)/2}$ is unchanged.

3.5.6 An Example

Except for the quantity a_α in the formula for r_2, the main terms in coverage probability expansions such as (3.40) and (3.42) are obtainable directly from the definition of the confidence interval. For example, as noted above, in the case of percentile-t intervals $s_1 = -q_{11} = q_1$ and $s_2 = q_{21}$. We shall illustrate the method of calculating a_α by considering the case of a confidence interval for a mean.

Adopt notation from Section 2.6 so that the sample is $\{W_1, \ldots, W_n\}$ with mean $\bar{W} = n^{-1} \sum W_i$ and variance $\hat{\beta}^2 = n^{-1} \sum (W_i - \bar{W})^2$. Let W denote a generic W_i, and write $m = E(W)$ and $\beta^2 = \mathrm{var}(W)$ for the population mean and variance, respectively. Take $\theta_0 = m$, $\hat{\theta} = \bar{W}$, and $\hat{\sigma}^2 = \hat{\beta}^2$. Then

$$
p_1(x) = -\tfrac{1}{6}\gamma(x^2 - 1), \qquad\qquad q_1(x) = \tfrac{1}{6}\gamma(2x^2 + 1),
$$

$$
p_2(x) = -x\left\{\tfrac{1}{24}\kappa(x^2 - 3) + \tfrac{1}{72}\gamma^2(x^4 - 10x^2 + 15)\right\}, \qquad (3.47)
$$

$$
q_2(x) = x\left\{\tfrac{1}{12}\kappa(x^2 - 3) - \tfrac{1}{18}\gamma^2(x^4 + 2x^2 - 3) - \tfrac{1}{4}(x^2 + 3)\right\},
$$

where $\gamma = E\{(W - m)/\beta\}^3$ and $\kappa = E\{(W - m)/\beta\}^4 - 3$; see (2.52), (2.53), (2.55), and (2.57). If s_1 denotes either of the polynomials p_1 or q_1 then $\hat{s}_1 - s_1 = (\hat{\gamma} - \gamma)\gamma^{-1} s_1$, and so calculation of a_α from formula (3.35) requires a tractable approximation to $\hat{\gamma} - \gamma$.

To obtain such an approximation, put

$$
U_j = n^{-1/2} \sum_{i=1}^{n} \left[\{(W_i - m)/\beta\}^j - E\{(W_i - m)/\beta\}^j\right]
$$

and note that

$\hat{\gamma} - \gamma$

$$
= \{\gamma + n^{-1/2}(U_3 - 3U_1) + O_p(n^{-1})\}\{1 + n^{-1/2}U_2 + O_p(n^{-1})\}^{-1} - \gamma
$$

$$
= n^{-1/2}(U_3 - 3U_1 - \tfrac{3}{2}\gamma U_2) + O_p(n^{-1}).
$$

Therefore

$$\Delta_n = n^{1/2}\{\hat{s}_1(z_\alpha) - s_1(z_\alpha)\} = n^{1/2}(\hat{\gamma} - \gamma)\gamma^{-1}s_1(z_\alpha)$$
$$= (U_3 - 3U_1 - \tfrac{3}{2}\gamma U_2)\gamma^{-1}s_1(z_\alpha) + O_p(n^{-1/2}). \qquad (3.48)$$

Similarly, in both the cases $T_n = n^{1/2}(\hat{\theta} - \theta_0)/\hat{\sigma}$ and $T_n = n^{1/2}(\hat{\theta} - \theta_0)/\sigma$ we have

$$T_n = U_1 + O_p(n^{-1/2}). \qquad (3.49)$$

Hence

$$E(T_n\Delta_n) = E\{U_1(U_3 - 3U_1 - \tfrac{3}{2}\gamma U_2)\gamma^{-1}s_1(z_\alpha)\} + O(n^{-1})$$
$$= (\kappa - \tfrac{3}{2}\gamma^2)\gamma^{-1}s_1(z_\alpha) + O(n^{-1}), \qquad (3.50)$$

where $\kappa = E\{(W - m)/\beta\}^4 - 3$. (That the remainder term here is $O(n^{-1})$, not simply $O(n^{-1/2})$, may be deduced by using more explicit forms of the remainder terms $O_p(n^{-1/2})$ in (3.48) and (3.49).) This proves that

$$a_\alpha = (\kappa - \tfrac{3}{2}\gamma^2)\gamma^{-1}s_1(z_\alpha). \qquad (3.51)$$

Using (3.40), (3.42), and (3.51) it is a simple matter to deduce formulae for coverage probabilities of the main types of confidence interval. We shall treat two cases, percentile and percentile-t. One- and two-sided percentile intervals were defined in Section 3.2; they are

$$\mathcal{I}_1 = \hat{I}_1 = \left(-\infty, \; \hat{\theta} - n^{-1/2}\hat{\sigma}\hat{u}_{1-\alpha}\right),$$
$$\mathcal{I}_2 = \hat{I}_2 = \left(\hat{\theta} - n^{-1/2}\hat{\sigma}\hat{u}_{(1+\alpha)/2}, \; \hat{\theta} - n^{-1/2}\hat{\sigma}\hat{u}_{(1-\alpha)/2}\right).$$

(The "other percentile intervals," $\hat{I}_{12}$ and $\hat{I}_{22}$, may be treated similarly.) Since

$$-\hat{u}_{1-\alpha} = -\{z_{1-\alpha} + n^{-1/2}\hat{p}_{11}(z_{1-\alpha}) + n^{-1}\hat{p}_{21}(z_{1-\alpha}) + \dots\}$$
$$= z_\alpha - n^{-1/2}\hat{p}_{11}(z_\alpha) + n^{-1}\hat{p}_{21}(z_\alpha) - \dots$$

then $s_1 = -p_{11} = p_1$ and $s_2(x) = p_{21}(x) = p_1(x)p_1'(x) - \tfrac{1}{2}xp_1(x)^2 - p_2(x)$. Hence by Proposition 3.1,

$$r_1(x) = p_1(x) - q_1(x) = -\tfrac{1}{2}\gamma x^2,$$
$$r_2(z_\alpha) = q_2(z_\alpha) - p_2(z_\alpha) - z_\alpha p_1(z_\alpha)^2 + p_1(z_\alpha)p_1'(z_\alpha)$$
$$\qquad\qquad + p_1(z_\alpha)\{z_\alpha q_1(z_\alpha) - q_1'(z_\alpha)\} - a_\alpha z_\alpha$$
$$= z_\alpha\{\tfrac{1}{24}\kappa(7z_\alpha^2 - 13) - \tfrac{1}{24}\gamma^2(3z_\alpha^4 + 6z_\alpha^2 - 11) - \tfrac{1}{4}(z_\alpha^2 + 3)\};$$

here we have used (3.47) and (3.51). In this notation the coverage probabilities are

$$P\big(m \in \widehat{I}_1\big) \;=\; \alpha + n^{-1/2}r_1(z_\alpha) + n^{-1}r_2(z_\alpha)\phi(z_\alpha) + \dots$$

and

$$P\big(m \in \widehat{I}_2\big) \;=\; \alpha + n^{-1}2r_2(z_{(1+\alpha)/2})\,\phi(z_{(1+\alpha)/2}) + \dots,$$

computed directly from (3.40) and (3.42).

One- and two-sided percentile-t intervals are

$$\mathcal{I}_1 \;=\; \widehat{J}_1 \;=\; \Big(-\infty,\; \hat\theta - n^{-1/2}\hat\sigma\hat v_{1-\alpha}\Big),$$

$$\mathcal{I}_2 \;=\; \widehat{J}_2 \;=\; \Big(\hat\theta - n^{-1/2}\hat\sigma\hat v_{(1+\alpha)/2},\; \hat\theta - n^{-1/2}\hat\sigma\hat v_{(1-\alpha)/2}\Big).$$

On this occasion, $s_1 = -q_{11} = q_1$ and $s_2(x) = q_1(x)q_1'(x) - \tfrac12 x q_1(x)^2 - q_2(x)$. Therefore, by Proposition 3.1, $r_1 \equiv 0$ and

$$\begin{aligned}
r_2(z_\alpha) &= -a_\alpha z_\alpha = -\big(\kappa - \tfrac32\,\gamma^2\big)\,\gamma^{-1}z_\alpha q_1(z_\alpha)\\
&= -\tfrac16\big(\kappa - \tfrac32\,\gamma^2\big)z_\alpha\big(2z_\alpha^2 + 1\big),
\end{aligned}$$

the last identity following from (3.47). Hence by (3.40) and (3.42),

$$P(m \in \widehat{J}_1) = \alpha - n^{-1}\tfrac16\big(\kappa - \tfrac32\gamma^2\big)z_\alpha\big(2z_\alpha^2 + 1\big)\phi(z_\alpha) + O(n^{-3/2}),$$
$$(3.52)$$
$$\begin{aligned}
P(m \in \widehat{J}_2) = \alpha &- n^{-1}\tfrac13\big(\kappa - \tfrac32\gamma^2\big)z_{(1+\alpha)/2}\big(2z_{(1+\alpha)/2}^2+1\big)\phi(z_{(1+\alpha)/2})\\
&+ O(n^{-2}).
\end{aligned}$$

3.6 Symmetric Bootstrap Confidence Intervals

Our description of bootstrap methods in Sections 3.3–3.5 conveys a particular view of the bootstrap in problems of distribution estimation: the bootstrap is a device for correcting an Edgeworth expansion or a Cornish-Fisher expansion for the first error term, due to the main effect of skewness. Of course, the bootstrap has important properties beyond Edgeworth correction; see for example Appendix V. However, its finesse at Edgeworth correction is one way of explaining the bootstrap's performance. When used correctly, the bootstrap approximation effectively removes the first error term in an Edgeworth expansion, and so its performance is generally an order of magnitude better than if only the "0th order" term, usually attributable to Normal approximation, had been accounted for. (However, note the caveats in the introduction to Section 3.10.)

Thus, one way of ensuring good performance from a bootstrap confidence interval is to pose the confidence interval problem in such a way that the terms after the first in an expansion of the distribution of the statistic on which the interval is based, are of as small an order as possible. Since the terms usually decrease in powers of $n^{-1/2}$ with the first term of size $n^{-1/2}$, this usually amounts to posing the problem so that the first term (representing the main effect of skewness) vanishes. That can be achieved in a variety of ways, for example by explicit correction to remove the term, or by transformation to minimize the effect of skewness. These methods will be discussed in Sections 3.8 and 3.9, respectively.

In the present section we point out that in the case of two-sided confidence intervals, the same objective may be achieved by constraining the interval to be symmetric, i.e., to be of the form $(\hat{\theta} - \hat{t}, \hat{\theta} + \hat{t})$. Naturally, we suggest choosing $\hat{t}$ by bootstrap methods. We shall consider only a pivotal (or percentile-t) approach, but there exist nonpivotal versions. The latter have coverage error of size n^{-1}, not n^{-2}, but are of interest because they are second-order correct for the theoretical (or "ideal") symmetric confidence interval J_{SY}, which we shall define shortly. This happens because, in view of parity properties of polynomials in Edgeworth expansions, Cornish-Fisher expansions of bootstrap quantiles are power series in n^{-1}, not $n^{-1/2}$, in the symmetric case. Result (3.56) below treats Cornish-Fisher expansions for the example of percentile-t; the percentile method and "other percentile method" are similar. This means that endpoints of both percentile and percentile-t symmetric confidence intervals differ from those of the theoretical interval in terms of order $n^{-1/2} \cdot n^{-1} = n^{-3/2}$. Thus, such confidence intervals are third-order correct. This property is perhaps of greater interest in the case of bootstrap multivariate confidence regions, which are often generalizations of symmetric univariate confidence intervals. For example, multivariate ellipsoidal regions of both the percentile and percentile-t type are second-order correct for their theoretical or "ideal" counterparts, as we shall show in Section 4.2.

We begin by defining theoretical and bootstrap forms of symmetric confidence intervals. Put $T = n^{1/2}(\hat{\theta} - \theta_0)/\hat{\sigma}$ and $T^* = n^{1/2}(\hat{\theta}^* - \theta_0)/\hat{\sigma}^*$, and let w_α and $\hat{w}_\alpha$ be the solutions of the equations

$$P\big(|T| \leq w_\alpha\big) \;=\; P\big(|T^*| \leq \hat{w}_\alpha \mid \mathcal{X}\big) \;=\; \alpha \, .$$

The theoretical symmetric interval $J_{\mathrm{SY}} = \big(\hat{\theta} - n^{-1/2}\,\hat{\sigma}\,w_\alpha \,,\; \hat{\theta} + n^{-1/2}\,\hat{\sigma}\,w_\alpha\big)$ has perfect coverage accuracy since

$$P\big(\theta_0 \in J_{\mathrm{SY}}\big) \;=\; P\big(|T| \leq w_\alpha\big) \;=\; \alpha \, .$$

However, J_{SY} is usually not computable since w_α is not known. The bootstrap interval $\widehat{J}_{\mathrm{SY}} = \big(\hat{\theta} - n^{-1/2}\,\hat{\sigma}\,\hat{w}_\alpha \,,\; \hat{\theta} + n^{-1/2}\,\hat{\sigma}\,\hat{w}_\alpha\big)$ is computable and has nominal coverage α. (In interpreting coverage formulae throughout this and later sections, the reader should bear in mind our convention that statistics such as T have a continuous distribution or at least have negligibly small atoms; note the last paragraph of subsection 3.5.1.)

Both w_α and $\hat{w}_\alpha$ converge to the Standard Normal critical point $z_{(1+\alpha)/2}$. We claim that

$$\hat{w}_\alpha - w_\alpha = O_p(n^{-3/2}), \qquad (3.53)$$

which immediately distinguishes symmetric bootstrap confidence intervals from their equal-tailed counterparts based on the quantile estimates $\hat{u}_\alpha$ or $\hat{v}_\alpha$. Recall from (3.23) and (3.24) that the differences $\hat{u}_\alpha - u_\alpha$ and $\hat{v}_\alpha - v_\alpha$ are of size n^{-1}, not $n^{-3/2}$.

Since $\hat{w}_\alpha - w_\alpha = O_p(n^{-3/2})$, the endpoints of $\widehat{J}_{\mathrm{SY}}$ agree with those of J_{SY} in terms of order $n^{-3/2}$; that is, $\widehat{J}_{\mathrm{SY}}$ is third-order correct for J_{SY}. This result implies that symmetric bootstrap confidence intervals will have greater coverage accuracy than their equal-tailed counterparts, which are only second-order correct. Indeed, since $T \mp \hat{w}_\alpha$ and $T \mp w_\alpha$ differ only in terms of $O_p(n^{-3/2})$, by (3.53), using the delta method (Section 2.7),

$$P\big(T \le \pm \hat{w}_\alpha\big) = P\big(T \le \pm w_\alpha\big) + O(n^{-3/2}).$$

Therefore

$$
\begin{aligned}
P\big(\theta_0 \in \widehat{J}_{\mathrm{SY}}\big) &= P\big(|T| \le \hat{w}_\alpha\big) \\
&= P\big(|T| \le w_\alpha\big) + O(n^{-3/2}) = \alpha + O(n^{-3/2}).
\end{aligned}
$$

The coverage error of $\widehat{J}_{\mathrm{SY}}$ is actually equal to $O(n^{-2})$, not simply $O(n^{-3/2})$. In this respect, symmetric intervals represent an improvement over equal-tailed intervals, which usually have coverage error of size n^{-1}; see subsection 3.5.5. The fact that $\widehat{J}_{\mathrm{SY}}$ has coverage error of order n^{-2} follows from the usual parity properties of polynomials in Edgeworth expansions. Indeed, the analogue of the coverage expansion (3.42) in the present context is, with $z = z_{(1+\alpha)/2}$,

$$
\begin{aligned}
P\big(\theta_0 &\in \widehat{J}_{\mathrm{SY}}\big) \\
&= \Phi(z) + n^{-3/2} r_3(z)\,\phi(z) + n^{-2} r_4(z)\,\phi(z) + \ldots \\
&\quad - \Big\{\Phi(-z) + n^{-3/2} r_3(-z)\,\phi(-z) + n^{-2} r_4(-z)\,\phi(-z) + \ldots \Big\} \\
&= \alpha + 2n^{-2} r_4(z)\,\phi(z) + O(n^{-3}),
\end{aligned}
$$

where $r_3, r_4, \ldots$ are polynomials (not the same as in Proposition 3.1), with r_j being an odd/even function for even/odd j respectively. Therefore $P\big(\theta_0 \in \widehat{J}_{\mathrm{SY}}\big) = \alpha + O(n^{-2})$.

In summary, symmetric bootstrap confidence intervals are third-order correct for their theoretical counterparts and have coverage error of order n^{-2}. They can actually be shorter than their equal-tailed counterparts, as we shall prove shortly. However, we should stress that symmetric confidence intervals are not a panacea for inaccuracies of the bootstrap. For one thing, our results about good performance are only asymptotic in character and are not necessarily available for all sample sizes. Furthermore, it is not always appropriate to constrain a confidence interval to be symmetric. The asymmetry of an equal-tailed confidence interval can convey important information about our uncertainty as to the location of the true parameter value, and it is not always prudent to ignore that information. This issue will be discussed in more detail in Appendix III.

In the remainder of this section we prove (3.53), from which follows our claim about third-order correctness. We also discuss interval length, and begin by developing a Cornish-Fisher expansion of w_α. Now,

$$
\begin{aligned}
\alpha &= P(T \leq w_\alpha) - P(T \leq -w_\alpha) \\
&= \Phi(w_\alpha) + n^{-1/2}q_1(w_\alpha)\phi(w_\alpha) + n^{-1}q_2(w_\alpha)\phi(w_\alpha) + \ldots \\
&\quad - \Big\{ \Phi(-w_\alpha) + n^{-1/2}q_1(-w_\alpha)\phi(-w_\alpha) + n^{-1}q_2(-w_\alpha)\phi(-w_\alpha) + \ldots \Big\}.
\end{aligned}
$$

(Here we have invoked the Edgeworth expansion (3.12).) Since q_j is an odd/even function for even/odd j, respectively, then terms of odd order in $n^{-1/2}$ vanish from this formula. Therefore

$$
\alpha = 2\Phi(w_\alpha) - 1 + 2\Big\{ n^{-1}q_2(w_\alpha)\phi(w_\alpha) + n^{-2}q_4(w_\alpha)\phi(w_\alpha) + \ldots \Big\}. \quad (3.54)
$$

Inverting, we derive a Cornish-Fisher expansion,

$$
w_\alpha = z_\xi + n^{-1}q_{13}(z_\xi) + n^{-2}q_{23}(z_\xi) + \ldots, \quad (3.55)
$$

where $\xi = \frac{1}{2}(1+\alpha)$,[3]

$$
\begin{aligned}
q_{13}(z) &= -q_2(z), \\
q_{23}(x) &= q_2(z)q_2'(z) - \tfrac{1}{2}xq_2(x)^2 - q_4(x).
\end{aligned}
$$

The bootstrap estimate of w_α is $\hat{w}_\alpha$, the solution of the equation $P(|T^*| \leq \hat{w}_\alpha \,|\, \mathcal{X}) = \alpha$ where $T^* = n^{1/2}(\hat{\theta}^* - \hat{\theta})/\hat{\sigma}^*$. By analogy with

---

[3] A note on notation is in order here. Given an Edgeworth expansion of a distribution function G in which the jth polynomial is r_j, we use r_{j1} to denote the jth polynomial in a Cornish-Fisher expansion of a G-quantile in terms of a Normal quantile, we let r_{j2} be the jth polynomial in a Cornish-Fisher expansion of a Normal quantile in terms of a G-quantile (see Section 2.5 for examples of the latter two notations), and we write r_{j3} for the jth polynomial in a Cornish-Fisher expansion of a quantile of the distribution function $G(x) - G(-x)$, $x \geq 0$.

Section 3.3, $\hat{w}_\alpha$ admits the obvious bootstrap analogue of expansion (3.55),

$$\hat{w}_\alpha = z_\xi + n^{-1}\hat{q}_{13}(z_\xi) + n^{-2}\hat{q}_{23}(z_\xi) + \dots, \qquad (3.56)$$

where $\hat{q}_{j3}$ is obtained from q_{j3} on replacing population moments by their sample estimates. (A formal proof is similar to that of Theorem 5.2.) In particular,

$$\hat{q}_{13}(x) = -\hat{q}_2(x), \quad \hat{q}_{23}(x) = \hat{q}_2(x)\hat{q}_2'(x) - \tfrac{1}{2}x\hat{q}_2(x)^2 - \hat{q}_4(x). \qquad (3.57)$$

Thus,

$$\begin{aligned}
\hat{w}_\alpha - w_\alpha &= n^{-1}\{\hat{q}_{13}(z_\xi) - q_{13}(z_\xi)\} + n^{-1}\{\hat{q}_{23}(z_\xi) - q_{23}(z_\xi)\} + \dots \\
&= -n^{-1}\{\hat{q}_2(z_\xi) - q_2(z_\xi)\} + O_p(n^{-5/2}) \\
&= O_p(n^{-3/2}), \qquad\qquad\qquad\qquad\qquad (3.58)
\end{aligned}$$

the last identity following from the fact that $\hat{q}_2 - q_2 = O_p(n^{-1/2})$. This proves (3.53).

In conclusion, we compare the length of the symmetric bootstrap confidence interval

$$\widehat{J}_{SY} = \left(\hat{\theta} - n^{-1/2}\hat{\sigma}\hat{w}_\alpha, \ \hat{\theta} + n^{-1/2}\hat{\sigma}\hat{w}_\alpha \right)$$

with that of its equal-tailed counterpart

$$\widehat{J}_2 = \left(\hat{\theta} - n^{-1/2}\hat{\sigma}\hat{v}_\xi, \ \hat{\theta} + n^{-1/2}\hat{\sigma}\hat{v}_{1-\xi} \right),$$

where $\xi = \tfrac{1}{2}(1 + \alpha)$. Let these lengths be denoted by

$$l_{SY} = 2n^{-1/2}\hat{\sigma}\hat{w}_\alpha, \qquad l_2 = n^{-1/2}\hat{\sigma}(\hat{v}_\xi - \hat{v}_{1-\xi}),$$

respectively. We claim that

$$l_{SY} - l_2 = n^{-3/2}\hat{\sigma}q_1(z_\xi)\left\{z_\xi q_1(z_\xi) - 2q_1'(z_\xi)\right\} + O_p(n^{-5/2}), \qquad (3.59)$$

where q_1 is the polynomial appearing in the coefficient of $n^{-1/2}$ in an Edgeworth expansion of the distribution of T. We shall derive (3.59) shortly. To appreciate the implications of (3.59), observe from Section 2.4 that if the first three cumulants of T equal $n^{-1/2}A_1$, 1, and $n^{-1/2}A_2$ to order $n^{-1/2}$, then

$$q_1(x) = -\left\{A_1 + \tfrac{1}{6}A_2(x^2 - 1)\right\};$$

see (2.36). In this event, and for $x > 0$, $q_1(x)\{xq_1(x) - 2q_1'(x)\} < 0$ if and only if

$$\left\{x^2 + (6A_1/A_2) - 1\right\}\left\{x^2 + (6A_1/A_2) - 5\right\} < 0.$$

Therefore a necessary and sufficient condition for

$$q_1(z_\xi)\left\{z_\xi q_1(z_\xi) - 2q_1'(z_\xi)\right\} < 0$$

is

$$1 - (6A_1/A_2) < z_\xi^2 < 5 - (6A_1/A_2), \tag{3.60}$$

or equivalently, assuming that the first three moments of T are $n^{-1/2}\mu_1$, 1, and $n^{-1/2}\mu_3$, to order $n^{-1/2}$,

$$\frac{\mu_3 - 9\mu_1}{\mu_3 - 3\mu_1} < z_\xi^2 < \frac{5\mu_3 - 21\mu_1}{\mu_3 - 3\mu_1}.$$

(Note that $A_1 = \mu_1$ and $A_2 = \mu_3 - 3\mu_1$; see Sections 2.3 and 2.4 for an account of the way in which cumulants, and moments, influence Edgeworth expansions.) When (3.60) holds, the probability that the symmetric bootstrap interval $\widehat{J}_{SY}$ has shorter length than its equal-tailed counterpart $\widehat{J}_2$ converges to one as $n \to \infty$.

Two examples involving the nonparametric bootstrap will serve to illustrate that these inequalities can be satisfied in cases of practical interest. This indicates that symmetric intervals do not necessarily have disadvantages, in terms of length, relative to equal-tailed intervals. Firstly, observe that when constructing a confidence interval for the mean m of a population, based on a sample $\{W_1, \ldots, W_n\}$, we would take $T = n^{1/2}(\hat{\theta} - \theta_0)/\hat{\sigma}$ where $\theta_0 = m$, $\hat{\theta} = \overline{W}$, and $\hat{\sigma}^2 = n^{-1}\sum(W_i - \overline{W})^2$. In this case, $A_1 = -\frac{1}{2}\gamma$ and $A_2 = -2\gamma$, where $\gamma = E\{(W - m)/\beta\}^3$ and $\beta^2 = \text{var}(W)$. See Example 2.1 in Section 2.6 for details. Therefore (3.60) reduces to $z_\xi^2 < \frac{7}{2}$, i.e., $0 \leq \alpha \leq 0.94$. It follows that symmetric 90% intervals tend to be shorter than equal-tailed 90% intervals, while symmetric 95% intervals tend to be slightly longer than their equal-tailed counterparts.

For the second example, consider constructing a confidence interval for population variance β^2, based on a sample $\{W_1, \ldots, W_n\}$. Here

$$\theta_0 = \beta^2, \qquad \hat{\theta} = n^{-1}\sum(W_i - \overline{W})^2, \qquad \hat{\sigma}^2 = n^{-1}\sum(W_i - \overline{W})^4 - \hat{\theta}^2,$$

$$A_1 = \tfrac{1}{2}(\nu_4 - 1)^{-3/2}(4\nu_3^2 + \nu_4 - \nu_6),$$

$$A_2 = 2(\nu_4 - 1)^{-3/2}(3\nu_3^2 + 3\nu_4 - \nu_6 - 2),$$

where $\nu_j = E\{(W - m)/\beta\}^j$. See Example 2.2 in Section 2.6; A_1, A_2 were denoted by C_1, C_2 there. Therefore, (3.60) becomes

$$\frac{-6\nu_3^2 - 3\nu_4 - \nu_6 + 4}{2(3\nu_3^2 + 3\nu_4 - \nu_6 - 2)} < z_\xi^2 < \frac{18\nu_3^2 + 27\nu_4 - 7\nu_6 - 20}{2(3\nu_3^2 + 3\nu_4 - \nu_6 - 2)}. \tag{3.61}$$

If the population happens to be Normal then $\nu_3 = 0$, $\nu_4 = 3$, $\nu_6 = 15$, and (3.61) reduces to $z_\xi^2 < \frac{11}{4}$, i.e., $0 \leq \alpha \leq 0.90$. Therefore, symmetric 90% intervals tend to be approximately the same length as equal-tailed 90% intervals, while symmetric 95% intervals tend to be longer than their equal-tailed counterparts.

Finally, we verify (3.59). Recall that

$$l_{\mathrm{SY}} = 2n^{-1/2}\hat{\sigma}\hat{w}_\alpha \qquad \text{and} \qquad l_2 = n^{-1/2}\hat{\sigma}(\hat{v}_\xi - \hat{v}_{1-\xi}),$$

whence

$$l_{\mathrm{SY}} - l_2 = n^{-1/2}\hat{\sigma}(2\hat{w}_\alpha - \hat{v}_\xi + \hat{v}_{1-\xi}).$$

By (3.55) and (3.56),

$$\hat{w}_\alpha = z_\xi - n^{-1}\hat{q}_2(z_\xi) + O_p(n^{-2});$$

and by (3.16) and (3.21),

$$\hat{v}_\xi = z_\xi + n^{-1/2}\hat{q}_{11}(z_\xi) + n^{-1}\hat{q}_{21}(z_\xi) + n^{-3/2}\hat{q}_{31}(z_\xi) + O_p(n^{-2}),$$

where

$$\hat{q}_{21}(x) = \hat{q}_1(x)\,\hat{q}_1'(x) - \tfrac{1}{2}x\hat{q}_1(x)^2 - \hat{q}_2(x)$$

and $\hat{q}_{11}$, $\hat{q}_{31}$ are even polynomials. Hence, noting that $z_{1-\xi} = -z_\xi$, we find that

$$l_{\mathrm{SY}} - l_2 = -2n^{-3/2}\hat{\sigma}\Big\{\hat{q}_2(z_\xi) + \hat{q}_{21}(z_\xi)\Big\} + O_p(n^{-5/2})$$

$$= n^{-3/2}\hat{\sigma}\Big\{z_\xi\hat{q}_1(z_\xi)^2 + 2\hat{q}_1(z_\xi)\hat{q}_1'(z_\xi)\Big\} + O_p(n^{-5/2}),$$

which establishes (3.59).

3.7 Short Bootstrap Confidence Intervals

Short confidence intervals were introduced in Section 1.3. We now elucidate their properties by applying techniques from Chapter 2. To recall the definition of short confidence intervals, note first that the quantiles v_α and $\hat{v}_\alpha$ are given by (3.6) and (3.8), respectively:

$$P\{n^{1/2}(\hat{\theta} - \theta_0)/\hat{\sigma} \leq v_\alpha\} = \alpha, \qquad P\{n^{1/2}(\hat{\theta}^* - \hat{\theta})/\hat{\sigma}^* \leq \hat{v}_\alpha \mid \mathcal{X}\} = \alpha.$$

If we knew v_α for a sufficiently wide range of α's then we could construct the theoretical (or "ideal") version of a short confidence interval,

$$J_{\mathrm{SH}} = \left(\hat{\theta} - n^{-1/2}\hat{\sigma}v_{\alpha+\beta}, \ \hat{\theta} - n^{-1/2}\hat{\sigma}v_\beta\right),$$

where $\beta = \delta$ minimizes $v_{\alpha+\delta} - v_\delta$ (which is proportional to interval length). In practice, J_{SH} is usually unobtainable, and so attention focuses on the sample (or bootstrap) version,

$$\widehat{J}_{\mathrm{SH}} = \left(\hat{\theta} - n^{-1/2}\hat{\sigma}\hat{v}_{\alpha+\hat{\beta}}, \ \hat{\theta} - n^{-1/2}\hat{\sigma}\hat{v}_{\hat{\beta}}\right),$$

where $\hat{\beta} = \delta$ minimizes $\hat{v}_{\alpha+\delta} - \hat{v}_\delta$.

We begin our study of J_{SH} and $\widehat{J}_{\text{SH}}$ by developing Cornish-Fisher expansions of the quantiles v_β and $v_{\alpha+\beta}$. Recall from (3.12) that the distribution function G of $n^{1/2}(\hat{\theta} - \theta_0)/\hat{\sigma}$ admits an Edgeworth expansion of the form

$$G(x) = P\{n^{1/2}(\hat{\theta} - \theta_0)/\hat{\sigma} \le x\} = \Phi(x) + \sum_{j \ge 1} n^{-j/2} q_j(x)\phi(x), \quad (3.62)$$

for polynomials q_j. Define $\xi = \frac{1}{2}(1 + \alpha)$ and

$$\psi_{ik} = (\partial/\partial x)^k \{q_i(x)\phi(x)\}\big|_{x=z_\xi},$$

and let $\sum_{(k,r)}$ denote summation over $j_1, \ldots, j_k \ge 1$ such that $j_1 + \ldots + j_k = r$. We claim that $v_{\alpha+\beta}$, v_β admit the Cornish-Fisher expansions

$$v_{\alpha+\beta} = z_\xi + \sum_{j \ge 1} n^{-j/2} a_j, \quad v_\beta = -\left\{ z_\xi + \sum_{j \ge 1} (-n^{-1/2})^j a_j \right\}, \quad (3.63)$$

where $a_1, a_2, \ldots$ are determined by minimizing $a_2, a_4, \ldots$ subject to

$$\psi_{2l,0} + \sum_{i=0}^{2l-1} \sum_{k=1}^{2l-1} \tfrac{1}{k!} \psi_{ik} \sum_{(k,2l-i)} a_{j_1} \ldots a_{j_k} = 0, \quad l \ge 1. \quad (3.64)$$

This prescription gives

$$a_1 = -\psi_{11}\psi_{02}^{-1} = z_\beta^{-1} q_1'(z_\xi) - q_1(z_\xi), \quad (3.65)$$

$$a_2 = \left(\tfrac{1}{2}\psi_{11}^2\psi_{02}^{-1} - \psi_{20}\right)\psi_{01}^{-1}$$

$$= -q_2(z_\xi) - \tfrac{1}{2} z_\xi \left\{ z_\xi^{-1} q_1'(z_\xi) - q_1(z_\xi) \right\}^2, \quad (3.66)$$

with more complex formulae for higher-order a_j's.

To check that our prescription is correct, observe that both $a = v_{\alpha+\beta}$ and $b = -v_\beta$ converge to z_ξ as $n \to \infty$, and in fact

$$a = z_\xi + \sum_{j \ge 1} n^{-j/2} a_j, \quad b = z_\xi + \sum_{j \ge 1} n^{-j/2} b_j$$

for quantities a_j, b_j that remain to be determined.

If $J_{\text{SH}} = \left(\hat{\theta} - n^{-1/2}\hat{\sigma}a, \hat{\theta} + n^{-1/2}\hat{\sigma}b\right)$ is the short confidence interval for θ then $\left(-\hat{\theta} - n^{-1/2}\hat{\sigma}b, -\hat{\theta} + n^{-1/2}\hat{\sigma}a\right)$ is the short confidence interval for $-\theta$. From this fact and the relation

$$P\left[n^{1/2}\{(-\hat{\theta}) - (-\theta)\}/\hat{\sigma} \le x\right] = \Phi(x) + \sum_{j \ge 1} (-n^{-1/2})^j q_j(x)\,\phi(x)$$

(which follows from (3.62) and the parity properties of q_j) we may deduce that

$$b = z_\xi + \sum_{j \ge 1} (-n^{-1/2})^j a_j,$$

which entails $a_j = (-1)^j b_j$. The coverage probability of the interval J_{SH} is therefore equal to

$$\alpha = G(a) - G(-b)$$

$$= \sum_{i \geq 0} n^{-i/2} \left[\phi_i \left(z_\xi + \sum_{j \geq 1} n^{-j/2} a_j \right) - \phi_i \left\{ - z_\xi - \sum_{j \geq 1} (-n^{-1/2})^j a_j \right\} \right]$$

$$= \sum_{i \geq 0} \sum_{k \geq 0} n^{-i/2} \tfrac{1}{k!} \left[\left(\sum_{j \geq 1} n^{-j/2} a_j \right) + (-1)^k \left\{ \sum_{j \geq 1} (-n^{-1/2})^j a_j \right\}^k \right] \psi_{ik} \,,$$

where $\phi_0(x) = \Phi(x)$ and $\phi_i(x) = q_i(x)\phi(x)$ for $i \geq 1$. In this formula, all coefficients of $n^{-i/2}$ for odd i vanish without further conditions. Equating to zero the coefficient of n^{-l} for $l \geq 1$, we obtain (3.64). Finally, the length of the interval J_{SH} equals $n^{-1/2}\hat{\sigma}$ multiplied by

$$a + b = 2 \left(z_\xi + \sum_{j \geq 1} n^{-j} a_{2j} \right),$$

and so is minimized by minimizing $a_2, a_4, \ldots$ subject to (3.64).

The quantity a_j may always be written as $a_j = t_j(z_\xi)$, where t_j is a polynomial whose properties are typical of those associated with Cornish-Fisher expansions: t_j is of degree at most $j+1$, and is an odd/even function for even/odd j, respectively. (It shares these features with the polynomials p_{j1}, q_{j1} introduced in Section 2.5.) In the case of estimating a population mean we may take $\hat{\theta}, \hat{\sigma}^2$ to equal sample mean, sample variance respectively, in which circumstance

$$q_1(x) = \tfrac{1}{6} \gamma(2x^2 + 1),$$

$$q_2(x) = x \left\{ \tfrac{1}{12} \kappa(x^2 - 3) - \tfrac{1}{18} \gamma^2(x^4 + 2x^2 - 3) - \tfrac{1}{4} (x^2 + 3) \right\},$$

where γ, κ denote skewness and kurtosis, respectively, in the parent population. See (2.55) and (2.57). It now follows from (3.65) and (3.66) that

$$a_1 = -\tfrac{1}{6} \gamma \left(2z_\xi^2 - 3 \right), \tag{3.67}$$

$$a_2 = z_\xi \left\{ - \tfrac{1}{12} \kappa \left(z_\xi^2 - 3 \right) + \tfrac{1}{72} \gamma^2 \left(20 z_\xi^2 - 21 \right) + \tfrac{1}{4} \left(z_\xi^2 + 3 \right) \right\}. \tag{3.68}$$

It is straightforward to deduce properties of short *bootstrap* confidence intervals from these results, as follows. Let $\hat{t}_j$ denote the version of t_j in which each population moment is replaced by the corresponding sample moment. (The relationship between $\hat{t}_j$ and t_j is exactly the same as that between $\hat{q}_{1j}$ and q_{1j}; see Section 3.3 for the latter.) Put $\hat{a}_j = \hat{t}_j(z_\xi)$. Then, in analogy to (3.63),

$$\hat{v}_{\alpha+\hat{\beta}} = z_\xi + \sum_{j \geq 1} n^{-j/2} \hat{a}_j, \qquad \hat{v}_{\hat{\beta}} = - \left\{ z_\xi + \sum_{j \geq 1} (-n^{-1/2})^j \hat{a}_j \right\}.$$

Of course, these Cornish-Fisher expansions are to be interpreted as asymptotic series. Thus, the lengths of confidence intervals J_{SH}, $\widehat{J}_{\mathrm{SH}}$ are, respectively,

$$n^{-1/2}\hat{\sigma}(v_{\alpha+\beta} - v_\beta) = 2n^{-1/2}\hat{\sigma}\big(z_\xi + n^{-1}a_2 + n^{-2}a_4\big) + O_p(n^{-7/2}),$$

$$n^{-1/2}\hat{\sigma}(\hat{v}_{\alpha+\beta} - \hat{v}_\beta) = 2n^{-1/2}\hat{\sigma}\big(z_\xi + n^{-1}\hat{a}_2 + n^{-2}\hat{a}_4\big) + O_p(n^{-7/2}).$$

Therefore, the lengths differ in terms of size n^{-2} (since $\hat{a}_j - a_j$ is of size $n^{-1/2}$), whereas the expected lengths differ in terms of size $n^{-5/2}$ (since $E(\hat{\sigma}\hat{a}_j) - E(\hat{\sigma})a_j$ is of size n^{-1}).

The coverage probability of the short bootstrap interval $\widehat{J}_{\mathrm{SH}}$ may be readily deduced from Proposition 3.1 in Section 3.5, which implies that

$$P\{\theta_0 \in (-\infty,\ \hat{\theta} - n^{-1/2}\hat{\sigma}\hat{v}_{\hat{\beta}})\}$$

$$= P\Big[\theta_0 \in \Big(-\infty,\ \hat{\theta} + n^{-1/2}\hat{\sigma}\{z_\xi - n^{-1/2}\hat{t}_1(z_\xi) + n^{-1}\hat{t}_2(z_\xi) - \ldots\}\Big)\Big]$$

$$= \xi + n^{-1/2}r_1(z_\xi)\phi(z_\xi) + n^{-1}r_2(z_\xi)\phi(z_\xi)$$
$$+ n^{-3/2}r_3(z_\xi)\phi(z_\xi) + O(n^{-2}), \qquad (3.69)$$

and

$$P\{\theta_0 \in (-\infty,\ \hat{\theta} - n^{-1/2}\hat{\sigma}\hat{v}_{\alpha+\hat{\beta}})\}$$

$$= 1 - \xi + n^{-1/2}r_1(z_\xi)\phi(z_\xi) - n^{-1}r_2(z_\xi)\phi(z_\xi)$$
$$+ n^{-3/2}r_3(z_\xi)\phi(z_\xi) + O(n^{-2}), \qquad (3.70)$$

where

$$r_2(z_\xi) = q_2(z_\xi) + t_2(z_\xi) - \tfrac{1}{2}z_\xi t_1(z_\xi)^2$$
$$- t_1(z_\xi)\{z_\xi q_1(z_\xi) - q_1'(z_\xi)\} - a_\xi z_\xi$$
$$= -a_\xi z_\xi$$

and a_ξ is defined by

$$-E\Big[n^{1/2}(\hat{\theta} - \theta_0)\hat{\sigma}^{-1} n^{1/2}\{\hat{t}_1(z_\xi) - t_1(z_\xi)\}\Big] = a_\xi + O(n^{-1}). \qquad (3.71)$$

Subtracting (3.69) and (3.70), and remembering that $\xi = \tfrac{1}{2}(1 + \alpha)$, we deduce that

$$P(\theta_0 \in J_{\mathrm{SH}}) = \alpha - n^{-1}2a_\xi z_\xi\, \phi(z_\xi) + O(n^{-2}). \qquad (3.72)$$

In the case of estimating a mean, where $\hat{\theta}$, $\hat{\sigma}^2$ denote sample mean and sample variance, respectively, we have by (3.67),

$$\hat{t}_1(z_\xi) - t_1(z_\xi) = \hat{a}_1 - a_1 = -\tfrac{1}{6}(\hat{\gamma} - \gamma)(2z_\xi^2 - 3).$$

We may now deduce from (3.50) and (3.71) that

$$a_\xi = \tfrac{1}{6}\left(\kappa - \tfrac{3}{2}\gamma^2\right)\left(2z_\xi^2 - 3\right),$$

and hence by (3.72),

$$\pi_{\text{SH}} = P\big(\theta_0 \in \widehat{J}_{\text{SH}}\big) = \alpha - n^{-1}\tfrac{1}{3}\left(\kappa - \tfrac{3}{2}\gamma^2\right)z_\xi\left(2z_\xi^2 - 3\right)\phi(z_\xi) + O(n^{-2}).$$

The equal-tailed, percentile-t counterpart of $\widehat{J}_{\text{SH}}$, the interval $\widehat{J}_2$ defined in Section 3.2, has coverage probability

$$\pi_2 = P\big(\theta_0 \in \widehat{J}_2\big) = \alpha - n^{-1}\tfrac{1}{3}\left(\kappa - \tfrac{3}{2}\gamma^2\right)z_\xi(2z_\xi^2 + 1)\phi(z_\xi) + O(n^{-2});$$

see (3.52). Therefore the ratio of coverage errors of short and equal-tailed percentile-t intervals is

$$\frac{\pi_{\text{SH}} - \alpha}{\pi_2 - \alpha} = \frac{2z_\xi^2 - 3}{2z_\xi^2 + 1} + O(n^{-1}).$$

The ratio on the right-hand side does not depend on population moments and is always positive for $z_\xi > \left(\tfrac{3}{2}\right)^{1/2}$, corresponding to nominal coverage $\alpha = 2\xi - 1 > 0.78$. For the important cases $\alpha = 0.90$, 0.95, and 0.99, the ratio equals 0.38, 0.54, and 0.72 respectively. Thus, in the case of estimating a mean, the coverage error of short bootstrap confidence intervals is typically of the same sign as, but smaller in magnitude than, the error of equal-tailed percentile-t intervals.

There exist multivariate versions of short confidence regions, of which one is the rectangular prism of smallest content with sides parallel to coordinate axes. The latter may be used as the basis for simultaneous confidence intervals for components of a parameter vector, and will be discussed at greater length in Section 4.2. Such "smallest prism" confidence regions are of theoretical interest because they are one of relatively few practicable types of multivariate bootstrap confidence regions that are not based on generalizing the concept of symmetric univariate confidence intervals.

Observe that the endpoints of short confidence intervals are proper power series in $n^{-1/2}$, not n^{-1}; see (3.63). Recall from the previous section that in the case of symmetric intervals, the analogous power series are in n^{-1}. If the endpoints of a short bootstrap confidence interval are constructed by the percentile method, rather than percentile-t, then the endpoints will differ from those of J_{SH} by terms of size n^{-1}; that is, they will be only first-order correct. In contrast, a percentile method symmetric bootstrap confidence interval is second-order correct for the "ideal" symmetric interval J_{SY}. This property persists for multivariate regions: percentile method regions of the "smallest prism" type are only first-order correct for their "ideal" (i.e., theoretical or nonbootstrap) counterpart, whereas percentile method symmetric regions (e.g., ellipsoidal regions) are second-order correct. Of course, percentile-t regions of both types are second-order correct for their "ideal" counterparts.

3.8 Explicit Corrections for Skewness

Let us assume that a confidence interval for θ_0 is to be based on the statistic $T = n^{1/2}(\hat{\theta} - \theta_0)/\hat{\sigma}$ and that the distribution of T admits the expansion

$$P(T \leq x) = \Phi(x) + n^{-1/2}q_1(x)\phi(x) + O(n^{-1}).$$

It may be proved via the delta method that

$$P\{T \leq x - n^{-1/2}\hat{q}_1(x)\} = \Phi(x) + O(n^{-1}),$$

where $\hat{q}_1$ is identical to q_1 except that population moments in coefficients are replaced by their sample counterparts. Then, the nominal α-level confidence interval,

$$\mathcal{I} = \left(-\infty, \ \hat{\theta} - n^{-1/2}\hat{\sigma}z_{1-\alpha} + n^{-1}\hat{\sigma}\hat{q}_1(z_{1-\alpha}) \right), \qquad (3.73)$$

has coverage error $O(n^{-1})$, not $O(n^{-1/2})$

$$P(\theta_0 \in \mathcal{I}) = P\left\{ n^{1/2}(\hat{\theta} - \theta_0)/\hat{\sigma} > z_{1-\alpha} - n^{-1/2}\hat{q}_1(z_{1-\alpha}) \right\}$$
$$= 1 - \Phi(z_{1-\alpha}) + O(n^{-1}) = \alpha + O(n^{-1}). \qquad (3.74)$$

The interval $\mathcal{I}$ is the result of a single Edgeworth correction.

We may correct the interval further, either by using the bootstrap or by applying another Edgeworth correction. To describe the bootstrap approach, let $\hat{q}_1^*$ denote the version of $\hat{q}_1$ constructed from the resample $\mathcal{X}^*$ rather than the sample $\mathcal{X}$ and define $\hat{y}_\alpha$ to be the solution of

$$P\left\{ n^{1/2}(\hat{\theta}^* - \hat{\theta})/\hat{\sigma}^* > z_{1-\alpha} - n^{-1/2}\hat{q}_1^*(z_{1-\alpha}) + \hat{y}_\alpha \mid \mathcal{X} \right\} = \alpha.$$

Then the interval

$$\mathcal{I}' = \left(-\infty, \ \hat{\theta} - n^{-1/2}\hat{\sigma}z_{1-\alpha} + n^{-1}\hat{\sigma}\hat{q}_1(z_{1-\alpha}) - n^{-1/2}\hat{\sigma}\hat{y}_\alpha \right)$$

has coverage error $O(n^{-3/2})$

$$P(\theta_0 \in \mathcal{I}') = P\left\{ n^{1/2}(\hat{\theta} - \theta_0)/\hat{\sigma} > z_{1-\alpha} - n^{-1/2}\hat{q}_1(z_{1-\alpha}) + \hat{y}_\alpha \right\}$$
$$= \alpha + O(n^{-3/2}).$$

The quantity $\hat{y}_\alpha$ is of order n^{-1} and amounts to a bootstrap correction for the first error term in an Edgeworth expansion of the coverage probability of $\mathcal{I}$.

That term may be corrected for explicitly, in much the same way that we introduced the original Edgeworth correction $n^{-1}\hat{\sigma}\hat{q}_1(z_{1-\alpha})$. To appreciate the method, observe that the coverage probability in (3.74) may be expressed as an Edgeworth expansion,

$$P(\theta_0 \in \mathcal{I}) = \alpha + n^{-1}r_1(z_{1-\alpha})\,\phi(z_{1-\alpha}) + O(n^{-3/2}),$$

where r_1 is a polynomial. Replace population moments in coefficients of r_1 by sample moments, obtaining an estimate $\hat{r}_1$ of r_1. We may use $\hat{r}_1$ as a subsidiary correction, to obtain the new confidence interval,

$$\mathcal{I}'' = \left(-\infty, \;\; \hat{\theta} - n^{-1/2}\hat{\sigma}z_{1-\alpha} + n^{-1}\hat{\sigma}\hat{q}_1(z_{1-\alpha}) - n^{-3/2}\hat{\sigma}\hat{r}_1(z_{1-\alpha}) \right),$$

which has coverage error $O(n^{-3/2})$:

$$P(\theta_0 \in \mathcal{I}'') = \alpha + n^{-3/2}r_2(z_{1-\alpha})\,\phi(z_{1-\alpha}) + O(n^{-2}),$$

where r_2 is a new polynomial. This procedure may be repeated, correcting for the term in r_2 and obtaining a confidence interval with coverage error $O(n^{-2})$.

In the illustrations above we applied Edgeworth and bootstrap corrections to one-sided confidence intervals, since this is the simplest case to describe. However, they may be used for two-sided confidence intervals, with only minor changes. The most important change is that each successive correction reduces coverage error by the factor n^{-1}, not $n^{-1/2}$. This derives from the fact that an Edgeworth expansion of coverage probability for a two-sided confidence interval is a power series in n^{-1}, not $n^{-1/2}$; see subsection 3.5.5. For example, suppose we wish to Edgeworth correct the Normal approximation interval

$$\mathcal{J} = \left(\hat{\theta} - n^{-1/2}\hat{\sigma}z_\xi, \;\; \hat{\theta} + n^{-1/2}\hat{\sigma}z_\xi \right),$$

where $\xi = \frac{1}{2}(1+\alpha)$. The coverage probability of $\mathcal{J}$ may be expanded as

$$
\begin{aligned}
P(\theta_0 \in \mathcal{J}) &= P(|T| \le z_\xi) \\
&= \alpha + 2\left\{ n^{-1}q_2(z_\xi)\phi(z_\xi) + n^{-2}q_4(z_\xi)\phi(z_\xi) + \ldots \right\},
\end{aligned}
$$

and so the first Edgeworth correction is based on $n^{-1}q_2$ rather than $n^{-1/2}q_1$. The coverage error of the corrected interval,

$$\mathcal{J}' = \left(\hat{\theta} - n^{-1/2}\hat{\sigma}z_\xi + n^{-3/2}\hat{\sigma}\hat{q}_2(z_\xi), \;\; \hat{\theta} + n^{-1/2}\hat{\sigma}z_\xi - n^{-3/2}\hat{\sigma}\hat{q}_2(z_\xi) \right),$$

is of order n^{-2}:

$$
\begin{aligned}
P(\theta_0 \in \mathcal{J}') &= P\{|T| \le z_\xi - n^{-1}\hat{q}_2(z_\xi)\} \\
&= \alpha + O(n^{-2}).
\end{aligned}
$$

This adjustment to two-sided intervals allows for skewness and kurtosis, not just skewness. A further adjustment, using either an explicit Edgeworth correction or a bootstrap correction, reduces coverage error to $O(n^{-3})$.

Techniques based on Edgeworth correction are sometimes considered to be more attractive than the bootstrap because they demand very little in the way of computation. However, this is achieved only at the expense of the algebraic effort, often considerable, of calculating the appropriate Edgeworth correction. There is a clear trade-off between algebraic and computational labour, and the latter is often more readily available.

Furthermore, Edgeworth corrections do not always perform particularly well. A second Edgeworth correction sometimes over-corrects an interval such as that defined in (3.73), resulting in inferior coverage probability. Even a single Edgeworth correction can produce an interval with poor coverage accuracy, either in small samples or in large samples with high nominal coverage level. This defect derives from the fact that a finite Edgeworth expansion is generally not a monotone function. For example, consider the one-term expansion

$$e(x) = \Phi(x) + n^{-1/2} \left(ax^2 + b \right) \phi(x),$$

for constants a and b. Since $1 - \Phi(x) \sim x^{-1}\phi(x)$ as $x \to \infty$, then if $a \neq 0$,

$$e(x) = \begin{cases} 1 + n^{-1/2}ax^2\phi(x)\{1 + O(x^{-2})\} & \text{as } x \to +\infty, \\ n^{-1/2}ax^2\phi(x)\{1 + O(x^{-2})\} & \text{as } x \to -\infty. \end{cases}$$

Hence if $a > 0$ then $e(x) > 1$ and $e(x) \downarrow 1$ as $x \uparrow \infty$, while if $a < 0$ then $e(x) < 0$ and $e(x) \uparrow 0$ as $x \downarrow -\infty$. Therefore e is not monotone; it does not even satisfy $0 \leq e \leq 1$. This fact makes it impossible to define the inverse function $e^{-1}(\alpha)$ over the entire range $0 < \alpha < 1$, except in an asymptotic sense as $n \to \infty$.

The Edgeworth-corrected interval

$$\mathcal{I} = \left(-\infty, \ \hat{\theta} - n^{-1/2}\hat{\sigma}z_{1-\alpha} + n^{-1}\hat{\sigma}\hat{q}_1(z_{1-\alpha}) \right)$$

is based on an asymptotic inversion of e, and it suffers from the drawbacks we might associate with such an approximation. For example, if $\hat{q}_1(x) = \hat{a}x^2 + \hat{b}$ then

$$\mathcal{I} = \left(-\infty, \ \hat{\theta} - n^{-1/2}\hat{\sigma}z_{1-\alpha} + n^{-1}\hat{\sigma}(\hat{a}z_{1-\alpha}^2 + \hat{b}) \right).$$

Should the value of $\hat{a}$ be negative then $\mathcal{I} \to \emptyset$ (the empty set) as $\alpha \to 1$. This obviously causes problems, with small samples or with large samples and large α's. If $\hat{a} > 0$ then the complementary one-sided interval, of the form $(\hat{\theta} - \hat{t}_\alpha, \ \infty)$ for an appropriate $\hat{t}_\alpha$, converges to the empty set as

$\alpha \to 1$. This perversity is behind the sometimes erratic performance of confidence intervals based on Edgeworth correction.

Most bootstrap methods avoid these problems by being based on inversions of *monotone* functions, such as distribution functions. For example, the quantile $\hat{v}_\alpha$, used for the percentile-t bootstrap (see Example 1.2 in Section 1.3), is given by $\widehat{G}^{-1}(\alpha)$ where $\widehat{G}(x) = P(T^* \le x | \mathcal{X})$ and T^* is the bootstrap version of T. However, bias correction methods for bootstrap confidence intervals are sometimes based in part on a form of Edgeworth correction, and they can suffer from the problems described in the previous paragraph. See Section 3.10.

3.9 Transformations that Correct for Skewness

In the previous section we showed that skewness could be corrected explicitly by adjusting the endpoints of a confidence interval, so as to allow for the first error term in an Edgeworth expansion of coverage probability. We pointed out that Edgeworth corrections do not always perform well, owing to the fact that Edgeworth expansions are not monotone. In the present section we suggest an alternative approach, based on skewness correction by transformation. Our argument is still founded on correcting for the first term in an Edgeworth expansion, but now the correction is carried out implicitly and the transformation is monotone.

We continue to construct percentile-t confidence intervals, based on the statistic $T = n^{1/2}(\hat{\theta} - \theta_0)/\hat{\sigma}$. Assume initially that the distribution of T admits an Edgeworth expansion of the form

$$P(T \le x) = \Phi(x) + n^{-1/2}\gamma(ax^2 + b)\,\phi(x) + O(n^{-1}), \qquad (3.75)$$

where a, b are known constants and γ is estimable. For example, in the case of estimating a mean, γ denotes skewness, and $a = \frac{1}{3}$ and $b = \frac{1}{6}$; see Example 2.1 in Section 2.6. Let $\hat{\gamma}$ be a $\sqrt{n}$-consistent estimate of γ. Then the transformed statistic

$$f(T) = T + n^{-1/2}a\hat{\gamma}T^2 + n^{-1/2}b\hat{\gamma}$$

admits an Edgeworth expansion in which the first term is of size n^{-1}, not $n^{-1/2}$,

$$P\{f(T) \le x\} = \Phi(x) + O(n^{-1}).$$

To see why, observe that if $\hat{\gamma} > 0$ then the event "$f(T) \le x$" is equivalent to

$$-n^{1/2}(2a\hat{\gamma})^{-1}\left[\{1 + 4n^{-1/2}a\hat{\gamma}(x - n^{-1/2}b\hat{\gamma})\}^{1/2} + 1\right]$$

$$\leq \ T \ \leq \ n^{1/2}(2a\hat{\gamma})^{-1}\left[\{1 + 4n^{-1/2}a\hat{\gamma}(x - n^{-1/2}b\hat{\gamma})\}^{1/2} - 1\right],$$

which may be expressed asymptotically as

$$-n^{1/2}(a\hat{\gamma})^{-1} + O_p(1) \ \leq \ T \ \leq \ x - n^{-1/2}\hat{\gamma}(ax^2 + b) + O_p(n^{-1}).$$

From this relation and a similar one for the case $\hat{\gamma} < 0$, we may deduce that if $\gamma \neq 0$,

$$\begin{aligned} P\{f(T) \leq x\} &= P\{-\infty < T \leq x - n^{-1/2}\hat{\gamma}(ax^2 + b)\} + O(n^{-1}) \\ &= P\{T \leq x - n^{-1/2}\gamma(ax^2 + b)\} + O(n^{-1}) \\ &= P(T \leq x) - n^{-1/2}\gamma(ax^2 + b)\,\phi(x) + O(n^{-1}) \\ &= \Phi(x) + O(n^{-1}), \end{aligned} \tag{3.76}$$

using the delta method.

Thus, the transformed statistic $f(T)$ suffers less from skewness than did T itself. We may base a confidence region $\mathcal{R}$ for θ_0 on this transformation, taking for example,

$$\mathcal{R} \ = \ \{\theta : f[n^{1/2}(\hat{\theta} - \theta)/\hat{\sigma}] \leq z_\alpha\}. \tag{3.77}$$

However, the quadratic function f is generally not one-to-one, which means that $\mathcal{R}$ has some undesirable features. If $a\hat{\gamma} > 0$ then for small values of α, $\mathcal{R}$ will equal the empty set; if $a\hat{\gamma} < 0$ then for large α, $\mathcal{R}$ will equal the entire real line $\mathbb{R}$; and for those α's such that $\mathcal{R}$ is a proper, nonempty subset of $\mathbb{R}$, $\mathcal{R}$ is always either a finite two-sided interval or a union of two disjoint one-sided intervals. In particular, $\mathcal{R}$ is never a one-sided interval $(\hat{\theta} - \hat{t}, \infty)$, as might be hoped from definition (3.77).

We may overcome these difficulties by converting the quadratic transformation f into a cubic,

$$\begin{aligned} g(T) &= f(T) + n^{-1}\tfrac{1}{3}a^2\hat{\gamma}^2T^3 \\ &= T + n^{-1/2}a\hat{\gamma}T^2 + n^{-1}\tfrac{1}{3}a^2\hat{\gamma}^2T^3 + n^{-1/2}b\hat{\gamma}. \end{aligned}$$

Since the added term is of order n^{-1}, it does not affect formula (3.76); we have

$$P\{g(T) \leq x\} \ = \ \Phi(x) + O(n^{-1}), \tag{3.78}$$

with much the same proof as (3.76). However, the addition of an appropriate term in T^3 converts f into a monotone, one-to-one function g. Note that

$$g'(T) \ = \ (1 + n^{-1/2}a\hat{\gamma}T)^2,$$

so that the function g is strictly monotone increasing from $g(-\infty) = -\infty$ to $g(+\infty) = +\infty$. The equation $g(T) = x$ always admits a unique solution,

$$T = g^{-1}(x) = n^{1/2}(a\hat{\gamma})^{-1}\left[\{1 + 3a\hat{\gamma}(n^{-1/2}x - n^{-1}b\hat{\gamma})\}^{1/3} - 1\right].$$

To construct a one-sided α-level confidence interval of the form $(-\infty, \hat{\theta} + \hat{t})$, take

$$\mathcal{I} = \{\theta : g[n^{1/2}(\hat{\theta} - \theta)/\hat{\sigma}] > z_{1-\alpha}\}$$
$$= \left(-\infty, \hat{\theta} - n^{-1/2}\hat{\sigma}g^{-1}(z_{1-\alpha})\right).$$

The coverage error of $\mathcal{I}$ equals $O(n^{-1})$, as may be deduced from (3.78). This represents an improvement on the error of $n^{-1/2}$ associated with the usual Normal approximation interval $(-\infty, \hat{\theta} - n^{-1/2}\hat{\sigma}z_{1-\alpha})$, and may be further improved to $O(n^{-3/2})$ by applying the bootstrap to the statistic $g(T)$, as follows. Let $T^* = n^{1/2}(\hat{\theta}^* - \hat{\theta})/\hat{\sigma}^*$, and write $\hat{y}_\alpha$ for the solution of

$$P\{g(T^*) \le \hat{y}_\alpha \mid \mathcal{X}\} = \alpha.$$

Put

$$\mathcal{I}' = \{\theta : g[n^{1/2}(\hat{\theta} - \theta)/\hat{\sigma}] > \hat{y}_{1-\alpha}\}$$
$$= \left(-\infty, \hat{\theta} - n^{-1/2}\hat{\sigma}g^{-1}(\hat{y}_{1-\alpha})\right).$$

Then

$$P(\theta_0 \in \mathcal{I}') = \alpha + O(n^{-3/2}). \tag{3.79}$$

A proof will be given shortly.

There are obvious two-sided analogues of these one-sided confidence intervals. Taking $\xi = \frac{1}{2}(1+\alpha)$ we see that the two-sided, equal-tailed version of $\mathcal{I}$ is

$$\mathcal{J} = \left(\hat{\theta} - n^{-1/2}\hat{\sigma}g^{-1}(z_\xi), \ \hat{\theta} - n^{-1/2}\hat{\sigma}g^{-1}(z_{1-\xi})\right).$$

Its bootstrap counterpart is

$$\mathcal{J}' = \left(\hat{\theta} - n^{-1/2}\hat{\sigma}g^{-1}(\hat{y}_\xi), \ \hat{\theta} - n^{-1/2}\hat{\sigma}g^{-1}(\hat{y}_{1-\xi})\right).$$

The coverage error of $\mathcal{J}$ equals $O(n^{-1})$, which does not represent an improvement on the order of coverage error of the common Normal approximation interval $(\hat{\theta} - n^{-1/2}\hat{\sigma}z_\xi, \ \hat{\theta} + n^{-1/2}\hat{\sigma}z_\xi)$. However, the coverage error of $\mathcal{J}'$ equals $O(n^{-2})$, i.e.,

$$P(\theta_0 \in \mathcal{J}') = \alpha + O(n^{-2}). \tag{3.80}$$

(A proof follows.) Therefore $\mathcal{J}'$ is competitive with the symmetric bootstrap confidence interval $\hat{\mathcal{J}}_{\mathrm{SY}}$ discussed in Section 3.6. Furthermore, $\mathcal{J}'$ is not subject to the artificial constraint of symmetry; the asymmetry of

$\mathcal{J}'$ can convey important information about our uncertainty as to the true value of θ_0, not available from $\hat{J}_{SY}$.

This feature underscores the philosophy behind our introduction of transformation-based confidence intervals: rather than impose the constraint of symmetry directly on the confidence interval, transform the statistic into another whose distribution is virtually symmetric, apply the bootstrap to this quantity, and then return to the original asymmetric problem by inverting the transform, yet retaining the high coverage accuracy conferred by bootstrapping a symmetric quantity.

Next we sketch proofs of (3.79) and (3.80). Let y_α be the solution of

$$P\{g(T) \leq y_\alpha\} = \alpha.$$

As noted earlier, an Edgeworth expansion of the distribution of T contains no term of order $n^{-1/2}$, and so

$$P\{g(T) \leq x\} = \Phi(x) + n^{-1}r(x)\phi(x) + O(n^{-3/2}),$$

where r is an odd polynomial of degree 5. Inverting this formula to derive the corresponding Cornish-Fisher expansion, we find that

$$y_\alpha = z_\alpha - n^{-1}r(z_\alpha) + O(n^{-3/2}).$$

Similarly,

$$\hat{y}_\alpha = z_\alpha - n^{-1}\hat{r}(z_\alpha) + O_p(n^{-3/2}),$$

where $\hat{r}$ is obtained from r on replacing population moments by sample moments in coefficients. Since $\hat{r} = r + O_p(n^{-1/2})$, $\hat{y}_\alpha = y_\alpha + O_p(n^{-3/2})$, whence

$$\begin{aligned} P(\theta_0 \in \mathcal{I}') &= P\{g(T) > \hat{y}_{1-\alpha}\} \\ &= P\{g(T) > y_{1-\alpha}\} + O(n^{-3/2}) \\ &= \alpha + O(n^{-3/2}), \end{aligned}$$

using the delta method. This gives (3.79). Similarly,

$$\begin{aligned} P(\theta_0 \in \mathcal{I}') &= P\{g(T) \leq \hat{y}_\xi\} - P\{g(T) \leq \hat{y}_{1-\xi}\} \\ &= \{\xi + O(n^{-3/2})\} - \{1 - \xi + O(n^{-3/2})\} \\ &= \alpha + O(n^{-3/2}). \end{aligned}$$

By making use of the parity properties of polynomials in Edgeworth expansions we may show that the last-written $O(n^{-3/2})$ is actually $O(n^{-2})$, thereby proving (3.80).

We next discuss transformations alternative to g, which achieve much the same end. The aim is to choose a function f_1 such that the transformation $g_1 = f + f_1$ has the following two properties:

(a) g_1 is monotone, one-to-one, and easily invertible,

(b) $f_1(T) = O_p(n^{-1})$.

The latter constraint is necessary to ensure that

$$P\{g_1(T) \le x\} = \Phi(x) + O(n^{-1});$$

compare (3.76) and (3.78). For ease of invertibility, given that we started with a quadratic transformation f, f_1 will have to be a polynomial. Since polynomials of even degree are never monotone, f_1 will have to be of odd degree greater than or equal to 3. However, polynomial equations of degrees $5, 7, \ldots$ are generally not solvable explicitly, and so we should take f_1 to be a cubic. Let the resulting cubic transformation g_1 be

$$g_1(T) = u + T + cT^2 + dT^3,$$

or perhaps a constant multiple of this T; the latter case may be treated similarly. Now,

$$g_1'(T) = 1 + 2cT + 3dT^2$$

and the discriminant of the quadratic equation $g_1'(T) = 0$ is $(2c)^2 - 4(3d)$. Therefore, g_1 is monotone if and only if $d \ge c^2/3$. We shall assume the latter condition. To solve the equation $g_1(T) = x$, first change variable from T to $U = T + (c/3d)$; this removes the quadratic term from $g_1(T)$. Then

$$g_1(T) - x = U^3 + pU + q,$$

where

$$p = \frac{1}{d} - \frac{c^2}{3d^2}, \qquad q = \frac{2}{27}\left(\frac{c}{d}\right)^3 - \frac{c}{3d^2} + v$$

and $v = (u - x)/d$. Since $d \ge c^2/3$, $p \ge 0$, and so $4p^3 + 27q^2 \ge 0$ for all values of x. This condition is necessary and sufficient to ensure that the equation $g_1(T) = x$ has a unique real solution for all x, that solution being given by

$$T = 2^{-1/3}\{(A - q)^{1/3} - (A + q)^{1/3}\} - \frac{c}{3d},$$

where $A = \left(\frac{4}{27}p^2 + q^2\right)^{1/2}$ (Littlewood 1950, p. 173ff). If we take $d = c^2/3$ then $p = 0$, and the complex inversion formula above reduces to

$$T = -q^{1/3} - \frac{c}{3d},$$

a much simpler expression. Our choice g of the transformation g_1 is of precisely this form, since in our case $c = n^{-1/2}a\hat\gamma$ and $d = \frac{1}{3}a^2\hat\gamma^2$, which entails $d = c^2/3$.

An alternative approach, with similar properties, is to construct the transformation g from scratch rather than try to adjust the quadratic transformation f. For example, we could take

$$g(T) = (2an^{-1/2}\hat{\gamma})^{-1}\left\{\exp(2an^{-1/2}\hat{\gamma}T) - 1\right\} + n^{-1}b\hat{\gamma},$$

which enjoys property (3.78) and has inverse

$$T = g^{-1}(x) = (2an^{-1/2}\hat{\gamma})^{-1}\log\left\{2an^{-1/2}\hat{\gamma}(x - n^{-1}b\hat{\gamma}) + 1\right\},$$

provided the argument of the logarithm is positive.

All of the work so far in this section is predicated on the assumption (3.75), that the polynomial q_1 in an Edgeworth expansion has the form $q_1(x) = \gamma(ax^2 + b)$ where a, b are known constants and γ is estimable. While several important cases are of this form (e.g., estimation of a mean and estimation of slope and intercept in regression; see Example 2.1 in Section 2.6 and Section 4.3), the property is not universal. In particular, it is not satisfied in the case where θ_0 is the population variance; see Example 2.2 in Section 2.6. However, we may put that problem and many others into the appropriate form by first correcting $T = n^{1/2}(\hat{\theta} - \theta_0)/\hat{\sigma}$ for the main effect of bias, and then applying our technique to the corrected statistic, as follows. Note that

$$E(T) = n^{-1/2}c_1 + O(n^{-3/2}), \qquad E(T^3) = n^{-1/2}c_2 + O(n^{-3/2}),$$

where the constants c_i depend on population moments; see Section 2.4. Let $\hat{c}_i$ denote the estimate of c_i that results from replacing those moments by sample moments, and put

$$T' = T - n^{-1/2}\hat{c}_1 = n^{1/2}(\hat{\theta} - \theta_0)\hat{\sigma}^{-1} - n^{-1/2}\hat{c}_1.$$

Then

$$E(T') = O(n^{-3/2}), \qquad E(T'^3) = n^{-1/2}\gamma' + O(n^{-3/2}),$$

where $\gamma' = c_2 - 3c_1$. It now follows from standard arguments for Edgeworth expansion in Chapter 2 that T' admits the version of formula (3.75) in which γ is replaced by γ' and $a = -1$, $b = 1$,

$$P(T' \le x) = \Phi(x) + n^{-1/2}\gamma'(1 - x^2)\phi(x) + O(n^{-1}).$$

Therefore we may apply the methods developed earlier to T' rather than T, taking $\hat{\gamma}' = \hat{c}_2 - 3\hat{c}_1$ as our estimate of γ', and thereby obtain confidence intervals for θ_0. In particular, redefining

$$g(T') = T' - n^{-1/2}\hat{\gamma}'T'^2 + n^{-1}\tfrac{1}{3}\hat{\gamma}'^2T'^3 + n^{-1/2}\hat{\gamma}'$$

and

$$g^{-1}(x) = n^{1/2}\hat{\gamma}'^{-1}\left[1 - \left\{1 + 3\hat{\gamma}'(n^{-1}\hat{\gamma}' - n^{-1/2}x)\right\}^{1/3}\right],$$

we see that a one-sided nominal α-level confidence interval for θ_0 is given by

$$\left(-\infty,\ \hat{\theta} - n^{-1/2}\hat{\sigma}\hat{g}^{-1}(z_{1-\alpha}) - n^{-1}\hat{\sigma}\hat{c}_1 \right),$$

with coverage error $O(n^{-1})$. Its bootstrap version has coverage error $O(n^{-3/2})$, and the bootstrap version of the two-sided interval has coverage error $O(n^{-2})$.

3.10 Methods Based on Bias Correction

3.10.1 Motivation for Bias Correction

Throughout our study of bootstrap methods for confidence intervals we have stressed the importance of pivoting. In most circumstances, pivoting amounts to "Studentizing" or correcting for scale. Our support of pivoting is difficult to sustain in problems where scale cannot be estimated in a stable way, i.e., with low variance,[4] and is impossible to sustain when no variance estimate is available. We have focussed on techniques that give good performance in relatively simple problems, such as estimation of a mean, and we have not, as yet, given due consideration to more complex problems where pivoting does not perform well. There are important examples of the latter, such as estimation of the correlation coefficient or (in some circumstances) of the ratio of two means. Here it can be difficult to construct an estimate of variance that is sufficiently stable for pivoting to be a reasonable option, particularly with small samples. This difficulty does not emerge clearly from our theoretical analysis based on Edgeworth expansions, but is readily apparent from simulation studies.

There is another drawback to pivotal methods, which some statisticians find quite dissatisfying: pivotal methods are generally not transformation-respecting. If $\mathcal{I} = (\hat{a}, \hat{b})$ is a confidence interval for an unknown quantity θ_0, and if f is a known monotone function (an increasing function, let us say), then we would ideally like $f(\mathcal{I}) = (f(\hat{a}), f(\hat{b}))$ to be the corresponding confidence interval for $f(\theta_0)$. Of course, $f(\mathcal{I})$ would be an appropriate confidence interval for $f(\theta_0)$, but it would not necessarily be produced by the method used to construct $\mathcal{I}$ for θ_0, and that is the point. It would

---

[4]When pivotal methods fail it is generally because the denominator of $(\hat{\theta}-\theta_0)/\hat{\sigma}$ fluctuates too erratically relative to the numerator. For example, in the case where θ_0 is a correlation coefficient, both $\hat{\theta}$ and θ_0 are restricted to the interval $[-1, 1]$, but typical standard deviation estimates (such as those based on the jackknife or on an asymptotic formula) are not so constrained. This problem persists when $(\hat{\theta}-\theta_0)/\hat{\sigma}$ is replaced by $(\hat{\theta}^*-\hat{\theta})/\hat{\sigma}^*$, and that is the key to poor performance of percentile-t in this problem.

usually not be produced if the method were percentile-t, except when f is a linear transformation.

The "other percentile method", which we have not viewed particularly favourably in our study of bootstrap methodology (see Example 1.2 in Section 1.3), does not require a scale estimate and is transformation-respecting. To check the latter assertion, observe that the endpoints of a nominal α-level confidence interval $(\hat{a}, \hat{b})$ for θ_0 have the property

$$P(\hat{a} \leq \hat{\theta}^* \leq \hat{b} \mid \mathcal{X}) = \alpha,$$

which entails

$$P\{f(\hat{a}) \leq f(\hat{\theta}^*) \leq f(\hat{b}) \mid \mathcal{X}\} = \alpha.$$

Therefore the "other percentile method" would automatically generate $(f(\hat{a}), f(\hat{b}))$ as the corresponding confidence interval for $f(\theta_0)$.

As we pointed out in Sections 3.4 and 3.5, percentile methods can have serious disadvantages in terms of accuracy (e.g., coverage accuracy) relative to percentile-t. One way of overcoming this problem is to adjust the endpoints of confidence intervals generated by the percentile method. If we think of those adjustments as corrections for bias in the position of the endpoints, then we might call the technique "bias correction". However, note that it is not the bias of $\hat{\theta}$ that is being corrected. The "bias correction" methods introduced in subsection 3.10.3 are transformation-respecting and do not demand the use of a pivotal statistic, although sometimes they require variance estimation as part of the bias correction.

Before passing to an account of bias correction, it is worth delineating the two quite different approaches to bootstrap methods that lead statisticians to advocate pivotal methods or bias correction. If we perceive that the major advantage of "standard" bootstrap methods is their ability to cope with particularly difficult problems (e.g., constructing a confidence interval for a correlation coefficient), and if we see respect of transformations as a major property, then percentile-t would not be favoured and bias correction might be advocated. If, on the other hand, we require the "standard" method to enjoy good accuracy in some of the simpler problems of classical statistics (e.g., constructing a confidence interval for a mean), and if transformation invariance is not uppermost in our minds, then the percentile-t method would be an obvious choice and bias correction might seem a little cumbersome.

3.10.2 Formula for the Amount of Correction Required

We begin by deriving an asymptotic formula for a one-sided confidence interval $\mathcal{I} = (-\infty, \hat{y}_\alpha)$, constructed by the "other percentile method"; then we write down the "ideal" version of $\mathcal{I}$, $J_1 = (-\infty,\ \hat{\theta}-n^{-1/2}\hat{\sigma}v_{1-\alpha})$, based on knowing the distribution of $(\hat{\theta} - \theta_0)/\hat{\sigma}$; and finally, by comparing these two formulae, we determine the amount of correction needed to convert $\mathcal{I}$ approximately into J_1.

The definitions of $\hat{y}_\alpha$ and v_α are in terms of solutions of the equations

$$P(\hat{\theta}^* \le \hat{y}_\alpha \mid \mathcal{X}) = P\{n^{1/2}(\hat{\theta} - \theta_0)/\hat{\sigma} \le v_\alpha\} = \alpha.$$

Now,

$$
\begin{aligned}
\hat{y}_\alpha &= \hat{\theta} + n^{-1/2}\hat{\sigma}\hat{u}_\alpha \\
&= \hat{\theta} + n^{-1/2}\hat{\sigma}\{z_\alpha + n^{-1/2}\hat{p}_{11}(z_\alpha) + O_p(n^{-1})\} \\
&= \hat{\theta} + n^{-1/2}\hat{\sigma}z_\alpha - n^{-1}\hat{\sigma}\hat{p}_1(z_\alpha) + O_p(n^{-3/2}),
\end{aligned}
\tag{3.81}
$$

using (3.15) and (3.20). Since $\hat{p}_1$ differs from p_1 by $O_p(n^{-1/2})$,

$$
\begin{aligned}
\mathcal{I} &= \left(-\infty,\ \hat{y}_\alpha\right) \\
&= \left(-\infty,\ \hat{\theta} + n^{-1/2}\hat{\sigma}z_\alpha - n^{-1}\hat{\sigma}p_1(z_\alpha) + O_p(n^{-3/2})\right).
\end{aligned}
\tag{3.82}
$$

Similarly, noting that $z_{1-\alpha} = -z_\alpha$, that q_1 is an even polynomial and that

$$
\begin{aligned}
v_\alpha &= z_\alpha + n^{-1/2}q_{11}(z_\alpha) + O(n^{-1}) \\
&= z_\alpha - n^{-1/2}q_1(z_\alpha) + O(n^{-1}),
\end{aligned}
$$

we see that

$$
\begin{aligned}
J_1 &= \left(-\infty,\ \hat{\theta} - n^{-1/2}\hat{\sigma}v_{1-\alpha}\right) \\
&= \left(-\infty,\ \hat{\theta} + n^{-1/2}\hat{\sigma}z_\alpha + n^{-1}\hat{\sigma}q_1(z_\alpha) + O_p(n^{-3/2})\right).
\end{aligned}
\tag{3.83}
$$

Comparing (3.82) and (3.83) we find that the accuracy of $\mathcal{I}$ relative to J_1 may be improved by shifting $\mathcal{I}$ to the right by an amount $n^{-1}\hat{\sigma}\{p_1(z_\alpha) + q_1(z_\alpha)\}$. In this event, $\mathcal{I}$ becomes second-order correct relative to J_1.

Our next task is to develop an explicit formula for $p_1 + q_1$. For this purpose it is beneficial to return to the "smooth function model" for general Edgeworth expansions, developed in Section 2.4. Assume that we may write $\theta = g(\boldsymbol{\mu})$ and $\hat{\theta} = g(\bar{\mathbf{X}})$, where $\bar{\mathbf{X}}$ is the mean of a random sample $\{\mathbf{X}_1, \dots, \mathbf{X}_n\}$ drawn from a d-variate population with mean $\boldsymbol{\mu}$. Write $v^{(i)} = (\mathbf{v})^{(i)}$ for the ith element of a vector $\mathbf{v}$, let $\mathbf{X}$ denote a generic $\mathbf{X}_i$, and put $A(\mathbf{x}) = g(\mathbf{x}) - g(\boldsymbol{\mu})$,

$$a_{i_1 \ldots i_j} = (\partial^j / \partial x^{(i_1)} \ldots \partial x^{(i_j)}) A(\mathbf{x}) \Big|_{\mathbf{x}=\boldsymbol{\mu}},$$

$$\mu_{i_1 \ldots i_j} = E\{(\mathbf{X} - \boldsymbol{\mu})^{(i_1)} \ldots (\mathbf{X} - \boldsymbol{\mu})^{(i_j)}\},$$

$$\sigma^2 = \sum_{i=1}^{d} \sum_{j=1}^{d} a_i a_j \mu_{ij}, \qquad A_1 = \frac{1}{2} \sum_{i=1}^{d} \sum_{j=1}^{d} a_{ij} \mu_{ij},$$

$$A_2 = \sum_{i=1}^{d} \sum_{j=1}^{d} \sum_{k=1}^{d} a_i a_j a_k \mu_{ijk} + 3 \sum_{i=1}^{d} \sum_{j=1}^{d} \sum_{k=1}^{d} \sum_{l=1}^{d} a_i a_j a_{kl} \mu_{ik} \mu_{jl}.$$

If necessary, adjoin additional elements to vectors $\mathbf{X}_i$ and $\boldsymbol{\mu}$ to ensure that $\sigma^2 = h^2(\boldsymbol{\mu})$ for a smooth function h; that this is possible was justified in Section 2.4, following Theorem 2.3. In this notation, $\hat{\sigma}^2 = h^2(\bar{\mathbf{X}})$. Put $B(\mathbf{x}) = A(\mathbf{x})/h(\mathbf{x})$,

$$b_{i_1 \ldots i_j} = (\partial^j / \partial x^{(i_1)} \ldots \partial x^{(i_j)}) B(\mathbf{x}) \Big|_{\mathbf{x}=\boldsymbol{\mu}}, \qquad c_i = (\partial / \partial x^{(i)}) h(\mathbf{x}) \Big|_{\mathbf{x}=\boldsymbol{\mu}},$$

$$B_1 = \frac{1}{2} \sum_{i=1}^{d} \sum_{j=1}^{d} b_{ij} \mu_{ij},$$

$$B_2 = \sum_{i=1}^{d} \sum_{j=1}^{d} \sum_{k=1}^{d} b_i b_j b_k \mu_{ijk} + 3 \sum_{i=1}^{d} \sum_{j=1}^{d} \sum_{k=1}^{d} \sum_{l=1}^{d} b_i b_j b_{kl} \mu_{ik} \mu_{jl}.$$

By Theorem 2.2,

$$p_1(x) = -\{A_1 \sigma^{-1} + \tfrac{1}{6} A_2 \sigma^{-3} (x^2 - 1)\},$$

$$q_1(x) = -\{B_1 + \tfrac{1}{6} B_2 (x^2 - 1)\}. \tag{3.84}$$

Using elementary rules for differentiating a product of two functions we see that

$$b_i = a_i \sigma^{-1}, \qquad b_{ij} = a_{ij} \sigma^{-1} - \tfrac{1}{2}(a_i c_j + a_j c_i) \sigma^{-3},$$

whence it follows that

$$B_2 = \sigma^{-3} A_2 - 3\sigma^{-3} \sum_{i=1}^{d} \sum_{j=1}^{d} a_i c_j \mu_{ij}. \tag{3.85}$$

Furthermore, $p(x)$ and $q_1(x)$ must agree when $x = 0$, since

$$P\{n^{1/2}(\hat{\theta} - \theta)/\sigma \le 0\} = P\{n^{1/2}(\hat{\theta} - \theta)/\hat{\sigma} \le 0\}.$$

Therefore, noting (3.84) and (3.85), we have

$$p_1(x) + q_1(x) = 2p_1(0) + \tfrac{1}{6} A \sigma^{-3} x^2, \tag{3.86}$$

where

$$A = -(A_2 + B_2\sigma^{-3}) = -\left(2A_2 - 3\sum_{i=1}^{d}\sum_{j=1}^{d} a_i c_j \mu_{ij}\right)$$

$$= 3\sum_{i=1}^{d}\sum_{j=1}^{d} a_i c_j \mu_{ij} - 2\sum_{i=1}^{d}\sum_{j=1}^{d}\sum_{k=1}^{d} a_i a_j a_k \mu_{ijk}$$

$$- 6\sum_{i=1}^{d}\sum_{j=1}^{d}\sum_{k=1}^{d}\sum_{l=1}^{d} a_i a_j a_{kl} \mu_{ik} \mu_{jl} . \qquad (3.87)$$

We claim that in many cases of practical interest, $n^{-1/2}\sigma^{-3}A$ equals the third moment, or skewness, of the first-order approximation to $n^{1/2}(\hat{\theta} - \theta)/\sigma$. To appreciate this point, observe that

$$n^{1/2}(\hat{\theta} - \theta)/\sigma = (n^{1/2}/\sigma)\sum_{i=1}^{d}(\overline{\mathbf{X}} - \boldsymbol{\mu})^{(i)} a_i + O_p(n^{-1/2}),$$

and so our claim is that

$$E\left\{ (n^{1/2}/\sigma)\sum_{i=1}^{d}(\overline{\mathbf{X}} - \boldsymbol{\mu})^{(i)} a_i \right\}^3 = n^{-1/2}A + O(n^{-1}).$$

Now, the left-hand side equals

$$n^{-1/2}\sigma^{-3}\sum_{i=1}^{d}\sum_{j=1}^{d}\sum_{k=1}^{d} a_i a_j a_k \mu_{ijk} ,$$

and so, in view of (3.87), it suffices to prove that

$$\sum_{i=1}^{d}\sum_{j=1}^{d} a_i c_j \mu_{ij} = 2\sum_{i=1}^{d}\sum_{j=1}^{d}\sum_{k=1}^{d}\sum_{l=1}^{d} a_i a_j a_{kl} \mu_{ik} \mu_{jl}$$

$$+ \sum_{i=1}^{d}\sum_{j=1}^{d}\sum_{k=1}^{d} a_i a_j a_k \mu_{ijk} . \qquad (3.88)$$

This formula will be verified in subsection 3.10.6, in the cases of nonparametric inference and of parametric inference in an exponential family model.

If we assume formula (3.88) then

$$A = \sum_{i=1}^{d}\sum_{j=1}^{d}\sum_{k=1}^{d} a_i a_j a_k \mu_{ijk} .$$

We may estimate a_i and μ_{ijk} as

$$\hat{a}_i = (\partial/\partial x^{(i)}) A(\mathbf{x})\big|_{\mathbf{x}=\overline{\mathbf{X}}} ,$$

$$\hat{\mu}_{ijk} = n^{-1} \sum_{l=1}^{n} (\mathbf{X}_l - \overline{\mathbf{X}})^{(i)} (\mathbf{X}_l - \overline{\mathbf{X}})^{(j)} (\mathbf{X}_l - \overline{\mathbf{X}})^{(k)} ,$$

respectively, and thereby compute an estimate of A:

$$\widehat{A} = \sum_{i=1}^{d} \sum_{j=1}^{d} \sum_{k=1}^{d} \hat{a}_i \hat{a}_j \hat{a}_k \hat{\mu}_{ijk} .$$

We already have an estimate of σ^2, $\hat{\sigma}^2 = h^2(\overline{\mathbf{X}})$. Combining these formulae we obtain an estimate of the coefficient $a = n^{-1/2}\frac{1}{6} A\sigma^{-3}$ of x^2 in $n^{-1/2}\{p_1(x) + q_1(x)\}$:

$$\hat{a} = \text{coefficient of } x^2 \text{ in } n^{-1/2}\{\hat{p}_1(x) + \hat{q}_1(x)\} = n^{-1/2}\frac{1}{6} \widehat{A}\hat{\sigma}^{-3} . \quad (3.89)$$

We call $\hat{a}$ the "acceleration constant". The value of $\hat{a}$ is a key ingredient in one form of bias correction, as we shall show in the next subsection. Note that, in view of the remarks in the previous paragraph, $\hat{a}$ is essentially a skewness estimate. Its incorporation into a confidence interval formula amounts to skewness correction.

3.10.3 The Nature of the Correction

We know from subsection 3.10.2 that in order to achieve performance commensurate with J_1, the "other percentile method" interval $\mathcal{I}$ should be shifted to the right by an amount $n^{-1}\hat{\sigma}\{p_1(z_\alpha) + q_1(z_\alpha)\}$. This end could be achieved by computing estimates $\hat{p}_1$ and $\hat{q}_1$ of p_1 and q_1, and adding the resulting estimate of $n^{-1}\{p_1(z_\alpha) + q_1(z_\alpha)\}$ to $\mathcal{I}$, obtaining the corrected interval

$$\mathcal{I}' = \mathcal{I} + n^{-1}\hat{\sigma}\{\hat{p}_1(z_\alpha) + \hat{q}_1(z_\alpha)\}$$

$$= \left(-\infty, \ \hat{y}_\alpha + n^{-1}\hat{\sigma}\{\hat{p}_1(z_\alpha) + \hat{q}_1(z_\alpha)\} \right) .$$

By construction, $\mathcal{I}'$ is second-order correct relative to the "ideal" interval $J_1 = (-\infty, \ \hat{\theta} - n^{-1/2}\hat{\sigma}v_{1-\alpha})$; see subsection 3.10.2.

This approach to correction is very much in the spirit of explicit skewness correction, discussed in Section 3.8. Efron (1982, 1987) suggested an alternative approach, which we describe in the present subsection. Efron's correction comes in two forms — a "bias correction" (BC) that corrects for the constant terms in $p_1 + q_1$, and an "accelerated bias correction" (ABC or BC_a) that corrects for all of $p_1 + q_1$. Accelerated bias correction is similar in several key respects to explicit skewness correction. In particular, it requires estimation of the constant $a = n^{-1/2} \frac{1}{6} A \sigma^{-3}$, which is a measure of skewness; it produces an interval that depends explicitly on the Standard Normal critical point z_α; and the interval does not increase monotonically with increasing α.

To describe the BC method, write $\widehat{H}(x) = P(\hat{\theta}^* \leq x \mid \mathcal{X})$ for the bootstrap distribution function of $\hat{\theta}^*$. In this notation,

$$\hat{y}_\alpha \;=\; \widehat{H}^{-1}(\alpha) \;=\; \widehat{H}^{-1}\{\Phi(z_\alpha)\}. \tag{3.90}$$

The sequence of quantiles $\{\hat{y}_\alpha, \, 0 < \alpha < 1\}$ would be "centred" empirically if $\hat{y}_{1/2}$ were equal to $\hat{\theta}$, i.e., if $\widehat{H}(\hat{\theta}) = \frac{1}{2}$. In general this identity does not hold, and we might wish to correct for centring error in formula (3.90), obtaining

$$\hat{y}_{\mathrm{BC},\alpha} \;=\; \widehat{H}^{-1}\big[\Phi\{\hat{m} + (\hat{m} + z_\alpha)\}\big] \;=\; \widehat{H}^{-1}\{\Phi(2\hat{m} + z_\alpha)\}, \tag{3.91}$$

where

$$\hat{m} \;=\; \Phi^{-1}\{\widehat{H}(\hat{\theta})\} \tag{3.92}$$

is the (estimate of the) centring correction or bias correction. A one-sided BC interval for θ_0, with nominal coverage α, is given by $(-\infty, \hat{y}_{\mathrm{BC},\,\alpha})$.

The accelerated bias-corrected version of this critical point is obtained by introducing a skewness adjustment to (3.91), yielding

$$\hat{y}_{\mathrm{ABC},\alpha} \;=\; \widehat{H}^{-1}\Big(\Phi\big[\hat{m} + (\hat{m} + z_\alpha)\{1 - \hat{a}(\hat{m} + z_\alpha)\}^{-1}\big]\Big). \tag{3.93}$$

Here $\hat{m}$ is defined by (3.92) and $\hat{a}$ is a small quantity of order $n^{-1/2}$ whose value is given by (3.89). We call $\hat{a}$ the (estimate of the) acceleration constant. A one-sided ABC interval with nominal coverage α is $(-\infty, \, \hat{y}_{\mathrm{ABC},\alpha})$. There is a variety of asymptotically equivalent forms of $\hat{y}_{\mathrm{ABC},\,\alpha}$ including

$$
\begin{aligned}
\hat{y}'_{\mathrm{ABC},\alpha} &\;=\; \widehat{H}^{-1}\Big[\Phi\big\{\hat{m} + (\hat{m} + z_\alpha) + \hat{a}(\hat{m} + z_\alpha)^2\big\}\Big] \\
&\;=\; \widehat{H}^{-1}\Big[\Phi\big\{2\hat{m} + z_\alpha + \hat{a}(\hat{m} + z_\alpha)^2\big\}\Big]
\end{aligned}
$$

and

$$
\begin{aligned}
\hat{y}''_{\mathrm{ABC},\alpha} &\;=\; \widehat{H}^{-1}\Big[\Phi\big\{\hat{m} + (\hat{m} + z_\alpha) + \hat{a}z_\alpha^2\big\}\Big] \\
&\;=\; \widehat{H}^{-1}\Big\{\Phi\big(2\hat{m} + z_\alpha + \hat{a}z_\alpha^2\big)\Big\}.
\end{aligned}
$$

The quantity $\hat{y}_{\mathrm{BC},\,\alpha}$ is a strictly monotone decreasing function of α, as it should be. Therefore the BC interval $(-\infty, \hat{y}_{\mathrm{BC},\,\alpha})$ has no obvious vices. However, the value of

$$\hat{y}_{\mathrm{ABC},\alpha} \;=\; \widehat{H}^{-1}\Big(\Phi\big[\hat{m} - \hat{a}^{-1} - \hat{a}^{-2}\{z_\alpha + (\hat{m} - \hat{a}^{-1})\}^{-1}\big]\Big)$$

has rather unsatisfactory properties as a function of α: it increases from $\widehat{H}^{-1}\{\Phi(\hat{m} - \hat{a}^{-1})\}$ to $+\infty$ as α increases from 0 to $1 - \Phi(\hat{m} - \hat{a}^{-1})$, and increases again from $-\infty$ to $\widehat{H}^{-1}\{\Phi(\hat{m} - \hat{a}^{-1})\}$ as α increases from $1 - \Phi(\hat{m} - \hat{a}^{-1})$ to 1.[5] In particular, it has an infinite jump discontinuity at $\alpha = 1 - \Phi(\hat{m} - \hat{a}^{-1})$. This causes irregularities in the coverage properties, which can be observed in simulation studies. (See also subsection 3.10.5.) The critical points $\hat{y}'_{\mathrm{ABC}, \alpha}$ and $\hat{y}''_{\mathrm{ABC}, \alpha}$ do not remedy this problem, since neither is a monotone function of α.

3.10.4 Performance of BC and ABC

It remains to determine the extent to which bias correction (BC) and accelerated bias-correction (ABC) achieve the goal of rendering the interval $\mathcal{I}$ second-order correct relative to the "ideal" interval J_1. Observe from subsection 3.10.2, particularly formula (3.86), that full correction demands that the critical point $\hat{y}_\alpha$ be transformed in the following way:

$$\hat{y}_\alpha \;\mapsto\; \hat{y}_\alpha + n^{-1}\hat{\sigma}\{p_1(z_\alpha) + q_1(z_\alpha)\} + O_p(n^{-3/2})$$
$$= \hat{y}_\alpha + n^{-1}\hat{\sigma}\{2p_1(0) + \tfrac{1}{6} A\sigma^{-3} z_\alpha^2\} + O_p(n^{-3/2}). \qquad (3.94)$$

We claim that ABC achieves the full correction, whereas BC only corrects by the "constant" part (i.e., the term $n^{-1}\hat{\sigma}2p_1(0)$) in formula (3.94).

First we examine the BC method. Define

$$\widehat{K}(x) \;=\; P\{n^{1/2}(\hat{\theta}^* - \hat{\theta})/\hat{\sigma} \le x \mid \mathcal{X}\}$$
$$= \; \Phi(x) + n^{-1/2}\hat{p}_1(x)\phi(x) + O_p(n^{-1})$$

(see equation (3.19)), and observe that

$$\widehat{H}(\hat{\theta}) \;=\; \widehat{K}(0) \;=\; \Phi(0) + n^{-1/2}\hat{p}_1(0)\phi(0) + O_p(n^{-1}).$$

Therefore

$$\hat{m} \;=\; \Phi^{-1}\{\Phi(0) + n^{-1/2}\hat{p}_1(0)\phi(0) + O_p(n^{-1})\}$$
$$= \; n^{-1/2}\hat{p}_1(0) + O_p(n^{-1}), \qquad (3.95)$$

whence

$$\Phi(2\hat{m} + z_\alpha) \;=\; \Phi(z_\alpha) + 2\hat{m}\phi(z_\alpha) + O_p(n^{-1})$$
$$= \; \alpha + n^{-1/2}2\hat{p}_1(0)\phi(z_\alpha) + O_p(n^{-1}).$$

---

[5]Strictly speaking, the values $\pm\infty$ here are actually finite, being respectively the largest and smallest values of the atoms of the bootstrap distribution. They diverge to $\pm\infty$ as $n \to \infty$.

In consequence,

$$\hat{y}_{\mathrm{BC},\,\alpha} = \widehat{H}^{-1}\big\{\Phi(2\hat{m}+z_\alpha)\big\}$$
$$= \widehat{H}^{-1}\big\{\alpha + n^{-1/2}2\hat{p}_1(0)\phi(z_\alpha) + O_p(n^{-1})\big\}$$
$$= \hat{y}_\alpha + n^{-1}\hat{\sigma}2p_1(0) + O_p(n^{-3/2}). \tag{3.96}$$

In proving (3.96) we have used the fact that

$$\hat{p}_1 = p_1 + O_p(n^{-1/2}), \qquad \widehat{H}^{-1}(\beta) = \hat{y}_\beta,$$

and, with $\beta = \alpha + n^{-1/2}2p_1(0)\phi(z_\alpha) + O_p(n^{-1})$,

$$\hat{y}_\beta = \hat{\theta} + n^{-1/2}\hat{\sigma}z_\beta - n^{-1}\hat{\sigma}\hat{p}_1(z_\beta) + O_p(n^{-3/2})$$
$$= \hat{\theta} + n^{-1/2}\hat{\sigma}\big\{z_\alpha + n^{-1/2}2p_1(0) + O_p(n^{-1})\big\}$$
$$\qquad\qquad\qquad - n^{-1}\hat{\sigma}\hat{p}_1(z_\alpha) + O_p(n^{-3/2})$$
$$= \hat{y}_\alpha + n^{-1}\hat{\sigma}2p_1(0) + O_p(n^{-3/2}).$$

(Note (3.81).)

Formula (3.96) demonstrates that bias correction adjusts only for the "constant" part in (3.94). In particular, the bias-corrected interval $(-\infty,\ \hat{y}_{\mathrm{BC},\,\alpha})$ is not second-order correct relative to J_1.

Next we show that accelerated bias correction achieves the full extent of the adjustment, which is necessary to ensure second-order correctness. Observe from (3.95) that

$$\hat{m} + (\hat{m}+z_\alpha)\big\{1 - \hat{a}(\hat{m}+z_\alpha)\big\}^{-1} = 2\hat{m} + z_\alpha + \hat{a}z_\alpha^2 + O_p(n^{-1}).$$

Therefore, by (3.93), and since

$$\hat{m} = n^{-1/2}p_1(0) + O_p(n^{-1}) \quad \text{and} \quad \hat{a} = n^{-1/2}\tfrac{1}{6}A\sigma^{-3} + O_p(n^{-1}),$$

$$\hat{y}_{\mathrm{ABC},\,\alpha} = \widehat{H}^{-1}\Big(\Phi\big[\hat{m} + (\hat{m}+z_\alpha)\{1 - \hat{a}(\hat{m}+z_\alpha)\}^{-1}\big]\Big)$$

$$= \widehat{H}^{-1}\big\{\alpha + (2\hat{m} + \hat{a}z_\alpha^2)\phi(z_\alpha) + O_p(n^{-1})\big\}$$

$$= \widehat{H}^{-1}\big[\alpha + n^{-1/2}\big\{2p_1(0) + \tfrac{1}{6}A\sigma^{-3}z_\alpha^2\big\} + O_p(n^{-1})\big]$$

$$= \hat{y}_\alpha + n^{-1}\hat{\sigma}\big\{2p_1(0) + \tfrac{1}{6}A\sigma^{-3}z_\alpha^2\big\} + O_p(n^{-3/2})$$

$$= \hat{y}_\alpha + n^{-1}\hat{\sigma}\big\{p_1(z_\alpha) + q_1(z_\alpha)\big\} + O_p(n^{-3/2}).$$

Here we have used (3.86). Comparing this expression with (3.94) we see that accelerated bias correction produces a confidence interval $(-\infty,\ \hat{y}_{\mathrm{ABC},\alpha})$, which is second-order correct relative to J_1.

Both the BC and ABC intervals are transformation-respecting. That is, if f is an increasing function, and if $(-\infty, \hat{y}_{BC,\alpha})$ and $(-\infty, \hat{y}_{ABC,\alpha})$ denote BC and ABC intervals for θ_0, constructed according to the methods described in subsection 3.10.3, then the same method applied in the same way produces $(-\infty, f(\hat{y}_{BC,\alpha}))$ and $(-\infty, f(\hat{y}_{ABC,\alpha}))$ as the BC and ABC intervals, respectively, for $f(\theta_0)$. To see why, write $\widehat{H}_f(x) = P\{f(\hat{\theta}^*) \leq x \mid \mathcal{X}\}$ for the bootstrap distribution function of $\hat{\theta}^*$. Since $\widehat{H} = \widehat{H}_f \circ f$,

$$\hat{m}_f = \Phi^{-1}\big[\widehat{H}_f\{f(\hat{\theta})\}\big] = \Phi^{-1}\{\widehat{H}(\hat{\theta})\} = \hat{m}.$$

Therefore, the version of $\hat{y}_{BC,\alpha}$ appropriate to inference about $f(\theta)$ equals

$$\begin{aligned}
\hat{y}_{f;BC,\alpha} &= \widehat{H}_f^{-1}\{\Phi(2\hat{m}_f + z_\alpha)\} \\
&= (\widehat{H} \circ f^{-1})^{-1}\{\Phi(2\hat{m} + z_\alpha)\} \\
&= f\big[\widehat{H}^{-1}\{\Phi(2\hat{m} + z_\alpha)\}\big] \\
&= f(\hat{y}_{BC,\alpha}).
\end{aligned} \tag{3.97}$$

A little manipulation with Taylor expansions shows that the value of $a = n^{-1/2}\frac{1}{6}A\sigma^{-3}$ does not alter if the confidence interval problem is changed from one of inference about θ_0 to that of inference about $f(\theta_0)$. Likewise, the value of $\hat{a}$ does not change, and so

$$\hat{y}_{f;ABC,\alpha} = \widehat{H}_f^{-1}\Big(\Phi\big[\hat{m} + (\hat{m} + z_\alpha)\{1 - \hat{a}(\hat{m} + z_\alpha)\}^{-1}\big]\Big) = f(\hat{y}_{ABC,\alpha}),$$

arguing as in (3.97).

3.10.5 Two-Sided Bias-Corrected Intervals

Two-sided BC and ABC intervals are readily constructed by taking intersections of the one-sided intervals suggested in subsection 3.10.3. In particular, defining $\xi = \frac{1}{2}(1 + \alpha)$ we obtain the following intervals having nominal coverage α:

$$\big(\hat{y}_{BC,\xi}, \hat{y}_{BC,1-\xi}\big), \qquad \big(\hat{y}_{ABC,\xi}, \hat{y}_{ABC,1-\xi}\big).$$

Note that as $\xi \uparrow 1$, the first of these two intervals increases steadily to the entire real line, whereas the second interval increases initially but then decreases, shrinking ultimately to the point $\widehat{H}^{-1}\{\Phi(\hat{m} - \hat{a}^{-1})\}$. The difficulties caused by this property can be observed in simulation studies, where for small samples and relatively large values of α the two-sided ABC interval has a tendency to undercover.

3.10.6 Proof of (3.88)

We consider separately the cases of nonparametric and parametric inference.

(A) NONPARAMETRIC INFERENCE

Recall from subsection 3.10.2 that

$$\sigma^2 = \sum_{i=1}^{d}\sum_{j=1}^{d} A_{(i)}(\boldsymbol{\mu})A_{(j)}(\boldsymbol{\mu})\{E(X^{(i)}X^{(j)}) - \mu^{(i)}\mu^{(j)}\},$$

where $A_{(i)}(\mathbf{x}) = (\partial/\partial x^{(i)})A(\mathbf{x})$ and $\mu^{(i)} = E(X^{(i)})$. If the products $X^{(i)}X^{(j)}$, for $1 \leq i, j \leq d$, are not components of $\mathbf{X}$ then we may adjoin them to $\mathbf{X}$. Therefore, we may assume that for each pair (i,j) such that $A(\mathbf{x})$ depends in a nontrivial way on both $x^{(i)}$ and $x^{(j)}$, there exists k such that $X^{(k)} \equiv X^{(i)}X^{(j)}$. Write this k as $\langle i,j \rangle$. In that notation,

$$\sigma^2 = h(\boldsymbol{\mu}) = \sum_{i=1}^{d}\sum_{j=1}^{d} A_{(i)}(\boldsymbol{\mu})A_{(j)}(\boldsymbol{\mu})(\mu^{(\langle i,j \rangle)} - \mu^{(i)}\mu^{(j)}), \qquad (3.98)$$

and our estimate of σ^2 is

$$\hat{\sigma}^2 = h(\bar{\mathbf{X}}) = \sum_{i=1}^{d}\sum_{j=1}^{d} A_{(i)}(\bar{\mathbf{X}})A_{(j)}(\bar{\mathbf{X}})(\bar{\mathbf{X}}^{(\langle i,j \rangle)} - \bar{\mathbf{X}}^{(i)}\bar{\mathbf{X}}^{(j)}).$$

Differentiate (3.98) to obtain a formula for $c_k = h_{(k)}(\boldsymbol{\mu})$,

$$\begin{aligned}
c_k = \sum_{i}\sum_{j} \Big[&\{A_{(i,k)}(\boldsymbol{\mu})A_{(j)}(\boldsymbol{\mu}) + A_{(i)}(\boldsymbol{\mu})A_{(j,k)}(\boldsymbol{\mu})\}(\mu^{(\langle i,j \rangle)} - \mu^{(i)}\mu^{(j)}) \\
&- A_{(i)}(\boldsymbol{\mu})A_{(j)}(\boldsymbol{\mu})(\delta_{ik}\mu^{(j)} + \delta_{jk}\mu^{(i)}) \Big] \\
&+ \sum_{i}\sum_{j}{}_{(k)} A_{(i)}(\boldsymbol{\mu})A_{(j)}(\boldsymbol{\mu}),
\end{aligned}$$

where $A_{(i,j)}(\mathbf{x}) = (\partial^2/\partial x^{(i)}\,\partial x^{(j)})A(\mathbf{x})$, δ_{ij} is the Kronecker delta, and $\sum_i\sum_{j\ (k)}$ denotes summation over i,j such that $\langle i,j \rangle = k$. (The sum contains zero or two summands.) Simplifying,

$$c_k = 2\sum_{i}\sum_{j} a_{ik}a_j\mu_{ij} - 2a_k \sum_{i} a_i\mu^{(i)} + \sum_{i}\sum_{j}{}_{(k)} a_i a_j.$$

Therefore

$$
\sum\sum a_i c_j \mu_{ij} = 2\sum_i\sum_j\sum_k\sum_l a_i a_j a_{kl}\mu_{ik}\mu_{jl}
$$
$$
- 2\sum_i\sum_j\sum_k a_i a_j a_k \mu_{ij}\mu^{(k)}
$$
$$
+ \sum_k\sum_l\sum_i\sum_{j\,(k)} a_i a_j a_l \mu_{kl}. \tag{3.99}
$$

Notice that $\mu_{kl} = E(X^{(k)}X^{(l)}) - \mu^{(k)}\mu^{(l)}$, and if $\langle i,j\rangle = k$,

$$
\mu_{ijl} = E(X^{(i)}X^{(j)}X^{(l)}) - \mu^{(i)}E(X^{(j)}X^{(l)}) - \mu^{(j)}E(X^{(i)}X^{(l)})
$$
$$
- \mu^{(l)}E(X^{(i)}X^{(j)}) + 2\mu^{(i)}\mu^{(j)}\mu^{(l)}
$$
$$
= E(X^{(k)}X^{(l)}) - \mu^{(i)}(\mu_{jl} + \mu^{(j)}\mu^{(l)}) - \mu^{(j)}(\mu_{il} + \mu^{(i)}\mu^{(l)})
$$
$$
- \mu^{(l)}\mu^{(k)} + 2\mu^{(i)}\mu^{(j)}\mu^{(l)}
$$
$$
= \mu_{kl} - \mu^{(i)}\mu_{jl} - \mu^{(j)}\mu_{il}.
$$

Using this result to eliminate μ_{kl} from the last-written term in (3.99), we obtain

$$
\sum_i\sum_j a_i c_j \mu_{ij} = 2\sum_i\sum_j\sum_k\sum_l a_i a_j a_{kl}\mu_{ik}\mu_{jl}
$$
$$
- 2\sum_i\sum_j\sum_k a_i a_j a_k \mu_{ij}\mu^{(k)}
$$
$$
+ \sum_i\sum_j\sum_l a_i a_j a_l\big(\mu_{ijl} + \mu^{(i)}\mu_{jl} + \mu^{(j)}\mu_{il}\big)
$$
$$
= 2\sum_i\sum_j\sum_k\sum_l a_i a_j a_{kl}\mu_{ik}\mu_{jl} + \sum_i\sum_j\sum_k a_i a_j a_k \mu_{ijk},
$$

which is identical to (3.88).

(B) Parametric Inference

Here we assume that the distribution of $\mathbf{X}$ is completely determined except for a vector $\boldsymbol{\lambda}$ of unknown parameters. This entails $\theta = f(\boldsymbol{\lambda})$, for some function f. Our estimate of $\boldsymbol{\lambda}$ would typically be the maximum likelihood estimate $\widehat{\boldsymbol{\lambda}}$, in which case $\hat{\theta}$ would be the maximum likelihood estimate $f(\widehat{\boldsymbol{\lambda}})$. Of course, there is no a priori reason for $\hat{\theta}$ to be expressible as a function of $\overline{\mathbf{X}}$. Since our theory is tailored to that case, we should concentrate on a parametric model under which we can write $\hat{\theta} = g(\overline{\mathbf{X}})$ for a known function g. This leads us to consider a simple exponential family model, where $\mathbf{X}$ has density

$$
h_\lambda(\mathbf{x}) = \exp\{\boldsymbol{\lambda}^{\mathrm{T}}\mathbf{x} - \psi(\boldsymbol{\lambda})\}h_0(\mathbf{x}),
$$

λ is a d-vector and ψ, h_0 are known real-valued functions of d variables.

Define

$$\psi_{(i_1,\ldots,i_j)} = (\partial^j/\partial x^{(i_1)} \ldots \partial x^{(i_j)})\, \psi(\mathbf{x})\,.$$

The maximum likelihood estimate $\widehat{\lambda}$ is the solution of the equations

$$\bar{X}^{(i)} = \psi_{(i)}(\widehat{\lambda})\,, \qquad 1 \le i \le d\,,$$

and so can be expressed as a function of $\bar{\mathbf{X}}$. Moments of $\mathbf{X}$ may be represented in terms of the derivatives of ψ, as

$$\mu^{(i)} = E(X^{(i)}) = \psi_{(i)}(\lambda)\,, \tag{3.100}$$

$$\mu_{ij} = E\{(X^{(i)} - \mu^{(i)})(X^{(j)} - \mu^{(j)})\} = \psi_{(i,j)}(\lambda)\,, \tag{3.101}$$

$$\mu_{ijk} = E\{(X^{(i)} - \mu^{(i)})(X^{(j)} - \mu^{(j)})(X^{(k)} - \mu^{(k)})\}$$
$$= \psi_{(i,j,k)}(\lambda)\,.$$

Therefore

$$\sigma^2 = \tau(\mu) = \sum_i \sum_j A_{(i)}(\mu) A_{(j)}(\mu) \psi_{(i,j)}(\lambda)\,,$$

and $\hat{\sigma}^2 = \tau(\bar{\mathbf{X}})$.

We wish to calculate $c_k = \tau_{(k)}(\mu)$. To this end, let $\mathbf{M} = (\mu_{ij})$ denote the $d \times d$ matrix, and put $\mathbf{N} = \mathbf{M}^{-1} = (\nu_{ij})$ say. By (3.100) and (3.101),

$$\frac{\partial \mu^{(i)}}{\partial \lambda^{(j)}} = \psi_{(i,j)}(\lambda) = \mu_{(i,j)}\,,$$

whence it follows that

$$\frac{\partial \lambda^{(i)}}{\partial \mu^{(j)}} = \nu_{ij}\,.$$

Hence

$$\frac{\partial}{\partial \mu^{(k)}}\, \psi_{(i,j)}(\lambda) = \sum_l \frac{\partial \lambda^{(l)}}{\partial \mu^{(k)}} \frac{\partial}{\partial \lambda^{(l)}}\, \psi_{(i,j)}(\lambda) = \sum_l \nu_{lk} \mu_{ijl} = \sum_l \nu_{kl} \mu_{ijl}\,,$$

since $\mathbf{M}$ (and hence $\mathbf{N}$) is symmetric. Therefore

$$c_k = \tau_{(k)}(\mu)$$

$$= \sum_i \sum_j \Big[\{A_{(i,k)}(\mu) A_{(j)}(\mu) + A_{(i)}(\mu) A_{(j,k)}(\mu)\} \psi_{(i,j)}(\lambda)$$
$$\quad + A_{(i)}(\mu) A_{(j)}(\mu) \frac{\partial}{\partial \mu^{(k)}}\, \psi_{(i,j)}(\lambda) \Big]$$

$$= 2 \sum_i \sum_j a_i a_{jk} \mu_{ij} + \sum_i \sum_j \sum_l a_i a_j \nu_{kl} \mu_{ijl}\,. \tag{3.102}$$

Now,

$$\sum_p \sum_k a_p \left(\sum_i \sum_j \sum_l a_i a_j \nu_{kl} \mu_{ijl} \right) \mu_{pk}$$

$$= \sum_i \sum_j \sum_l \sum_p a_i a_j a_p \mu_{ijl} \left(\sum_k \mu_{pk} \nu_{kl} \right)$$

$$= \sum_i \sum_j \sum_l a_i a_j a_l \mu_{ijl} . \qquad (3.103)$$

Multiplying (3.102) by $a_p \mu_{pk}$, adding over p and k, and using (3.103) we deduce that

$$\sum_i \sum_j a_i c_j \mu_{ij} = 2 \sum_i \sum_j \sum_k \sum_l a_i a_j a_{kl} \mu_{ik} \mu_{jl}$$

$$+ \sum_i \sum_j \sum_k a_i a_j a_k \mu_{ijk} ,$$

which is identical to (3.88).

3.11 Bootstrap Iteration

3.11.1. Pros and Cons of Bootstrap Iteration

We introduced bootstrap iteration in Section 1.4. In broad terms, bootstrap iteration enhances the accuracy of a bootstrap method by estimating an error term, such as coverage error of a confidence interval, and adjusting the bootstrap method so as to reduce that error. One advantage of bootstrap iteration is that it can substantially improve the performance of "naive" bootstrap methods, such as the "other percentile method," which are not in themselves good performers. A disadvantage is that iteration is highly computer intensive; in the nonparametric case it involves drawing resamples from resamples drawn from the sample, with the number of operations typically increasing as the square of the number for the ordinary bootstrap.

The reader might wonder why one would consider using a highly computer intensive method to improve the performance of a percentile method, when we have argued that techniques such as percentile-t are generally to be preferred over percentile. However, recall from subsection 3.10.1 that percentile-t only performs well when the variance of $\hat{\theta}$ can be estimated reasonably well. In other cases, such as that where $\hat{\theta}$ is the sample correlation coefficient, percentile-t can give disastrous results. Percentile methods will often perform better in those circumstances, although they will still give the poor coverage accuracy typically associated with uncorrected percentile techniques. Bootstrap iteration offers the prospect of retaining the excellent stability properties of percentile methods but enhancing their coverage accuracy by adjusting the nominal confidence level or the interval endpoints. Thus, the iterated percentile method interval is a highly computer-intensive hybrid, and in this hybridization lies its strength. If we had to recommend a utilitarian technique, which could be applied to a wide range of problems with satisfactory results, it would be hard to go past an iterated percentile method. Either of the two percentile methods could be used, although the "other percentile method" seems to give better results in simulations, for reasons that are not clear to us.

It may be that a significant part of the future of bootstrap methods lies in this direction, using techniques such as the iterated bootstrap to marry the best features of different methods, and thereby obtaining new techniques with widespread applicability. That is not to say that bootstrap iteration cannot be used to good effect in other ways, for example to enhance the already satisfactory performance of methods such as percentile-t in problems like estimating a population mean. However, the relative gain in performance when iteration is applied to methods that work well is not so great as when it is applied to methods whose performance is substandard; yet the computational cost is much the same.

3.11.2 Bootstrap Iteration in the Case of Confidence Intervals

As we showed in Section 1.4, bootstrap iteration involves selecting (i.e., estimating) a tuning parameter $t = t(n)$, so as to render an error term closer to zero. Assume that the problem has been parametrized so that $t(n) \to 0$ as $n \to \infty$. For example, t might be an additive correction to the endpoint of a nominal α-level confidence interval of the form $(-\infty, \hat{x}_\alpha)$ or $(\hat{x}_{(1-\alpha)/2}, \hat{x}_{(1+\alpha)/2})$. In this case we might ask that t be chosen such that

$$P\Big\{\theta_0 \in (-\infty, \hat{x}_\alpha + t)\Big\} = \alpha$$

or

$$P\Big\{\theta_0 \in \big(\hat{x}_{(1-\alpha)/2} - t, \ \hat{x}_{(1+\alpha)/2} + t\big)\Big\} = \alpha.$$

Alternatively, t might be an additive correction to the nominal level, in which case we would want

$$P\Big\{\theta_0 \in \big(-\infty, \hat{x}_{\alpha+t}\big)\Big\} = \alpha$$

or

$$P\Big\{\theta_0 \in \big(\hat{x}_{(1-\alpha-t)/2}, \ \hat{x}_{(1+\alpha+t)/2}\big)\Big\} = \alpha.$$

For either of these two approaches to correction, t may be expressed as an asymptotic power series in $n^{-1/2}$ in the case of correcting a one-sided interval and an asymptotic series in n^{-1} in the case of correcting a two-sided interval. Subsection 3.11.3 will illustrate this point. The expansions of t are obtainable directly from Edgeworth expansions of the coverage error of the uncorrected interval, and so the coefficients of terms in the expansions depend on population moments. The bootstrap estimate of t may be expressed in an identical manner, by an expansion, except that the population moments are now replaced by their sample counterparts.

The main interest in these expansions lies in their usefulness for explaining why bootstrap iteration works. However, they could in principle be used to bypass the second level of bootstrapping in bootstrap iteration. We could compute the first term in the theoretical expansion of t, replace unknowns in that formula by sample estimates, and use the resulting quantity $\tilde{t}$ in place of the $\hat{t}$ that would be obtained by bootstrap iteration. This technique is very much in the spirit of explicit skewness correction, discussed in Section 3.8. Indeed, if the original interval $\mathcal{I} = (-\infty, \hat{x}_\alpha)$ were not adequately corrected for skewness (so that $\mathcal{I}$ had coverage error of size $n^{-1/2}$), and if the correction to the endpoint were made additively, then the corrected interval $(-\infty, \hat{x}_\alpha + \tilde{t})$ would be exactly of the form considered in Section 3.8. The iterated bootstrap approach to confidence intervals may be viewed as a method of correcting for skewness and higher-order cumulants, although the correction is carried out in a monotone way that avoids the problems described in Section 3.8.

It is instructive to spell out the mechanics of bootstrap iteration in a special case. (The method was described in general terms in Section 1.4.) For that purpose we shall use the example of the "other percentile method," in the context of equal-tailed, two-sided confidence intervals and with the correction applied to the level of nominal coverage. Given $\frac{1}{2} < \alpha < 1$, define $\xi = \frac{1}{2}(1 + \alpha)$ and let $\hat{y}_\beta$ be the solution of the equation

$$P(\hat{\theta}^* \leq \hat{y}_\beta \mid \mathcal{X}) = \beta.$$

Then $(\hat{y}_{1-\xi}, \hat{y}_\xi)$ is the nominal α-level confidence interval for θ_0 and

$$P\left\{\theta_0 \in (\hat{y}_{1-\xi}, \hat{y}_\xi)\right\} = \alpha + O(n^{-1}).$$

See subsection 3.5.5. If we adjust α to the new level $\alpha + t$ then ξ changes to $\xi + \frac{1}{2}t$. In principle it is possible to select t so that

$$P\left\{\theta_0 \in (\hat{y}_{1-\xi-(t/2)}, \hat{y}_{\xi+(t/2)})\right\} = \alpha,$$

although of course this t depends on the unknown sampling distribution. To estimate t in the case of the nonparametric bootstrap, let $\mathcal{X}^{**}$ denote a resample drawn randomly, with replacement, from $\mathcal{X}^*$, let $\hat{\theta}^{**}$ denote the version of $\hat{\theta}$ computed from $\mathcal{X}^{**}$ instead of $\mathcal{X}$, and let $\hat{y}_\beta^*$ be the solution of the equation $P\big(\hat{\theta}^{**} \leq \hat{y}_\beta^* \mid \mathcal{X}^*\big) = \beta$. Then

$$\hat{\pi}(t) = P\Big\{\hat{\theta} \in \big(\hat{y}_{1-\xi-(t/2)}^*, \, \hat{y}_{\xi+(t/2)}^*\big) \mid \mathcal{X}\Big\}$$

is an estimate of

$$\pi(t) = P\Big\{\theta_0 \in \big(\hat{y}_{1-\xi-(t/2)}, \, \hat{y}_{\xi+(t/2)}\big)\Big\}.$$

The solution $\hat{t}$ of $\hat{\pi}(\hat{t}) = \alpha$ is our estimate of the solution of $\pi(t) = \alpha$. The iterated bootstrap interval is

$$\mathcal{I} = \big(\hat{y}_{1-\xi-(\hat{t}/2)}, \, \hat{y}_{\xi+(\hat{t}/2)}\big). \tag{3.104}$$

In practice, computation of $\hat{t}$ demands repeated resampling from resamples, as follows. Draw independent resamples $\mathcal{X}_1^*, \ldots, \mathcal{X}_{B_1}^*$ by sampling randomly, with replacement, from $\mathcal{X}$. For each $1 \leq b \leq B_1$, draw independent resamples $\mathcal{X}_{b1}^{**}, \ldots, \mathcal{X}_{bB_2}^{**}$ by sampling randomly, with replacement, from $\mathcal{X}_b^*$. Write $\hat{\theta}_b^*$, $\hat{\theta}_{bc}^{**}$ for values of $\hat{\theta}$ computed from $\mathcal{X}_b^*$, $\mathcal{X}_{bc}^{**}$ respectively, instead of $\mathcal{X}$. Rank $\{\hat{\theta}_b^*, \, 1 \leq b \leq B_1\}$ as $\hat{\theta}_{(1)}^* \leq \ldots \leq \hat{\theta}_{(B_1)}^*$ and rank $\{\hat{\theta}_{bc}^{**}, \, 1 \leq c \leq B_2\}$ as $\hat{\theta}_{b,(1)}^* \leq \ldots \leq \hat{\theta}_{b,(B_2)}^*$. Let $\nu_i(\beta)$, for $i = 1, 2$, be monotone functions taking values in the sequence $1, \ldots, B_i$ and such that $\nu_i(\beta)/B_i \simeq \beta$ for $i = 1, 2$ and $0 \leq \beta \leq 1$. Then $\hat{\theta}_{(\nu_1(\beta))}^*$ is our approximation to $\hat{y}_\beta$ and

$$\hat{\pi}_{B_1 B_2}(t) = B_1^{-1} \sum_{b=1}^{B_1} I\Big\{\hat{\theta} \in \big(\hat{\theta}_{b,(\nu_2(1-\xi-(t/2)))}^{**}, \, \hat{\theta}_{b,(\nu_2(\xi+(t/2)))}^{**}\big)\Big\}$$

is our approximation to $\hat{\pi}(t)$. Choose $\hat{t}_{B_1 B_2}$ to solve the equation

$$\hat{\pi}_{B_1 B_2}\big(\hat{t}_{B_1 B_2}\big) = \alpha$$

as nearly as possible, bearing in mind the discrete nature of the function $\hat{\pi}_{B_1 B_2}$. Then

$$\mathcal{I}_{B_1 B_2} = \Big(\hat{\theta}_{(\nu_1(1-\xi-(\hat{t}_{B_1 B_2}/2)))}^*, \, \hat{\theta}_{(\nu_1(\xi+(\hat{t}_{B_1 B_2}/2)))}^*\Big)$$

is our computable approximation to the interval $\mathcal{I}$ defined at (3.104). As ν_1, ν_2, B_1, B_2 diverge in such a way that $\nu_i(\beta)/B_i \to \beta$, for $i = 1, 2$ and $0 \leq \beta \leq 1$, we have $\hat{\pi}_{B_1, B_2} \to \hat{\pi}$ and $\hat{\theta}_{(\nu_1(\beta))}^* \to \hat{y}_\beta$ with probability one conditional on $\mathcal{X}$. This ensures that $\mathcal{I}_{B_1, B_2} \to \mathcal{I}$ as $B_1, B_2 \to \infty$, at least within the limitations imposed by discreteness in the definition of the latter interval.

Our account of the mechanics of bootstrap iteration, leading to the interval $\mathcal{I}$ defined at (3.104), was tailored to the case of the nonparametric bootstrap. To treat the case of the parametric bootstrap, let us assume that the population Π_λ is completely known up to a vector λ of unknown parameters. Write $\hat{\lambda}$ for an estimate of λ, perhaps the maximum likelihood estimate; let $\mathcal{X}^*$ denote a sample drawn randomly from $\Pi_{\hat{\lambda}}$; write $\hat{\lambda}^*$ for the version of $\hat{\lambda}$ computed for $\mathcal{X}^*$ instead of $\mathcal{X}$; and let $\mathcal{X}^{**}$ be a sample drawn randomly from $\Pi_{\hat{\lambda}^*}$. The prescription that produced $\mathcal{I}$ in the nonparametric case may be used exactly as before, except that the new definitions of $\mathcal{X}^*$ and $\mathcal{X}^{**}$ should replace the "resampling" definitions given earlier. Likewise, the argument that gave the computable approximation $\mathcal{I}_{B_1 B_2}$ is still applicable, provided the samples $\mathcal{X}_b^*$ and $\mathcal{X}_{bc}^{**}$ are generated in the manner described above.

3.11.3 Asymptotic Theory for the Iterated Bootstrap

For the sake of definiteness we shall develop theory in the case of confidence intervals constructed by the "other percentile method," when correction is carried out at the level of nominal coverage. This is the context described in subsection 3.11.2.

Let $\hat{y}_\beta$ be the solution of the equation

$$P(\hat{\theta}^* \leq \hat{y}_\beta \mid \mathcal{X}) = \beta,$$

and put $\xi = \frac{1}{2}(1 + \alpha)$, where $0 < \alpha < 1$. Then $(-\infty, \hat{y}_\alpha)$, $(\hat{y}_{1-\xi}, \hat{y}_\xi)$ are nominal α-level confidence intervals for θ_0. Ideally we would like to choose tuning parameters t_1, t_2 such that

$$P\{\theta_0 \in (-\infty, \hat{y}_{\alpha+t_1})\} = \alpha,$$

$$P\{\theta_0 \in (\hat{y}_{1-\xi-(t_2/2)}, \hat{y}_{\xi+(t_2/2)})\} = \alpha. \tag{3.105}$$

The iterated bootstrap produces estimates $\hat{t}_1, \hat{t}_2$. We shall derive asymptotic expressions for t_1 and t_2, in powers of $n^{-1/2}$ and n^{-1} respectively, and with first terms of sizes $n^{-1/2}$ and n^{-1}. Then we shall show that $\hat{t}_1$ and $\hat{t}_2$ admit identical expansions, except that population moments in coefficients are replaced by sample moments. Since these sample moments are $O_p(n^{-1/2})$ away from their corresponding population versions,

$$\hat{t}_1 - t_1 = O_p(n^{-1/2} \cdot n^{-1/2}) = O_p(n^{-1})$$

and

$$\hat{t}_2 - t_2 = O_p(n^{-1} \cdot n^{-1/2}) = O_p(n^{-3/2}).$$

We shall also prove that the coverage errors of the empirically corrected intervals $(-\infty, \hat{y}_{\alpha+\hat{t}_1})$, $(\hat{y}_{1-\xi-(\hat{t}_2/2)}, \hat{y}_{\xi+(\hat{t}_2/2)})$ are $O(n^{-1})$, $O(n^{-2})$ respectively. This compares with errors of size $n^{-1/2}, n^{-1}$ in the case of the uncorrected intervals $(-\infty, \hat{y}_\alpha), (\hat{y}_{1-\xi}, \hat{y}_\xi)$; see Section 3.5 for an account of the latter.

To begin our theoretical development, let $\hat{u}_\beta$ be the solution of the equation

$$P\{n^{1/2}(\hat{\theta}^* - \hat{\theta})/\hat{\sigma} \le \hat{u}_\beta \mid \mathcal{X}\} = \beta,$$

and observe that

$$\hat{y}_\beta = \hat{\theta} + n^{-1/2}\hat{\sigma}\hat{u}_\beta$$

$$= \hat{\theta} + n^{-1/2}\hat{\sigma}\{z_\beta + n^{-1/2}\hat{p}_{11}(z_\beta) + n^{-1}\hat{p}_{21}(z_\beta) + \dots\}.$$

(Here we have used (3.20).) Hence by Proposition 3.1 in Section 3.5,

$$P\{\theta_0 \in (-\infty, \hat{y}_\beta)\}$$

$$= P\Big[\theta_0 \in \big(-\infty, \hat{\theta} + n^{-1/2}\hat{\sigma}\{z_\beta + n^{-1/2}\hat{p}_{11}(z_\beta) + n^{-1}\hat{p}_{21}(z_\beta) + \dots\}\big)\Big]$$

$$= \beta + n^{-1/2}r_1(z_\beta)\phi(z_\beta) + n^{-1}r_2(z_\beta)\phi(z_\beta) + \dots,$$

where r_j is an odd/even polynomial for even/odd j, respectively. Thus,

$$P\{\theta_0 \in (\hat{y}_{1-\beta}, \hat{y}_\beta)\}$$

$$= P\{\theta_0 \in (-\infty, \hat{y}_\beta)\} - P\{\theta_0 \in (-\infty, \hat{y}_{1-\beta})\}$$

$$= 2\beta - 1 + n^{-1}2r_2(z_\beta)\phi(z_\beta) + n^{-2}2r_4(z_\beta)\phi(z_\beta) + \dots.$$

The "ideal" tuning parameters t_1, t_2 are given by equations (3.105), which are here seen to be equivalent to

$$t_1 = -n^{-1/2}r_1(z_{\alpha+t_1})\phi(z_{\alpha+t_1})$$

$$- n^{-1}r_2(z_{\alpha+t_1})\phi(z_{\alpha+t_1}) + \dots, \tag{3.106}$$

$$t_2 = -n^{-1}2r_2(z_{\xi+(t_2/2)})\phi(z_{\xi+(t_2/2)})$$

$$- n^{-2}2r_4(z_{\xi+(t_2/2)})\phi(z_{\xi+(t_2/2)}) - \dots. \tag{3.107}$$

From these formulae and the Taylor expansion

$$z_{\beta+\epsilon} = z_\beta + \epsilon\phi(z_\beta)^{-1} + \tfrac{1}{2}\epsilon^2 z_\beta\phi(z_\beta)^{-2} + \dots,$$

one may deduce that t_1, t_2 admit the asymptotic expansions

$$t_1 = n^{-1/2}\pi_{11}(z_\alpha)\phi(z_\alpha) + n^{-1}\pi_{21}(z_\alpha)\phi(z_\alpha) + \dots, \tag{3.108}$$

$$t_2 = n^{-1}\pi_{12}(z_\xi)\phi(z_\xi) + n^{-1}\pi_{22}(z_\xi)\phi(z_\xi) + \dots, \tag{3.109}$$

where π_{j1} is an odd/even polynomial for even/odd j, respectively, and each π_{j2} is an odd polynomial. In particular,

$$\pi_{11} = -r_1, \quad \pi_{21}(z) = r_1(z)r_1'(z) - zr_1(z)^2 - r_2(z),$$

$$\pi_{12} = -r_2, \quad \pi_{22}(z) = r_2(z)r_2'(z) - zr_2(z)^2 - r_4(z).$$

By analogy with (3.106) and (3.107),

$$\hat{t}_1 = -n^{-1/2}\hat{r}_1(z_{\alpha+\hat{t}_1})\phi(z_{\alpha+\hat{t}_1}) - n^{-1}\hat{r}_2(z_{\alpha+\hat{t}_1})\phi(z_{\alpha+\hat{t}_1}) + \ldots$$

$$\hat{t}_2 = -n^{-1}2\hat{r}_2(z_{\xi+(\hat{t}_2/2)})\phi(z_{\xi+(\hat{t}_2/2)})$$
$$- n^{-2}2\hat{r}_2(z_{\xi+(\hat{t}_2/2)})\phi(z_{\xi+(\hat{t}_2/2)}) - \ldots,$$

where $\hat{r}_j$ is identical to r_j except that sample moments in coefficients are replaced by population moments. Inverting these expansions we deduce the analogues of (3.108) and (3.109), which differ only in that (t_i, π_{ji}) is replaced by $(\hat{t}_i, \hat{\pi}_{ji})$, where $\hat{\pi}_{ji}$ results from π_{ji} on replacing population moments by sample moments.

To obtain a formula for the coverage error of the interval $(-\infty, \hat{y}_{\alpha+\hat{t}_1})$, note that since $\hat{t}_1 = t_1 + O_p(n^{-1})$, by the delta method,

$$P\{\theta_0 \in (-\infty, \hat{y}_{\alpha+\hat{t}_1})\}$$
$$= P[\theta_0 \le \hat{\theta} + n^{-1/2}\hat{\sigma}\{z_{\alpha+\hat{t}_1} + n^{-1/2}\hat{p}_{11}(z_{\alpha+\hat{t}_1}) + \ldots\}]$$
$$= P\{n^{1/2}(\hat{\theta} - \theta_0)/\hat{\sigma} > -z_{\alpha+\hat{t}_1} - n^{-1/2}\hat{p}_{11}(z_{\alpha+\hat{t}_1}) - \ldots\}$$
$$= P\{n^{1/2}(\hat{\theta} - \theta_0)/\hat{\sigma} > -z_{\alpha+t_1} - n^{-1/2}\hat{p}_{11}(z_{\alpha+t_1}) + O_p(n^{-1})\}$$
$$= P\{n^{1/2}(\hat{\theta} - \theta_0)/\hat{\sigma} > -z_{\alpha+t_1} - n^{-1/2}\hat{p}_{11}(z_{\alpha+t_1}) - \ldots\} + O(n^{-1})$$
$$= P\{\theta_0 \in (-\infty, \hat{y}_{\alpha+t_1})\} + O(n^{-1})$$
$$= \alpha + O(n^{-1}).$$

That is, the coverage error of $(-\infty, \hat{y}_{\alpha+t_1})$ equals $O(n^{-1})$.

Derivation of the coverage error of $(\hat{y}_{1-\xi-(\hat{t}_2/2)}, \hat{y}_{\xi+(\hat{t}_2/2)})$ is similar although more complex. Writing $\Delta = \frac{1}{2}(\hat{t}_2 - t_2) = O_p(n^{-3/2})$, we have

$$P\{\theta_0 \in (-\infty, \hat{y}_{\beta\pm(\hat{t}_2/2)})\}$$
$$= P\{n^{1/2}(\hat{\theta} - \theta_0)/\hat{\sigma} > -z_{\beta\pm(\hat{t}_2/2)} - n^{-1/2}\hat{p}_{11}(z_{\beta\pm(\hat{t}_2/2)}) - \ldots\}$$
$$= P\{n^{1/2}(\hat{\theta} - \theta_0)/\hat{\sigma} > -z_{\beta\pm(t_2/2)} - n^{-1/2}\hat{p}_{11}(z_{\beta\pm(t_2/2)})$$
$$\mp \Delta\phi(z_\beta)^{-1} + O_p(n^{-2})\}$$
$$= P\{n^{1/2}(\hat{\theta} - \theta_0)/\hat{\sigma} > -z_{\beta\pm(t_2/2)} - n^{-1/2}\hat{p}_{11}(z_{1-\beta\pm(t_2/2)}) - \ldots$$
$$\mp \Delta\phi(z_\beta)^{-1}\} + O(n^{-2})$$
$$= P\{\theta_0 \in (-\infty, \hat{y}_{\beta\pm t_2})\} \pm n^{-3/2}bz_\beta\phi(z_\beta) + O(n^{-2}),$$

where b is the constant such that

$$E\{n^{1/2}(\hat{\theta} - \theta_0)\hat{\sigma}^{-1} \cdot n^{3/2}\tfrac{1}{2}(\hat{t}_2 - t_2)\} = b + O(n^{-1}).$$

(Use the argument leading to (3.36).) Therefore

$$
\begin{aligned}
P\{\theta_0 &\in (\hat{y}_{1-\xi-(\hat{t}_2/2)}, \; \hat{y}_{\xi+(\hat{t}_2/2)})\} \\
&= P\{\theta_0 \in (-\infty, \; \hat{y}_{\xi+(\hat{t}_2/2)})\} - P\{\theta_0 \in (-\infty, \; \hat{y}_{1-\xi-(\hat{t}_2/2)})\} \\
&= P\{\theta_0 \in (-\infty, \; \hat{y}_{\xi+(t_2/2)})\} + n^{-3/2} b z_\xi \phi(z_\xi) + O(n^{-2}) \\
&\quad - \left[P\{\theta_0 \in (-\infty, \; \hat{y}_{1-\xi-(t_2/2)})\} - n^{-3/2} b z_{1-\xi} \phi(z_{1-\xi}) + O(n^{-2}) \right] \\
&= P\{\theta_0 \in (\hat{y}_{1-\xi-(t_2/2)}, \; \hat{y}_{\xi+(t_2/2)})\} + O(n^{-2}) \\
&= \alpha + O(n^{-2}).
\end{aligned}
$$

That is, the coverage error of $(\hat{y}_{1-\xi-(\hat{t}_2/2)}, \; \hat{y}_{\xi+(\hat{t}_2/2)})$ equals $O(n^{-2})$.

3.12 Hypothesis Testing

There is a well known and established duality between confidence intervals and hypothesis testing: if $\mathcal{I}$ is a confidence interval for an unknown parameter θ, with coverage probability α, then a $(1 - \alpha)$-level test of the null hypothesis $H_0 : \theta = \theta_0$ against $H_1 : \theta \neq \theta_0$ is to reject H_0 if $\theta_0 \notin \mathcal{I}$. In the case of this two-sided test it would be appropriate for $\mathcal{I}$ to be a two-sided interval. (The level of a test equals the probability of committing a Type I error, i.e., of rejecting H_0 when H_0 is true.) More generally, to test a composite hypothesis $H_0 : \theta \in \Omega$ against $H_1 : \theta \in \Theta \backslash \Omega$, where Θ denotes the parameter space and Ω is a subset of Θ, reject H_0 if $\mathcal{I} \cap \Omega = \emptyset$ (the empty set). Thus, many of the issues that arise in the context of confidence intervals, such as the debate over pivotal and nonpivotal methods, also have a place in the development of bootstrap methods for hypothesis tests.

However, there is a point of conceptual difference, which in simpler problems emerges in terms of centring a parameter estimate $\hat{\theta}$. Let $\hat{\theta}^*$ denote the bootstrap version of $\hat{\theta}$, computed from a resample $\mathcal{X}^*$ instead of the entire sample $\mathcal{X}$. Suppose we wish to test $H_0 : \theta = \theta_1$ against $H_1 : \theta \neq \theta_1$. In most circumstances we would base the test on the difference $\hat{\theta} - \theta_1$, rejecting H_0 in favour of H_1 if $|\hat{\theta} - \theta_1|$ were too large. However, we should bootstrap the difference $\hat{\theta}^* - \hat{\theta}$ (or rather, a pivotal version of it), not $\hat{\theta}^* - \theta_1$. For example, a percentile-t, nominal $(1 - \alpha)$-level test of H_0 versus H_1 would be to reject H_0 if $|\hat{\theta} - \theta_1| > n^{-1/2} \hat{\sigma} \hat{w}_\alpha$, where $\hat{w}_\alpha$ is defined by

$$
P\big(n^{1/2} |\hat{\theta}^* - \hat{\theta}| / \hat{\sigma}^* \leq \hat{w}_\alpha \mid \mathcal{X}\big) = \alpha,
$$

and $\hat{\sigma}^{*2}$ is the resample version of an estimate $\hat{\sigma}^2$ of the variance of $n^{1/2} \hat{\theta}$. This is equivalent to rejecting H_0 if θ_1 lies outside the symmetric percentile-t confidence interval $(\hat{\theta} - n^{-1/2} \hat{\sigma} \, \hat{w}_\alpha, \hat{\theta} + n^{-1/2} \hat{\sigma} \, \hat{w}_\alpha)$; see Section 3.6

for an account of the latter. An alternative approach, which involves centring at θ_1 rather than $\hat{\theta}$, would be to compute the value $\hat{w}_{\alpha 1}$ such that

$$P\left(n^{1/2}|\hat{\theta}^* - \theta_1|/\hat{\sigma}^* \leq \hat{w}_{\alpha 1} \mid \mathcal{X}\right) = \alpha, \qquad (3.110)$$

and reject H_0 if $|\hat{\theta} - \theta_1| > n^{-1/2}\,\hat{\sigma}\,\hat{w}_{\alpha 1}$. This technique is occasionally advocated, but it has low power.

To appreciate the point about low power, let us assume that the true value of θ is a fixed number θ_0, and take $\theta_1 = \theta_0 - n^{-1/2}\,c\sigma$, where c is a constant and σ^2 equals the asymptotic variance of $n^{1/2}\,\hat{\theta}$. Define

$$\pi_n(c) = P\left(|\hat{\theta} - \theta_1| > n^{-1/2}\,\hat{\sigma}\,\hat{w}_\alpha\right),$$
$$\pi_{n1}(c) = P\left(|\hat{\theta} - \theta_1| > n^{-1/2}\,\hat{\sigma}\,\hat{w}_{\alpha 1}\right),$$

which represent the powers of nominal $(1 - \alpha)$-level tests based on resampling $\hat{\theta}^* - \hat{\theta}$ and $\hat{\theta}^* - \theta_1$ respectively. Denote the asymptotic powers by

$$\pi_\infty(c) = \lim_{n\to\infty} \pi_n(c), \qquad \pi_{\infty 1}(c) = \lim_{n\to\infty} \pi_{n1}(c).$$

We claim that as $c \to \infty$, and for $\alpha > \frac{1}{2}$, the asymptotic power of the first test converges to 1 but that of the second test tends to 0,

$$\pi_\infty(c) \longrightarrow 1, \qquad \pi_{\infty 1}(c) \longrightarrow 0. \qquad (3.111)$$

The latter result is clearly unsatisfactory, demonstrating the superiority of bootstrapping $\hat{\theta}^* - \hat{\theta}$ rather than $\hat{\theta}^* - \theta_1$.

We now prove (3.111), assuming that $Z_n = n^{1/2}(\hat{\theta} - \theta_0)/\hat{\sigma}$ is asymptotically Normal $N(0, 1)$ and that, conditional on $\mathcal{X}$, the same is true of $Z_n^* = n^{1/2}(\hat{\theta}^* - \hat{\theta})/\hat{\sigma}^*$. These assumptions entail

$$\hat{w}_\alpha = z_\xi + o_p(1) \quad \text{and} \quad \hat{w}_{\alpha,1} = \zeta_\alpha(Z_n + c) + o_p(1), \qquad (3.112)$$

where $\xi = \frac{1}{2}(1 + \alpha)$, $\zeta_\alpha(x)$ is defined by

$$P\{|Z + x| \leq \zeta_\alpha(x)\} = \alpha,$$

and Z is a Standard Normal random variable. (To obtain the second part of (3.112), observe that

$$n^{1/2}|\hat{\theta}^* - \theta_1|/\hat{\sigma}^* = |Z_n^* + (\hat{\sigma}/\hat{\sigma}^*)\,Z_n + (\sigma/\hat{\sigma}^*)c|,$$

and note the definition (3.110) of $\hat{w}_{\alpha 1}$.) From (3.112) and the fact that $n^{1/2}|\hat{\theta} - \theta_1|/\hat{\sigma} = |Z_n + c|$ we may deduce that

$$\pi_\infty(c) = P\left(|Z + c| > z_\xi\right),$$
$$\pi_{\infty 1}(c) = P\{|Z + c| > \zeta_\alpha(Z + c)\}. \qquad (3.113)$$

The first part of (3.111) is an immediate consequence of the first part of (3.113). To obtain the second part of (3.111), note that as $x \to \infty$,

$$\zeta_\alpha(x) - x \longrightarrow z_\alpha.$$

Therefore as $c \to \infty$,

$$
\begin{aligned}
\pi_{\infty 1}(c) &= P\{|Z + c| > \zeta_\alpha(Z + c)\} \\
&= P\{|Z + c| > Z + c + z_\alpha + o_p(1)\} \\
&= P(Z + c > Z + c + z_\alpha) + o(1) \quad \longrightarrow \quad 0.
\end{aligned}
$$

The effect of pivoting is to enhance the level accuracy of a bootstrap test. Just as in the case of confidence intervals, the effect of pivoting is most obvious in the one-sided case. A one-sided test based on a Normal approximation typically has level error of order $n^{-1/2}$, whereas the percentile-t bootstrap method can reduce the error to order n^{-1}. A two-sided test based on the same Normal approximation has level error of order n^{-1}, which is reduced to order n^{-2} by using a two-sided symmetric test of the type described above (i.e., reject H_0 if $|\hat\theta - \theta_1| > n^{1/2} \hat\sigma \hat w_\alpha$). However, the analogous two-sided equal-tailed test, where H_0 is rejected if $\hat\theta - \theta_1 > n^{1/2} \hat\sigma \hat v_\xi$ or $\hat\theta - \theta_1 < n^{-1/2} \hat\sigma \hat v_{1-\xi}$, and $\hat v_\beta$ is defined by

$$P\{n^{1/2}(\hat\theta^* - \hat\theta)/\hat\sigma^* \le \hat v_\beta \mid \mathcal{X}\} = \beta,$$

has level error of size n^{-1} rather than n^{-2}. All these results follow immediately from the duality between confidence intervals and hypothesis tests, noted earlier. (Recall from Sections 3.5 and 3.6 that equal-tailed two-sided percentile-t intervals have coverage error of size n^{-1}, whereas their symmetric counterparts have coverage error of size n^{-2}.)

We should stress that our results about the efficacy of pivoting are asymptotic in nature, and do not apply in all circumstances. For problems such as testing a hypothesis about a correlation coefficient, where it is not possible to estimate variance sufficiently accurately for pivotal methods to be viable, alternative techniques are called for. Bias correction and bootstrap iteration are both candidates.

Of course, the duality between hypothesis testing and confidence intervals allows all the confidence interval technology to be brought to bear. In view of our earlier work on confidence intervals, this duality has been a convenient device for introducing hypothesis testing. However, many would rightly object to this approach, pointing out that an important aspect of significance testing is the actual level of significance attained by the observed data, not some artificial threshold imposed by the experimenter or statistician. Of course, bootstrap methods may be used to estimate the actual significance level. For example, if the test is of $H_0 : \theta = \theta_0$ against

$H_1 : \theta \neq \theta_0$, then the significance level for a symmetric test is estimated to be

$$P\big(|\hat{\theta}^* - \hat{\theta}|/\hat{\sigma}^* > |\hat{\theta} - \theta_0|/\hat{\sigma} \mid \mathcal{X}\big).$$

If the alternative hypothesis is $\theta > \theta_0$, then the estimate of significance for a one-sided test is

$$P\big\{(\hat{\theta}^* - \hat{\theta})/\hat{\sigma}^* > (\hat{\theta} - \theta_0)/\hat{\sigma} \mid \mathcal{X}\big\}.$$

3.13 Bibliographical Notes

The use of Edgeworth expansions to elucidate properties of the bootstrap dates from Singh's (1981) seminal paper. See also Bickel and Freedman (1980). Singh, and Bickel and Freedman, developed one-term Edgeworth expansions to describe the skewness-correcting role of bootstrap methods in the case of the non-Studentized sample mean. In the context of this work, population variance was assumed known and so $n^{1/2}(\hat{\theta} - \theta_0)$ was asymptotically pivotal. Babu and Singh (1983, 1984, 1985) and Hall (1986a, 1988a) later treated the case of a Studentized statistic, and brought out some of the differences between pivotal and nonpivotal applications of the bootstrap. See also Beran (1982). Singh (1981) addressed the problem of bootstrap approximations to the distribution of the mean of a sample from a lattice population, and Hall (1987a) treated continuity corrections for enhancing those approximations.

Hall (1986a) derived asymptotic formulae for coverage errors of percentile-t confidence intervals, and subsequently (1988a) developed a more general theory encompassing a full range of bootstrap confidence intervals. Theoretical support for pivotal, rather than nonpivotal, methods was established in a sequence of papers by Babu and Singh (1984), Hall (1986a, 1988a), Beran (1987), and Liu and Singh (1987), all of which call upon tools of Edgeworth expansion.

Methods based on symmetric bootstrap confidence intervals were suggested by Hall (1988b). The "short" bootstrap confidence intervals described in Section 3.7 were proposed by Hall (1988a); see also Buckland (1980, 1983). Explicit Edgeworth corrections for skewness and higher-order effects were discussed by Pfanzagl (1979), Hall (1983), Withers (1983), and Abramovitch and Singh (1985). In the latter paper, second-order Edgeworth corrections are compared with bootstrap adjustments of first-order corrections. Transformations for correcting for skewness were described by Johnson (1978), Mallows and Tukey (1982), and Hall (1992), among others. These techniques are distantly related to the transformations of Fisher, and Wilson and Hilferty, for improving approximations to the distribution of the chi-squared statistic; see Kendall and Stuart (1977, p. 399) for an account of the latter.

The "other percentile method" dates back to Efron's (1979) original paper. Efron (1981a, 1982) noted that this technique does not perform as well as might be hoped, and proposed the bias correction (BC) method. See Buckland (1983, 1984, 1985) for applications of and algorithms for both the percentile method and its BC version. Schenker (1985) argued cogently that neither the percentile method nor its bias-corrected version is adequate for many purposes, and prompted Efron (1987) to develop accelerated bias correction (ABC). It was here that the term "second-order correct" was introduced.

Bootstrap iteration in the context of confidence intervals was introduced by Hall (1986a) and Beran (1987), with the correction being made additively to the interval endpoint in the former case and additively to nominal coverage level in the latter. Hall and Martin (1988) developed a general framework allowing a variety of different types of correction. See also Martin (1989, 1990). Loh (1987) introduced the notion of bootstrap calibration, a device for improving the coverage accuracy of bootstrap confidence intervals. If calibration is applied to bootstrap confidence intervals, it amounts to bootstrap iteration; if used in conjunction with intervals that have been Edgeworth-corrected for skewness, it produces an adjustment of the type considered by Abramovitch and Singh (1985).

The problem of constructing a confidence interval for a correlation coefficient is the "smoking gun" of bootstrap methods. When used without a variance-stabilizing transformation, percentile-t fails spectacularly, producing intervals that can have poor coverage accuracy and fail to respect the range of the coefficient. This seems to be the case regardless of choice of the variance estimate $\hat{\sigma}^2$ — versions include the jackknife estimate and an estimate based on substituting sample moments for population moments in the asymptotic variance formula. Percentile methods, including bias-corrected versions, also suffer from poor coverage accuracy, as do techniques such as empirical likelihood (Owen 1988). A lively debate on this problem has developed in the psychology literature (e.g., Lunneborg 1985, Rasmussen 1987, and Efron 1988), even calling into question the ability of any bootstrap technique to solve this problem. However, several techniques perform respectably well, including coverage correction (by bootstrap iteration) of the percentile-method interval, and application of percentile-t to a transformed version of the correlation coefficient, the transformation being chosen to stabilize variance. See Hall, Martin, and Schucany (1989). Young (1988) has discussed bootstrap smoothing in the context of the correlation coefficient; see also Silverman and Young (1987).

In general, variance-stabilization is a wise prelude to percentile-t methods. For example, log transformations are appropriate in the problem of constructing confidence intervals for a variance or standard deviation. Tibshirani (1988) has introduced an empirical device for obtaining an "automatic" variance-stabilizing transformation in a wide variety of problems, using a double bootstrap argument.

Related work on bootstrap confidence intervals includes J.A. Robinson (1983), Swanepoel, van Wyk, and Venter (1983), J. Robinson (1986), Rothe (1986), Tibshirani (1986), Schemper (1987b), DiCiccio and Tibshirani (1987), Laird and Louis (1987), Babu and Bose (1988), Beran (1988a, 1988b), Bose (1988) and DiCiccio and Romano (1989). The effect of tail weight on the bootstrap approximation has been discussed by Athreya (1987), Knight (1989), and Hall (1990d).

General theory for bootstrap hypothesis testing is discussed briefly by Hinkley (1988) during a survey of bootstrap methods, and at greater length by Hinkley (1989). Bunke and Riemer (1983) describe bootstrap hypothesis testing in linear models, Beran and Srivastava (1985) suggest bootstrap tests and confidence regions for functions of a covariance matrix, Young (1986) raises some of the issues of hypothesis testing in the setting of geometric statistics, Beran and Millar (1987) develop methods for stochastic estimation and testing, Beran (1988c) discusses pivoting in the context of hypothesis testing, Romano (1988) describes bootstrap versions of distance-based tests (such as tests of goodness of fit), Chen (1990) proposes bootstrap hypothesis tests that are based on transforming the data so that the null hypothesis is satisfied, Fisher and Hall (1990) treat bootstrap hypothesis testing via the example of analysis of variance, and Hall and Wilson (1991) illustrate the two guidelines of "pivoting" and "sampling under the null hypothesis" by applying bootstrap tests to specific data sets. The innovative work of Silverman (1981) on testing for multimodality should be mentioned. See also Mammen (1991).

4

Bootstrap Curve Estimation

4.1 Introduction

Our purpose in this chapter is to develop the ideas and techniques introduced in Chapter 3, this time along somewhat less conventional lines. We are particularly concerned with problems of curve estimation, which include parametric regression, nonparametric regression, and density estimation. Some of the ideas in Chapter 3 extend relatively straightforwardly to these new areas. In problems where variance can be estimated reasonably accurately (i.e., with low relative error), pivotal methods have a distinct advantage over nonpivotal ones, and this comparison may be clearly delineated by using arguments based on Edgeworth expansions.

However, there are distinctly different sides to some of the problems considered here, and it is those features that we stress. In the case of parametric regression (Section 4.3), confidence intervals for slope may be constructed with particularly good coverage accuracy, the errors being of order $n^{-3/2}$ and n^{-2} in the cases of one-sided and two-sided percentile-t intervals, respectively. Both these errors are of size n^{-1} in the more conventional circumstances described in Chapter 3. The reason that intervals for slope enjoy such a high level of coverage accuracy is that the presence of design points confers an extra degree of symmetry specifically on the problem of slope. This is true even if design is asymmetric. Estimation of an intercept in regression, or of a regression mean, is quite mundane in comparison and has properties that follow closely those described in Chapter 3.

In nonparametric curve estimation, for example nonparametric density or regression estimation (Sections 4.4 and 4.5, respectively), Edgeworth expansions are no longer power series in $n^{-1/2}$. This reflects the "infinite parameter" nature of such problems and the fact that nonparametric estimation of a curve must rely heavily on local properties. Only data points in the close vicinity of x convey useful information about the value of the curve at x, and so the sample size is effectively reduced. In the case of kernel estimation of a univariate density or regression function the effective sample size is roughly nh where h denotes the bandwidth of the estimator. Therefore the Edgeworth expansion is basically a power series in $(nh)^{-1/2}$, rather than $n^{-1/2}$. However, there are other complicating factors. In the case of nonparametric regression, the variance of the estimate can be estimated $\sqrt{n}$-consistently. Therefore the Studentized statistic is a ratio of one quantity whose Edgeworth expansion goes down in powers of $(nh)^{-1/2}$, and another whose expansion is in powers of $n^{-1/2}$. The interaction between these two quantities produces a most unusual overall expansion, and ensures that the case of nonparametric regression is particularly interesting.

Another distinctive feature of nonparametric curve estimation is the problem of bias. The "usual" bootstrap method applied to nonparametric curve estimation produces a confidence band for the expected value of the estimate, not for the curve itself. This can be rather awkward, since most prescriptions for curve estimation produce an estimate whose distance from its expected value is of approximately the same size as the distance between the true curve and the expected value. Therefore, bias is significant and must be allowed for. In the case of nonparametric density estimation we investigate this problem in some detail, comparing the effect on coverage error of different approaches to bias correction. However, the context of nonparametric regression is already rather complex, given the particularly unusual nature of its Edgeworth expansion theory, and so for the sake of simplicity we do not dwell on bias correction there.

In practical nonparametric curve estimation the bandwidth h would usually be estimated from the data, so that h would be a function of the sample values. Let us write h as $\hat{h}$ to indicate this data dependence. Most bandwidth estimation methods produce a function that is asymptotically constant, in the sense that $\hat{h}/c_n \to 1$ in probability for a sequence of constants c_n. However, that does not alter the fact that in detailed second-order arguments about coverage accuracy, such as those in Sections 4.4 and 4.5, we should really take account of the precise way in which $\hat{h}$ depends on the data; we do not, and in fact we assume that h is nonrandom. A more comprehensive account is well beyond the scope of this monograph. The differences probably do not have much impact if, in the bootstrap part of the algorithm, we replace $\hat{h}$ by the appropriate function of the resample values rather than sample values. That is, if $\hat{h} = A_n(X_1, \ldots, X_n)$ is the bandwidth used for the curve estimate then $\hat{h}^* = A_n(X_1^*, \ldots, X_n^*)$ is the bandwidth appropriate for the bootstrap version of the curve estimate. However, this approach causes other problems. For one thing, the common prescriptions for calculating $\hat{h}$ are often highly computer intensive, and sometimes involve a degree of subjectivity, such as selection of the bandwidth for a pilot curve estimate. Both of these features can rule out the use of such methods in the bootstrap case, where many Monte Carlo simulations are usually involved. Furthermore, the popular bandwidth selection method of cross-validation is affected seriously by ties among sample values, and this can be a problem when it is used with a bootstrap resample.

It is appropriate here to mention the issue of "smoothing" the bootstrap, which has been publicized in some quarters. So far in this monograph, in our accounts of the nonparametric bootstrap, we have recommended resampling directly from the sample. That is, we have suggested drawing the resample $\mathcal{X}^*$ by sampling randomly, with replacement, from $\mathcal{X}$. This procedure may be generalized by computing an estimate $\hat{f}$ of the unknown population density f, and resampling from the distribution with density $\hat{f}$. The shape and properties of $\hat{f}$ are governed by choice of smoothing parameter (the bandwidth in the case of a kernel estimator) used to construct $\hat{f}$. If there is no smoothing then $\hat{f}$ is simply the Dirac delta density of the sample values, and so the procedure reduces to the usual method of resampling with replacement. More generally, it would often be theoretically possible to improve the performance of the bootstrap algorithm by smoothing a little. This problem is a little like that of shrinkage, encountered in Example 1.4 of Section 1.3: the performance of a point estimate can often be improved by employing a small amount of shrinkage, although the extent of improvement will typically be a second-order effect. The same is generally true in the case of smoothing and the bootstrap, at least in the circumstances encountered in Chapter 3. Smoothing can improve performance a little, but the amount of improvement will be small, and if too large a level of smoothing is applied then performance can deteriorate sharply. The practical problem is to determine how much to smooth.

To appreciate why smoothing usually cannot have a first-order effect, recall our argument in Section 3.1, which uses Edgeworth expansions to elucidate properties of the bootstrap. The bootstrap works because it produces a distribution function estimate in which the bootstrap version of the effect due to skewness is a good approximation to the true effect of skewness. For example, if the true skewness is the third population moment, as in the case of Edgeworth expansion for the sample mean, then its bootstrap estimate is the third sample moment. The distance apart of these quantities is $O(n^{-1/2})$, and so the efficacy of the approximation is improved by a factor of $n^{-1/2}$ over that of the Normal approximation, from $n^{-1/2}$ to $n^{-1/2} \cdot n^{-1/2} = n^{-1}$. If we resampled from a smoothed empirical distribution (i.e., the distribution with density $\hat{f}$), then the third sample moment would be replaced by the third moment of the smoothed empiric. There would be no improvement in the rate of convergence, since arguments based on the Cramér-Rao lower bound can be used to show that the third population moment cannot, in general, be estimated at a faster rate than $n^{-1/2}$.

Nevertheless, it is likely that some reduction in the distance between the true distribution function and its estimate can be achieved. For example, if we use the L_1 metric $\int E|\widehat{H} - H|$ to measure the distance between the true distribution function H of a pivotal statistic and its bootstrap estimate $\widehat{H}$, then attention centres on the formula

$$\int E|\widehat{H} - H| = C n^{-1} + o(n^{-1}),$$

where $C > 0$ is a constant. It is usually possible to reduce the size of the $o(n^{-1})$ term, although not that of the constant C, by smoothing appropriately when constructing $\widehat{H}$. However, the amount of smoothing that achieves this end is generally much less than that which would normally be used for optimal point estimation of the density of the population from which the sample $\mathcal{X}$ was drawn.

In summary, smoothing usually has only a second-order effect, and the amount of smoothing that leads to an improvement in performance of the bootstrap is generally quite small. Similar remarks apply to other problems of point estimation. For example, smoothing usually does not have an effect on the first-order accuracy of a bootstrap estimate of bias or variance, although it can reduce the size of second- and higher-order terms. Proofs of all these results in the context of the "smooth function model," where the statistic in question can be expressed as a smooth function of a vector sample mean (see Section 2.4), are straightforward. They amount to noting that, by Taylor expansion, the bootstrap is being used (to first order) to estimate a fixed function of a multivariate population mean (e.g., skewness in the case of distribution function estimation) and that the first-order performance (e.g., mean squared error) of the sample mean as an estimate of the population mean cannot be improved by smoothing. (In close analogy, the first-order performance of the sample mean as an estimate of the population mean cannot be improved by shrinkage.) However, in cases outside the smooth function model, smoothing can have an impact on the rate of convergence; see Sections 4.4 and 4.5, and Appendix IV.

Section 4.2 treats multivariate confidence regions, as a prelude to the discussion of multivariate, multiparameter regression in Section 4.3. Nonparametric curve estimation is analyzed in Sections 4.4 and 4.5, although only in the case of pointwise confidence intervals, not confidence bands.

4.2 Multivariate Confidence Regions

The main issue that distinguishes multivariate from univariate confidence regions is that of configuration. In one dimension, an interval is by far the most easily interpreted confidence region. The position of the centre of the interval relative to a point estimate of the unknown parameter can convey important information about our uncertainty as to the true parameter value, but there is seldom any reason for taking the confidence region to be a disjoint union of two or more intervals.

Of course, some of the appeal of a confidence *interval* lies in the fact that an interval is the only type of convex set available in one dimension. In two or more dimensions there is a much richer variety of convex sets. Even if we constrain a multivariate region to be convex, there are important questions of shape and orientation to be answered, and the bootstrap does not automatically provide a response to any of them.

The classical Normal approximation method would construct a two-dimensional confidence region to be an ellipse (an ellipsoid in higher dimensions), since level sets of a bivariate Normal distribution function are ellipses. Let $\widehat{\boldsymbol{\theta}}$ be an estimate of an unknown d-vector $\boldsymbol{\theta}_0$, let $n^{-1}\widehat{\boldsymbol{\Sigma}}$ be an estimate of the variance matrix of $\widehat{\boldsymbol{\theta}}$, and assume that $\mathbf{T} = n^{1/2}\widehat{\boldsymbol{\Sigma}}^{-1/2}(\widehat{\boldsymbol{\theta}} - \boldsymbol{\theta}_0)$ is approximately Normal $N(\mathbf{O}, \mathbf{I})$, where $\mathbf{I}$ is the $d \times d$ identity matrix. If $\mathcal{S}$ is a d-variate sphere, centred at the origin and with radius chosen such that $P(\mathbf{N} \in \mathcal{S}) = \alpha$, where $\mathbf{N}$ has the $N(\mathbf{O}, \mathbf{I})$ distribution, then the set

$$\mathcal{R} = \left\{ \widehat{\boldsymbol{\theta}} - n^{-1/2}\widehat{\boldsymbol{\Sigma}}^{1/2}\mathbf{x} : \mathbf{x} \in \mathcal{S} \right\} \tag{4.1}$$

is an ellipsoidal confidence region for $\boldsymbol{\theta}_0$ with nominal coverage α. The actual coverage of $\mathcal{R}$ converges to α as $n \to \infty$. The rate of convergence may be determined from multivariate Edgeworth expansions, as we shall shortly relate.

Occasionally the shape of a confidence region is given by the nature of a statistical problem. For example, if our brief is to construct simultaneous confidence intervals for the individual components of $\boldsymbol{\theta}_0$, then we would ask that the confidence region be a rectangular prism with sides parallel to the coordinate axes. If $\mathcal{R}$ is the rectangular prism of smallest content subject to the probability of a Normal $N(\widehat{\boldsymbol{\theta}}, n^{-1}\widehat{\boldsymbol{\Sigma}})$ random variable lying within $\mathcal{R}$ being equal to α, then the boundaries of $\mathcal{R}$ define a set of simultaneous confidence intervals for the components of $\boldsymbol{\theta}_0$, with nominal coverage α.

The coverage inaccuracies of these confidence regions arise from errors in the Normal approximation. We showed in Chapter 3 that in a univariate setting, the bootstrap can provide significantly more accurate approximations than the Normal approximation, in that it automatically corrects for the main effects of departure from Normality. The same is true in higher dimensions, as we shall see shortly. We begin by describing bootstrap alternatives to the ellipsoidal and rectangular Normal approximation regions described in the two previous paragraphs.

Let $\widehat{\boldsymbol{\theta}}^*$ and $\widehat{\boldsymbol{\Sigma}}^*$ denote versions of $\widehat{\boldsymbol{\theta}}$ and $\widehat{\boldsymbol{\Sigma}}$ computed for a bootstrap resample $\mathcal{X}^*$ rather than the original sample $\mathcal{X}$. Put

$$\mathbf{T}^* = n^{1/2}\widehat{\boldsymbol{\Sigma}}^{*-1/2}(\widehat{\boldsymbol{\theta}}^* - \widehat{\boldsymbol{\theta}}).$$

Then the distribution of $\mathbf{T}^*$ conditional on $\mathcal{X}$ is a bootstrap approximation to the distribution of $\mathbf{T}$. To construct a bootstrap ellipsoidal region $\mathcal{R}$, choose $\mathcal{S}$ to be the d-variate sphere centred at the origin and such that $P(\mathbf{T}^* \in \mathcal{S} \mid \mathcal{X}) = \alpha$, and define $\mathcal{R}$ by (4.1). To construct a set of simultaneous confidence intervals, take $\mathcal{P}$ to be the rectangular prism with sides parallel to coordinate axes and of smallest content subject to the constraint

$$P(\mathbf{T}^* \in \widehat{\mathbf{\Sigma}}^{-1/2}\mathcal{P} \mid \mathcal{X}) = \alpha,$$

where

$$\widehat{\mathbf{\Sigma}}^{-1/2}\mathcal{P} = \{\widehat{\mathbf{\Sigma}}^{-1/2}\mathbf{x} : \mathbf{x} \in \mathcal{P}\}.$$

Then the boundaries of the prism

$$\mathcal{R} = \{\widehat{\boldsymbol{\theta}} - n^{-1/2}\mathbf{x} : \mathbf{x} \in \mathcal{P}\}$$

define the desired simultaneous confidence intervals. There are of course other ways of constructing confidence prisms, or simultaneous confidence intervals; for example, there is an equal-tailed version in which the marginal coverage probabilities are constrained to be equal.

This method of constructing confidence regions is of course a multivariate version of the percentile-t technique described in Chapters 1 and 3. The percentile method also has a multivariate analogue, although adjustments such as bias correction are cumbersome in two or more dimensions. The advantages of percentile-t over percentile may be explained in terms of skewness corrections, just as in the univariate case treated in Chapter 3. However, while the percentile-t method automatically corrects for skewness errors in the Normal approximation, it does little to help us select the shape of the confidence region. As the examples above show, shape is still subject to the whim of the experimenter.

In many practical problems it is desirable to have the data themselves determine region shape. One way of achieving this end is to take the confidence region $\mathcal{R}$ to be an estimate of a level set of the density f of $\mathbf{T}$. Now, the conditional distribution of $\mathbf{T}^*$ given $\mathcal{X}$ is an approximation to the distribution of $\mathbf{T}$, and so we could use Monte Carlo methods and density estimation to approximate f. Specifically, draw B resamples of size n, and let $\mathbf{T}_1^*, \ldots, \mathbf{T}_b^*$ denote the corresponding values of $\mathbf{T}^*$; using the $\mathbf{T}_b^*$'s, calculate a d-variate nonparametric density estimate $\hat{f}_B$, say, of f; construct a level set $\mathcal{S}$ of $\hat{f}_B$ that contains a proportion α of the B values $T_1^*, \ldots, T_B^*$; and take

$$\mathcal{R} = \{\widehat{\boldsymbol{\theta}} - n^{-1/2}\widehat{\mathbf{\Sigma}}^{1/2}\mathbf{x} : \mathbf{x} \in \mathcal{S}\}.$$

One difficulty with this prescription lies in calculating the estimate $\hat{f}_B$. Typically $\hat{f}_B$ depends on a "smoothing parameter" whose choice is up to the experimenter, and so there may still be a degree of subjectivity about construction of the confidence region $\mathcal{R}$. For example, if $\hat{f}_B$ is calculated by the kernel method (see for example Silverman (1986, Chapter 4), and also Section 4.4 below), then the smoothing parameter takes the form of a bandwidth h, in the formula

$$\hat{f}_B(\mathbf{x}) = (Bh^d)^{-1} \sum_{b=1}^{B} K\{(\mathbf{x} - \mathbf{T}_b^*)/h\},$$

where the kernel function K is typically a known d-variate population density function such as the Standard Normal density. Selecting too large a value of h results in an overly smooth and flat density estimate $\hat{f}_B$, whereas choosing h too small can produce an estimate with a mode at each data point. Standard methods of bandwidth selection, such as cross-validation, can be unworkable in this context because of the presence of ties among values $\mathbf{T}_b^*$. (If there are ties, cross-validation produces the bandwidth estimate $\hat{h} = 0$.) Furthermore, techniques such as cross-validation, which are based on global criteria, are difficult to justify when our aim is to estimate a contour that characterizes the extreme 5% of a distribution (in the case of a 95% confidence region). One alternative approach is to choose h as the smallest bandwidth for which the confidence region $\mathcal{R}$ is convex.

We should mention here the technique of empirical likelihood, introduced by Owen (1988, 1990). Empirical likelihood places arbitrary probabilities on the data points, say weight p_i on the ith data value, in contrast with the bootstrap, which assigns equal probabilities to all data values. The p_i's are chosen to maximize a constrained multinomial form of "likelihood," and empirical likelihood confidence regions are constructed by contouring this "likelihood." An attractive feature of empirical likelihood is that it produces confidence regions whose shape and orientation are determined entirely by the data, and that have coverage accuracy properties at least comparable with those of bootstrap confidence regions.

It remains to elucidate the performance of Normal approximation, percentile, and percentile-t methods for constructing multivariate confidence regions. There are important differences between univariate and multivariate cases, owing to the fact that many multivariate bootstrap confidence regions, such as confidence ellipsoids, are generalizations of symmetric univariate confidence intervals. Expansions associated with the boundaries of such regions are power series in n^{-1}, not $n^{-1/2}$, and so even the percentile method can produce second-order correct regions. (Compare Section 3.6. In the multivariate context, one region is jth order correct for another if the boundaries of the two regions agree in terms of order $n^{-j/2}$; that is, if the boundaries are order $n^{-(j+1)/2}$ apart.) The "smallest content" confidence prism described earlier is an exception. It is a generalization of the short confidence intervals introduced in Section 3.7, and so the position of its boundary is affected by the ability of the bootstrap method to correctly allow for skewness errors. However, there is no natural multivariate analogue of equal-tailed bootstrap confidence intervals. The latter are perhaps the most often used in univariate problems, and as noted in Chapter 3 they are susceptible to skewness errors arising from using percentile rather than percentile-t techniques.

As in the univariate case, Edgeworth expansions are a convenient tool for studying multivariate confidence regions. However, on the present occasion it is particularly convenient to work with so-called "density versions" of those expansions, introduced in Section 2.8 of Chapter 2. To recapitulate that work, recall that in the univariate case the expansion of a statistic $T = n^{1/2}(\hat{\theta} - \theta_0)/\hat{\sigma}$ typically has the form

$$P(T \leq x) = \Phi(x) + n^{-1/2} q_1(x)\, \phi(x) + n^{-1} q_2(x)\, \phi(x) + \ldots$$

for polynomials q_j. Therefore, for each $-\infty < x < y < \infty$,

$$P(x < T \leq y)$$

$$= P(T \leq y) - P(T \leq x)$$

$$= \int_{(x,y]} \left\{ 1 + n^{-1/2} s_1(u) + n^{-1} s_2(u) + \ldots \right\} \phi(u)\, du\,, \quad (4.2)$$

where

$$s_j(x) = (d/dx)\, \{q_j(x)\, \phi(x)\}\, \phi(x)^{-1}\,.$$

The s_j's are polynomials in an Edgeworth expansion of the density of T. Since q_j is of degree $3j - 1$ and has the opposite parity to j, s_j is of degree $3j$ and has the same parity as j.

More generally, if a set S is expressible as a union of a finite number of convex sets (i.e., a union of intervals) then

$$P(T \in S) = \int_S \left\{ 1 + s_1(x) + s_2(x) + \ldots \right\} \phi(x)\, dx\,,$$

as is immediately apparent from (4.2). In the d-variate case, if $S \subseteq \mathbb{R}^d$ is a finite union of convex sets then

$$P(\mathbf{T} \in S) = \int_S \left\{ 1 + n^{-1/2} s_1(\mathbf{x}) + n^{-1} s_2(\mathbf{x}) + \ldots \right\} \phi(\mathbf{x})\, d\mathbf{x}\,, \quad (4.3)$$

where $\mathbf{T} = n^{1/2}\, \widehat{\boldsymbol{\Sigma}}^{-1/2}(\widehat{\boldsymbol{\theta}} - \boldsymbol{\theta}_0)$ and on this occasion, s_j is a polynomial in the components of $\mathbf{x}$, of degree $3j$ and an odd/even function for odd/even j respectively, and ϕ is the d-variate Standard Normal density. If $n^{-1}\boldsymbol{\Sigma}$ denotes the asymptotic variance of $\widehat{\boldsymbol{\theta}}$, and if we define $\mathbf{S} = n^{1/2}\, \boldsymbol{\Sigma}^{-1/2}(\widehat{\boldsymbol{\theta}} - \boldsymbol{\theta}_0)$, then the non-Studentized version of (4.3) is

$$P(\mathbf{S} \in S) = \int_S \left\{ 1 + n^{-1/2} r_1(\mathbf{x}) + n^{-1} r_2(\mathbf{x}) + \ldots \right\} \phi(\mathbf{x})\, d\mathbf{x} \quad (4.4)$$

for polynomials $r_1, r_2, \ldots$ enjoying the properties ascribed to the s_j's above.

Bootstrap analogues of (4.4) and (4.3) are

$$P(\mathbf{S}^* \in \mathcal{S} \mid \mathcal{X}) = \int_{\mathcal{S}} \left\{ 1 + n^{-1/2}\,\hat{r}_1(\mathbf{x}) + n^{-1}\,\hat{r}_2(\mathbf{x}) + \ldots \right\} \phi(\mathbf{x})\,d\mathbf{x},$$

$$P(\mathbf{T}^* \in \mathcal{S} \mid \mathcal{X}) = \int_{\mathcal{S}} \left\{ 1 + n^{-1/2}\,\hat{s}_1(\mathbf{x}) + n^{-1}\,\hat{s}_2(\mathbf{x}) + \ldots \right\} \phi(\mathbf{x})\,d\mathbf{x},$$

respectively, where $\mathbf{S}^* = n^{1/2}\,\widehat{\Sigma}^{-1/2}(\hat{\boldsymbol{\theta}}^* - \hat{\boldsymbol{\theta}})$, $\mathbf{T}^* = n^{1/2}\,\widehat{\Sigma}^{*-1/2}(\hat{\boldsymbol{\theta}}^* - \hat{\boldsymbol{\theta}})$, and the polynomials $\hat{r}_j$, $\hat{s}_j$ are obtained from r_j, s_j on replacing population moments by sample moments. As explained in Chapter 2, formulae such as these must be interpreted as *asymptotic* expansions. In particular, under appropriate smoothness and moment conditions on the sampling distributions, we have for $j \geq 1$,

$$P(\mathbf{T} \in \mathcal{S}) = \int_{\mathcal{S}} \left\{ 1 + n^{-1/2}\,s_1(\mathbf{x}) + \ldots + n^{-j/2} s_j(\mathbf{x}) \right\} \phi(\mathbf{x})\,d\mathbf{x}$$
$$+ o(n^{-j/2}),$$

$$P(\mathbf{T}^* \in \mathcal{S} \mid \mathcal{X}) = \int_{\mathcal{S}} \left\{ 1 + n^{-1/2}\,\hat{s}_1(\mathbf{x}) + \ldots + n^{-j/2}\hat{s}_j(\mathbf{x}) \right\} \phi(\mathbf{x})\,d\mathbf{x}$$
$$+ o_p(n^{-j/2}),$$

uniformly in sets $\mathcal{S}$ that are unions of no more than m (any fixed number) of convex subsets of $\mathbb{R}^d$. Taking account of the fact that $\hat{s}_j = s_j + O_p(n^{-1/2})$ for $j \geq 1$, we see that

$$P(\mathbf{T}^* \in \mathcal{S} \mid \mathcal{X}) = \int_{\mathcal{S}} \left\{ 1 + n^{-1/2}\,\hat{s}_1(\mathbf{x}) \right\} \phi(\mathbf{x})\,d\mathbf{x} + O_p(n^{-1})$$

$$= \int_{\mathcal{S}} \left\{ 1 + n^{-1/2}\,s_1(\mathbf{x}) \right\} \phi(\mathbf{x})\,d\mathbf{x} + O_p(n^{-1})$$

$$= P(\mathbf{T} \in \mathcal{S}) + O_p(n^{-1}). \qquad (4.5)$$

Thus, the percentile-t bootstrap estimate of the distribution of $\mathbf{T}$ is in error by only $O_p(n^{-1})$, whereas the Normal approximation is in error by terms of size $n^{-1/2}$. A similar argument shows that if the polynomials r_1 and s_1 are distinct then $P(\mathbf{T} \in \mathcal{S})$ and $P(\mathbf{S} \in \mathcal{S} \mid \mathcal{X})$ differ in terms of size $n^{-1/2}$; that is, the percentile approximation of the distribution of $\mathbf{T}$ is in error by terms of size $n^{-1/2}$.

As noted earlier, the implications of this result for multivariate confidence regions are generally not so great as in the univariate case. To illustrate this point we shall consider the ellipsoidal regions introduced earlier. Let $\mathcal{S}_1, \ldots, \mathcal{S}_4$ denote d-dimensional spheres centred at the origin and such that

$$P(\mathbf{T} \in \mathcal{S}_1) = P(\mathbf{N} \in \mathcal{S}_2) = P(\mathbf{S}^* \in \mathcal{S}_3 \mid \mathcal{X}) = P(\mathbf{T}^* \in \mathcal{S}_4 \mid \mathcal{X}) = \alpha,$$

where $\mathbf{N}$ is d-variate Standard Normal. Of course, the radii of $\mathcal{S}_3$ and $\mathcal{S}_4$ are random variables. Define

$$\mathcal{R}_i = \{\widehat{\boldsymbol{\theta}} - n^{-1/2}\,\widehat{\boldsymbol{\Sigma}}^{1/2}\,\mathbf{x} : \mathbf{x} \in \mathcal{S}_i\}, \qquad 1 \le i \le 4.$$

Then $\mathcal{R}_1$ is an "ideal" confidence region for $\boldsymbol{\theta}_0$, with perfect coverage accuracy; $\mathcal{R}_2$ is the region based on Normal approximation; and $\mathcal{R}_3$, $\mathcal{R}_4$ are percentile, percentile-t regions, respectively. If π is an odd polynomial then for any sphere $\mathcal{S}$ centred at the origin,

$$\int_{\mathcal{S}} \pi(\mathbf{x})\,\phi(\mathbf{x})\,d\mathbf{x} = 0. \tag{4.6}$$

In particular, since s_{2j+1}, $\hat{r}_{2j+1}$, and $\hat{s}_{2j+1}$ are odd polynomials,

$$P(\mathbf{T} \in \mathcal{S}_1) = \int_{\mathcal{S}_1} \left\{1 + n^{-1}\,s_2(\mathbf{x}) + n^{-2}\,s_4(\mathbf{x}) + \ldots\right\}\phi(\mathbf{x})\,d\mathbf{x},$$

$$P(\mathbf{S}^* \in \mathcal{S}_3 \mid \mathcal{X}) = \int_{\mathcal{S}_3} \left\{1 + n^{-1}\,\hat{r}_2(\mathbf{x}) + n^{-2}\,\hat{r}_4(\mathbf{x}) + \ldots\right\}\phi(\mathbf{x})\,d\mathbf{x},$$

$$P(\mathbf{T}^* \in \mathcal{S}_4 \mid \mathcal{X}) = \int_{\mathcal{S}_4} \left\{1 + n^{-1}\,\hat{s}_2(\mathbf{x}) + n^{-2}\,\hat{s}_4(\mathbf{x}) + \ldots\right\}\phi(\mathbf{x})\,d\mathbf{x}.$$

It follows that the radii of $\mathcal{S}_1$, $\mathcal{S}_3$, and $\mathcal{S}_4$ may be represented as power series in n^{-1} (rather than $n^{-1/2}$), in which the leading term equals the radius of $\mathcal{S}_2$. Therefore, the boundaries of $\mathcal{S}_2$, $\mathcal{S}_3$, and $\mathcal{S}_4$ are all $O_p(n^{-1})$ away from the boundary of $\mathcal{S}_1$, and so the boundaries of $\mathcal{R}_2$, $\mathcal{R}_3$, and $\mathcal{R}_4$ are $O_p(n^{-3/2})$ away from the boundary of the "ideal" region $\mathcal{R}_1$. That is, $\mathcal{R}_2$, $\mathcal{R}_3$, and $\mathcal{R}_4$ are all second-order correct for $\mathcal{R}_1$.

In the event that the polynomials r_1 and s_1 are identical, we have in place of (4.5) the result

$$P(\mathbf{S}^* \in \mathcal{S} \mid \mathcal{X}) = \int_{\mathcal{S}} \left\{1 + n^{-1/2}\,\hat{r}_1(\mathbf{x})\right\}\phi(\mathbf{x})\,d\mathbf{x} + O_p(n^{-1})$$

$$= \int_{\mathcal{S}} \left\{1 + n^{-1/2}\,r_1(\mathbf{x})\right\}\phi(\mathbf{x})\,d\mathbf{x} + O_p(n^{-1})$$

$$= \int_{\mathcal{S}} \left\{1 + n^{-1/2}\,s_1(\mathbf{x})\right\}\phi(\mathbf{x})\,d\mathbf{x} + O_p(n^{-1})$$

$$= P(\mathbf{T} \in \mathcal{S}) + O_p(n^{-1}).$$

This situation is encountered only infrequently, although we shall meet with such a case in subsection 4.3.5.

The reader will recall from Sections 3.5 and 3.6 that two-sided confidence intervals have coverage error of at most $O(n^{-1})$, and that the error is only $O(n^{-2})$ in the case of symmetric intervals. These results followed from symmetry properties of terms in Edgeworth expansions and have straightforward analogues in a multivariate setting. For example, if $\mathcal{R}_2$, $\mathcal{R}_3$, and $\mathcal{R}_4$ denote the Normal approximation, percentile, and percentile-t ellipsoidal confidence regions defined two paragraphs above, then

$$
\begin{aligned}
P(\boldsymbol{\theta}_0 \in \mathcal{R}_2) &= \alpha + O(n^{-1}), \\
P(\boldsymbol{\theta}_0 \in \mathcal{R}_3) &= \alpha + O(n^{-1}), \\
P(\boldsymbol{\theta}_0 \in \mathcal{R}_4) &= \alpha + O(n^{-2}).
\end{aligned}
\tag{4.7}
$$

(Note that $\mathcal{R}_4$ is a multivariate version of the symmetric percentile-t confidence intervals defined in Section 3.6.) The rectangular confidence regions introduced earlier in this section, using either the Normal approximation or the percentile-t method, have coverage error $O(n^{-1})$.

We shall sketch proofs of the first and last of the identities in (4.7). To treat the first, observe that

$$
\begin{aligned}
P(\boldsymbol{\theta}_0 \in \mathcal{R}_2) &= P(\mathbf{T} \in \mathcal{S}_2) \\
&= \int_{\mathcal{S}_2} \left\{ 1 + n^{-1/2} s_1(\mathbf{x}) + n^{-1} s_2(\mathbf{x}) + \dots \right\} \phi(\mathbf{x})\, d\mathbf{x} \\
&= \int_{\mathcal{S}_2} \left\{ 1 + n^{-1} s_2(\mathbf{x}) + n^{-2} s_4(\mathbf{x}) + \dots \right\} \phi(\mathbf{x})\, d\mathbf{x} \\
&= \alpha + O(n^{-1}).
\end{aligned}
\tag{4.8}
$$

Terms of odd order in $n^{-1/2}$ have vanished completely from (4.8), owing to the fact that the polynomials s_{2j+1}, $j \geq 0$, are odd; see (4.6).

To check the last identity in (4.7), recall from three paragraphs earlier that $\mathcal{S}_1$ is a sphere centred at the origin and of radius r such that

$$
\begin{aligned}
\alpha &= P(\mathbf{T} \in \mathcal{S}_1) \\
&= \int_{\mathcal{S}_1} \left\{ 1 + n^{-1/2} s_1(\mathbf{x}) + n^{-1} s_2(\mathbf{x}) + \dots \right\} \phi(\mathbf{x})\, d\mathbf{x} \\
&= \int_{\mathcal{S}_1} \left\{ 1 + n^{-1} s_1(\mathbf{x}) + n^{-2} s_2(\mathbf{x}) + \dots \right\} \phi(\mathbf{x})\, d\mathbf{x},
\end{aligned}
\tag{4.9}
$$

the last identity following from (4.6). Therefore,

$$
r = r_0 + n^{-1} c_1 + n^{-2} c_2 + \dots,
$$

where r_0 denotes the radius of the sphere $\mathcal{S}_2$ centred at the origin and such that

$$
P(\mathbf{N} \in \mathcal{S}) = \int_{\mathcal{S}_2} \phi(\mathbf{x})\, d\mathbf{x} = \alpha,
$$

and where $c_1, c_2, \dots$ are constants depending on population moments (from the coefficients of polynomials s_{2j}). Let $\hat{c}_j$ denote the version of c_j obtained by replacing population moments by sample moments, and note that

$\hat{c}_j - c_j = O_p(n^{-3/2})$. Applying the argument at (4.9) to $P(\mathbf{T}^* \in \mathcal{S}_4 \mid \mathcal{X})$ instead of $P(\mathbf{T} \in \mathcal{S}_1)$, we deduce that $\mathcal{S}_4$ is a sphere centred at the origin and of random radius $\hat{r}$, where

$$\hat{r} = r_0 + n^{-1}\hat{c}_1 + n^{-2}\hat{c}_2 + \dots .$$

Therefore,

$$
\begin{aligned}
P(\boldsymbol{\theta}_0 \in \mathcal{R}_4) &= P(\mathbf{T} \in \mathcal{S}_4) = P(\mathbf{T} \in \hat{r}r^{-1}\mathcal{S}_1) \\
&= P\Big[\mathbf{T} \in \{1 + (nr_0)^{-1}(\hat{c}_1 - c_1) + O_p(n^{-2})\}\,\mathcal{S}_1\Big] \\
&= P\Big[\{1 - (nr_0)^{-1}(\hat{c}_1 - c_1)\}\,\mathbf{T} \in \mathcal{S}_1\Big] + O(n^{-2}), \qquad (4.10)
\end{aligned}
$$

the last identity following by the delta method (Section 2.7). The distribution of the d-variate statistic

$$\mathbf{T}_1 = \{1 - (nr_0)^{-1}(\hat{c}_1 - c_1)\}\mathbf{T}$$

admits an Edgeworth expansion of the form

$$P(\mathbf{T}_1 \in \mathcal{S}) = \int_{\mathcal{S}} \{1 + n^{-1/2}t_1(\mathbf{x}) + n^{-1}t_2(\mathbf{x}) + \dots\}\,\phi(\mathbf{x})\,d\mathbf{x}, \qquad (4.11)$$

where t_j is a polynomial of degree $3j$ and is an odd/even function for odd/even j, respectively, and $\mathcal{S}$ is any union of a finite number of convex sets. In fact, $t_j = s_j$ for $j = 1, 2$, as follows from the fact that $\mathbf{T}$ and $\mathbf{T}_1$ differ only in terms of $O_p(n^{-3/2})$. When $\mathcal{S}$ is a sphere centred at the origin, terms of odd order in $n^{-1/2}$ vanish from the expansion (4.11) (note (4.6)), and so

$$
\begin{aligned}
P(\mathbf{T}_1 \in \mathcal{S}_1) &= \int_{\mathcal{S}_1} \{1 + n^{-1}t_2(\mathbf{x}) + n^{-2}t_4(\mathbf{x}) + \dots\}\,\phi(\mathbf{x})\,d\mathbf{x} \\
&= \int_{\mathcal{S}_1} \{1 + n^{-1}t_2(\mathbf{x})\}\,\phi(\mathbf{x})\,d\mathbf{x} + O(n^{-2}) \\
&= \int_{\mathcal{S}_1} \{1 + n^{-1}s_2(\mathbf{x})\}\,\phi(\mathbf{x})\,d\mathbf{x} + O(n^{-2}) \\
&= \int_{\mathcal{S}_1} \{1 + n^{-1/2}s_1(\mathbf{x}) + n^{-1}s_2(\mathbf{x}) + \dots\}\,\phi(\mathbf{x})\,d\mathbf{x} + O(n^{-2}) \\
&= P(\mathbf{T} \in \mathcal{S}_1) + O(n^{-2}) \\
&= \alpha + O(n^{-2}).
\end{aligned}
$$

It now follows from (4.10) that $P(\boldsymbol{\theta}_0 \in \mathcal{R}_4) = \alpha + O(n^{-2})$, as had to be shown.

We should stress one caveat concerning the performance of percentile-t methods in multivariate problems. Percentile-t cannot be expected to produce good performance unless the variance matrix Σ can be estimated reasonably accurately, with only moderate variance. That requirement becomes increasingly difficult to ensure as the dimension of the problem increases, and so percentile-t bootstrap methods generally cannot be recommended in high-dimensional problems unless sample size is very large. This is one aspect of the so-called "curse of dimensionality" — inference becomes increasingly difficult as dimension grows, since the complexity of different plausible data and parameter configurations increases rapidly, and the data become more sparse. Another aspect of the "curse" is that Monte Carlo approximation of bootstrap confidence regions demands a rapidly increasing order of simulation as dimension increases, to overcome problems of sparseness of the simulated values.

4.3 Parametric Regression

4.3.1 Introduction

In Chapter 3 we drew attention to several important properties of the bootstrap in a wide range of statistical problems. We stressed the advantages of pivoting, and pointed out that ordinary percentile method confidence intervals (as distinct from those constructed by the pivotal percentile-t method) usually require some sort of correction if they are to achieve good coverage accuracy. For example, the coverage error of a one-sided percentile-t confidence interval is of size n^{-1}, but the coverage error of an uncorrected one-sided percentile interval is of size $n^{-1/2}$.

The good performance of a percentile-t interval is available in problems where the variance of a parameter estimate may be estimated accurately. Many regression problems are of this type. Thus, we might expect the endearing properties of percentile-t to go over without change to the regression case. In a sense, this is true; one-sided percentile-t confidence intervals for regression mean, intercept or slope all have coverage error at most $O(n^{-1})$, whereas their percentile method counterparts generally only have coverage error of size $n^{-1/2}$. However, this generalization conceals several very important differences in the case of slope estimation. One-sided percentile-t confidence intervals for slope have coverage error $O(n^{-3/2})$, not $O(n^{-1})$; and the error is only $O(n^{-2})$ in the case of two-sided intervals. Furthermore, one-sided confidence intervals for slope constructed by the ordinary percentile method (but not by the "other percentile method";

see Example 1.2 in Section 1.3, and Section 3.4, for accounts of the distinction between these two methods) have coverage error $O(n^{-1})$, not just $O(n^{-1/2})$. They are, in fact, second-order equivalent to the accelerated bias-corrected version of the "other percentile method" (see Section 3.10), and so are second-order correct. This is quite unusual for percentile intervals.

These exceptional properties apply only to estimates of slope, not to estimates of intercept or means. However, slope parameters are particularly important in the study of regression, and our interpretation of slope is quite general. For example, in the polynomial regression model

$$Y_i = c + x_i\, d_1 + \ldots + x_i^m\, d_m + \epsilon_i, \qquad 1 \le i \le n, \qquad (4.12)$$

we regard each d_j as a slope parameter. A one-sided percentile-t confidence interval for d_j has coverage error $O(n^{-3/2})$, although a one-sided percentile-t interval for c or for

$$E(Y \mid x = x_0) = c + x_0\, d_1 + \ldots + x_0^m\, d_m$$

has coverage error of size n^{-1}.

The reason that slope parameters have this distinctive property is that the design points x_i confer a significant amount of extra symmetry. Note that we may rewrite the model (4.12) as

$$Y_i = c' + (x_i - \xi_1)\, d_1 + \ldots + (x_i^m - \xi_m)\, d_m + \epsilon_i, \qquad 1 \le i \le n,$$

where $\xi_j = n^{-1} \sum x_i^j$ and $c' = c + \xi_1 d_1 + \ldots + \xi_m d_m$. The extra symmetry arises from the fact that

$$\sum_{i=1}^{n} \left(x_i^j - \xi_j\right) = 0, \qquad 1 \le j \le m. \qquad (4.13)$$

For example, (4.13) implies that random variables $\sum_i \left(x_i^j - \xi_j\right) \epsilon_i^k$ and $\sum \epsilon_i^l$ are uncorrelated for each triple (j, k, l) of nonnegative integers, and this symmetry property is enough to establish the claimed performance of percentile-t confidence intervals.

Subsection 4.3.2 draws a distinction between the "regression" and "correlation" linear models, and points out that the method of resampling should be determined by a decision as to which model is more appropriate. We argue that regression is inherently concerned with inference conditional on the design points. This definition dictates that the residuals $\hat{\epsilon}_i$, and not the pairs (x_i, Y_i), be resampled in regression problems. On the other hand, resampling the pairs is appropriate under the correlation model. In subsection 4.3.5 we point out that our conclusions about unusually good performance of percentile-t methods in problems involving slope remain true in the case of the correlation model with independent errors ϵ_i. However, they are not valid for the general correlation model.

Subsections 4.3.3 and 4.3.4 treat simple linear regression, looking first at the case of slope. Subsection 4.3.6 examines multivariate, multiparameter regression, and points out that the unusual properties of percentile-t bootstrap confidence intervals continue to apply in that context. Finally, subsection 4.3.7 discusses bootstrap methods for constructing simultaneous confidence bands for a regression mean. In all these instances, methods based on Edgeworth expansions have an important role to play in elucidating properties.

As in many other applications of the bootstrap, one of its more important features is its success as a method for correcting skewness. In particular, the percentile-t bootstrap captures and corrects for the first-order term in an Edgeworth expansion, representing the main effect of skewness. Now, the skewness term in the case of regression includes contributions from both the design points and the error distribution. In the case of slope, the overall error due to skewness is proportional to $n^{-1/2} \gamma_x \gamma$, where

$$\gamma_x = n^{-1} \sum_{i=1}^{n} (x_i - \bar{x})^3$$

is the skewness of the design and $\gamma = E(\epsilon_i^3)$ is the skewness of the error distribution. If the design points x_i are regularly spaced, say $x_i = u\,n^{-1}i + v$ for constants u and v, then we have $\gamma_x = O(n^{-1/2})$ as $n \to \infty$, and so the skewness term will be of order $n^{-1/2} \cdot n^{-1/2} = n^{-1}$. In this case, even the simple Normal approximation can perform exceptionally well, and even for relatively highly skewed error distributions. When the design skewness is of order $n^{-1/2}$, the "other percentile method" may be virtually indistinguishable from the ordinary percentile method; in fact, the acceleration constant $\hat{a}$ and the bias correction constant $\hat{m}$ are both of order n^{-1}, rather than $n^{-1/2}$, indicating that the extent of any bias correction is very small. (See Section 3.10.)

Throughout our work in this section we employ the nonparametric bootstrap, for which resampling involves sampling with replacement from the set of residuals (in the case of the regression model) or from the set of all data pairs (for the correlation models). Similar conclusions may be drawn in the case of the parametric bootstrap, where the bootstrap errors ϵ_i^* are drawn from a specified distribution with estimated parameters. In particular, percentile-t confidence intervals for slope have the unusual orders of coverage accuracy reported above.

4.3.2 Resampling in Regression and Correlation Models

Consider a linear model of the form

$$\mathbf{Y}_i = \mathbf{c} + \mathbf{x}_i \mathbf{d} + \boldsymbol{\epsilon}_i, \tag{4.14}$$

where $\mathbf{Y}_i$, $\mathbf{c}$, $\boldsymbol{\epsilon}_i$ denote $q \times 1$ vectors, $\mathbf{d}$ is a $p \times 1$ vector, and $\mathbf{x}_i$ is a $q \times p$ matrix. Data are observed in the form $\{(\mathbf{x}_i, \mathbf{Y}_i), 1 \leq i \leq n\}$, and we call $\{\mathbf{x}_i, 1 \leq i \leq n\}$ the set of design points of the model. We say that (4.14) is a regression model if analysis is carried out conditional on the design points. The design points could be genuinely fixed (for example, they could be regularly spaced) or they could have a random source but be conditioned upon. In a regression model the $\boldsymbol{\epsilon}_i$'s would be regarded as independent random vectors with zero mean, usually assumed to be identically distributed. The case where $(\mathbf{x}_i, Y_i)$, $1 \leq i \leq n$, are independent and identically distributed pairs of random vectors is, in our nomenclature, a correlation model. There, inference would be carried out unconditionally, and we would usually write $\mathbf{X}_i$ instead of $\mathbf{x}_i$ to indicate that the randomness of design is being taken into account. A third model, that of correlation with independent errors, arises when the pairs $(\mathbf{X}_i, \boldsymbol{\epsilon}_i)$ are assumed independent and identically distributed, with $\mathbf{X}_i$ and $\boldsymbol{\epsilon}_i$ independent.

The regression and correlation approaches to linear models demand quite different bootstrap algorithms. Recall from Chapter 1 that the idea behind the bootstrap is to replace the true distribution function F, in a formula for an unknown quantity, by its empirical estimate $\widehat{F}$. Under the regression model, F is the distribution function of the errors $\boldsymbol{\epsilon}_i$, and $\widehat{F}$ is the empirical distribution function of the sequence of residuals

$$\widehat{\boldsymbol{\epsilon}}_i = \mathbf{Y}_i - \widehat{\mathbf{c}} - \mathbf{x}_i \widehat{\mathbf{d}}, \qquad 1 \leq i \leq n, \tag{4.15}$$

where $\widehat{\mathbf{c}}$ and $\widehat{\mathbf{d}}$ are the usual least-squares estimates of $\mathbf{c}$ and $\mathbf{d}$. In the case of the correlation model, F is the distribution function of the pairs $(\mathbf{X}_i, \mathbf{Y}_i)$, $1 \leq i \leq n$. Under the model of correlation with independent errors, F is the distribution function of the pairs $(\mathbf{X}_i, \boldsymbol{\epsilon}_i)$ and $\widehat{F} = \widehat{F}_1 \widehat{F}_2$ where $\widehat{F}_1$ is the empirical distribution function of the observed design points $\{\mathbf{X}_i, 1 \leq i \leq n\}$ and $\widehat{F}_2$ is the empirical distribution function of the sequence of residuals

$$\widehat{\boldsymbol{\epsilon}}_i = \mathbf{Y}_i - \widehat{\mathbf{c}} - \mathbf{X}_i \widehat{\mathbf{d}}, \qquad 1 \leq i \leq n. \tag{4.16}$$

These differences may be clearly seen in the Monte Carlo methods used to construct numerical approximations. First we treat the case of the regression model. Here, the residuals $\widehat{\boldsymbol{\epsilon}}_i$ are resampled; note that the residuals are centred, in the sense that $\sum \widehat{\boldsymbol{\epsilon}}_i = \mathbf{0}$. We take $\{\boldsymbol{\epsilon}_1^*, \ldots, \boldsymbol{\epsilon}_n^*\}$ to be a resample drawn randomly, with replacement, from the set $\{\widehat{\boldsymbol{\epsilon}}_1, \ldots, \widehat{\boldsymbol{\epsilon}}_n\}$ and

define
$$\mathbf{Y}_i^* = \widehat{\mathbf{c}} + \mathbf{x}_i \widehat{\mathbf{d}} + \boldsymbol{\epsilon}_i^*, \qquad 1 \le i \le n. \tag{4.17}$$
The joint distribution of $\{\mathbf{Y}_i^*, 1 \le i \le n\}$, conditional on

$$\mathcal{X} = \{(\mathbf{x}_1, \mathbf{Y}_1), \dots, (\mathbf{x}_n, \mathbf{Y}_n)\},$$

is the bootstrap estimate of the joint distribution of $\{\mathbf{Y}_i, 1 \le i \le n\}$, conditional on $\mathbf{x}_1, \dots, \mathbf{x}_n$. To appreciate how this estimate is utilized, consider the example where $p = q = 1$, and put $\hat{\epsilon}_i = Y_i - \bar{Y} - (x_i - \bar{x})\hat{d}$, $\bar{x} = n^{-1} \sum x_i$,

$$\hat{\sigma}^2 = n^{-1} \sum_{i=1}^n \hat{\epsilon}_i^2, \qquad \sigma_x^2 = n^{-1} \sum_{i=1}^n (x_i - \bar{x})^2. \tag{4.18}$$

Then $\hat{\sigma}^2$ estimates σ^2, and in this notation,

$$\hat{d} = \sigma_x^{-2} n^{-1} \sum_{i=1}^n (x_i - \bar{x})(Y_i - \bar{Y}), \qquad \hat{c} = \bar{Y} - \bar{x}\hat{d}. \tag{4.19}$$

Since $\hat{d}$ has variance $n^{-1} \sigma_x^{-2} \sigma^2$, $n^{1/2} (\hat{d} - d) \sigma_x / \hat{\sigma}$ is (asymptotically) pivotal, and the distribution of this quantity is estimated by the distribution of $n^{1/2}(\hat{d}^* - \hat{d}) \sigma_x / \hat{\sigma}^*$, where $\hat{d}^*$, $\hat{\sigma}^*$ have the same formulae as $\hat{d}$, $\hat{\sigma}$, respectively, except that Y_i is replaced by Y_i^* throughout.

In the case of the correlation model, the pairs $(\mathbf{X}_i, \mathbf{Y}_i)$ are resampled directly. That is, we draw a resample

$$\mathcal{X}^* = \{(\mathbf{X}_1^*, \mathbf{Y}_1^*), \dots, (\mathbf{X}_n^*, \mathbf{Y}_n^*)\}$$

randomly, with replacement, from

$$\mathcal{X} = \{(\mathbf{X}_1, \mathbf{Y}_1), \dots, (\mathbf{X}_n, \mathbf{Y}_n)\}.$$

The joint distribution of $\mathcal{X}^*$ conditional on $\mathcal{X}$ is the bootstrap estimate of the unconditional joint distribution of $\mathcal{X}$. Under the model of correlation with independent errors, the variables $\mathbf{X}_i$ and $\hat{\epsilon}_i$ are resampled independently. That is, we conduct two totally independent resampling operations in which a random sample $\{\mathbf{X}_1^*, \dots, \mathbf{X}_n^*\}$ is drawn with replacement from $\{\mathbf{X}_1, \dots, \mathbf{X}_n\}$ and a random sample $\{\epsilon_1^*, \dots, \epsilon_n^*\}$ is drawn with replacement from $\{\hat{\epsilon}_1, \dots, \hat{\epsilon}_n\}$. (We define $\hat{\epsilon}_i$ by (4.16).) Put

$$\mathbf{Y}_i^* = \widehat{\mathbf{c}} + \mathbf{X}_i^* \widehat{\mathbf{d}} + \boldsymbol{\epsilon}_i^*.$$

The joint distribution of $\{(\mathbf{X}_i^*, \mathbf{Y}_i^*), 1 \le i \le n\}$, conditional on

$$\mathcal{X} = \{(\mathbf{X}_1, \mathbf{Y}_1), \dots, (\mathbf{X}_n, \mathbf{Y}_n)\},$$

is the bootstrap estimate of the unconditional joint distribution of $\mathcal{X}$, under the model of correlation with independent errors.

4.3.3 Simple Linear Regression: Slope Parameter

In this subsection we describe bootstrap methods for constructing confidence intervals for the slope parameter in simple linear regression and sketch proofs of the main properties of those intervals, noted in subsection 4.3.1.

The simple linear regression model is

$$Y_i \;=\; c + x_i\,d + \epsilon_i, \qquad 1 \le i \le n,$$

where c, d, x_i, Y_i, ϵ_i are scalars, c and d are unknown constants representing intercept and slope, respectively, the ϵ_i's are independent and identically distributed random variables with zero mean and variance σ^2, and the x_i's are fixed design points. The usual least-squares estimates of c, d, and σ^2, and the method of bootstrap resampling for regression, are given in subsection 4.3.2. In particular,

$$Y_i^* \;=\; \hat{c} + x_i\,\hat{d} + \epsilon_i^*, \qquad 1 \le i \le n$$

(see (4.17)), where the ϵ_i^*'s are generated by resampling randomly from the residuals $\hat{\epsilon}_i$. Furthermore, $\hat{c}^*$, $\hat{d}^*$, and $\hat{\sigma}^*$ have the same formulae as $\hat{c}$, $\hat{d}$, and $\hat{\sigma}$, except that Y_i is replaced by Y_i^* at each appearance of the former. Note too that $\hat{d}$ has variance $n^{-1}\sigma_x^{-2}\sigma^2$.

Quantiles u_α and v_α of the distributions of

$$n^{1/2}\,(\hat{d}-d)\,\sigma_x/\sigma \qquad \text{and} \qquad n^{1/2}\,(\hat{d}-d)\,\sigma_x/\hat{\sigma}$$

may be defined by

$$P\{n^{1/2}\,(\hat{d}-d)\,\sigma_x/\sigma \le u_\alpha\} = P\{n^{1/2}\,(\hat{d}-d)\,\sigma_x/\hat{\sigma} \le v_\alpha\}$$

$$= \alpha, \tag{4.20}$$

and their bootstrap estimates $\hat{u}_\alpha$ and $\hat{v}_\alpha$ by

$$P\{n^{1/2}\,(\hat{d}^*-\hat{d})\,\sigma_x/\hat{\sigma} \le \hat{u}_\alpha \mid \mathcal{X}\} = P\{n^{1/2}\,(\hat{d}^*-\hat{d})\,\sigma_x/\hat{\sigma}^* \le \hat{v}_\alpha \mid \mathcal{X}\}$$

$$= \alpha, \tag{4.21}$$

where $\mathcal{X}$ denotes the sample of pairs $\{(x_1, Y_1), \dots, (x_n, Y_n)\}$. In this notation, one-sided percentile and percentile-t bootstrap confidence intervals for d are given by

$$\hat{I}_1 \;=\; \left(-\infty, \;\; \hat{d} - n^{-1/2}\,\sigma_x^{-1}\,\hat{\sigma}\,\hat{u}_{1-\alpha}\right),$$

$$\hat{J}_1 \;=\; \left(-\infty, \;\; \hat{d} - n^{-1/2}\,\sigma_x^{-1}\,\hat{\sigma}\,\hat{v}_{1-\alpha}\right) \tag{4.22}$$

respectively; compare (3.10). The analogous "other percentile method" confidence interval is

$$\widehat{I}_{12} = \left(-\infty, \ \hat{d} + n^{-1/2} \sigma_x^{-1} \hat{\sigma} \hat{u}_\alpha \right) = \left(-\infty, \ \widehat{H}^{-1}(\alpha) \right), \qquad (4.23)$$

where $\widehat{H}(x) = P(\hat{d}^* \le x \mid \mathcal{X})$ is the conditional distribution function of $\hat{d}^*$; see (1.15). Each of these three confidence intervals has nominal coverage α. The percentile-t interval $\widehat{J}_1$ is the bootstrap version of an "ideal" interval

$$J_1 = \left(-\infty, \ \hat{d} - n^{-1/2} \sigma_x^{-1} \hat{\sigma} v_{1-\alpha} \right). \qquad (4.24)$$

Of course, each of the intervals has a two-sided counterpart, which may be defined using the "equal-tailed" approach (subsection 3.5.5), the symmetric method (Section 3.6), or the "shortest" method (Section 3.7).

Recall our definition that one confidence interval is second-order correct relative to another if the (finite) endpoints of the intervals agree up to and including terms of order $(n^{-1/2})^2 = n^{-1}$; see Section 3.4. It comes as no surprise to find that $\widehat{J}_1$ is second-order correct for J_1, given what was learned in Section 3.4 about bootstrap confidence intervals in more conventional problems. However, on the present occasion $\widehat{I}_1$ is also second-order correct for J_1, and that property is quite unusual. It arises because Edgeworth expansions of the distributions of $n^{1/2}(\hat{d} - d)\,\sigma_x/\sigma$ and $n^{1/2}(\hat{d} - d)\,\sigma_x/\hat{\sigma}$ contain identical terms of size $n^{-1/2}$; that is, Studentizing has no effect on the first term in the expansion. This is a consequence of the extra symmetry conferred by the presence of the design points x_i, as we shall show in the next paragraph. The reason why second-order correctness follows from identical formulae for the $n^{-1/2}$ terms in expansions was made clear in Section 3.4.

Assume that σ_x^2 is bounded away from zero as $n \to \infty$, and that $\max_{1 \le i \le n} |x_i - \bar{x}|$ is bounded as $n \to \infty$. (In refined versions of the proof below, this boundedness condition may be replaced by a moment condition on the design points x_i, such as $\sup_n n^{-1} \sum (x_i - \bar{x})^4 < \infty$.) Put $\bar{\epsilon} = n^{-1} \sum \epsilon_i$, and observe that

$$\hat{\sigma}^2 = n^{-1} \sum_{i=1}^{n} \hat{\epsilon}_i^2 = n^{-1} \sum_{i=1}^{n} \left\{ \epsilon_i - \bar{\epsilon} - (x_i - \bar{x})(\hat{d} - d) \right\}^2$$

$$= \sigma^2 + n^{-1} \sum_{i=1}^{n} (\epsilon_i^2 - \sigma^2) + O_p(n^{-1}).$$

Therefore, defining $S = n^{1/2}(\hat{d} - d)\,\sigma_x/\sigma$, $T = n^{1/2}(\hat{d} - d)\,\sigma_x/\hat{\sigma}$, and

$$\Delta = \tfrac{1}{2} n^{-1} \sigma^{-2} \sum_{i=1}^{n} (\epsilon_i^2 - \sigma^2),$$

we have

$$T = S(1 - \Delta) + O_p(n^{-1}) = S + O_p(n^{-1/2}). \qquad (4.25)$$

By making use of the fact that $\sum (x_i - \bar{x}) = 0$ (this is where the extra symmetry conferred by the design comes in) and of the representation

$$S = n^{-1/2} \sigma_x^{-1} \sigma^{-1} \sum_{i=1}^{n} (x_i - \bar{x}) \epsilon_i,$$

we may easily prove that $E\{S(1 - \Delta)\}^j - E(S^j) = O(n^{-1})$ for $j = 1, 2, 3$. Therefore the first three cumulants of S and $S(1 - \Delta)$ agree up to and including terms of order $n^{-1/2}$. Higher-order cumulants are of size n^{-1} or smaller, as may be proved by methods similar to those used to establish Theorem 2.1 in Section 2.4. It follows that Edgeworth expansions of the distributions of S and $S(1 - \Delta)$ differ only in terms of order n^{-1}. In view of (4.25), the same is true for S and T,

$$P(S \le w) = P(T \le w) + O(n^{-1}).$$

(This step uses the delta method.) Therefore, Studentizing has no effect on the first term in the expansion, as had to be proved.

The terms in Edgeworth expansions of S and T may be calculated explicitly using methods developed in Chapter 2: first compute the cumulants of S, or of an appropriate approximant to T (here the delta method is needed); then expand the characteristic functions of S and T as power series in $n^{-1/2}$; and finally, invert the resulting Fourier-Stieltjes transforms. Arguing thus we may prove that

$$P\{n^{1/2}(\hat{d} - d)\,\sigma_x/\sigma \le w\}$$
$$= \Phi(w) + n^{-1/2} p_1(w)\,\phi(w) + n^{-1} p_2(w)\,\phi(w) + \dots, \qquad (4.26)$$

$$P\{n^{1/2}(\hat{d} - d)\,\sigma_x/\hat{\sigma} \le w\}$$
$$= \Phi(w) + n^{-1/2} q_1(w)\,\phi(w) + n^{-1} q_2(w)\,\phi(w) + \dots, \qquad (4.27)$$

where

$$p_1(w) = q_1(w) = -\tfrac{1}{6}\,\gamma\gamma_x(w^2 - 1), \qquad (4.28)$$

$$p_2(w) = -w\left\{\tfrac{1}{24}\,\kappa(\kappa_x + 3)(w^2 - 3) + \tfrac{1}{72}\,\gamma^2\gamma_x^2(w^4 - 10w^2 + 15)\right\},$$

$$q_2(w) = -w\left\{2 + \tfrac{1}{24}\,(\kappa\kappa_x + 6)(w^2 - 3) + \tfrac{1}{72}\,\gamma^2\gamma_x^2(w^4 - 10w^2 + 15)\right\},$$

$$\gamma = E(\epsilon/\sigma)^3, \qquad \gamma_x = n^{-1} \sum_{i=1}^{n} \{(x_i - \bar{x})/\sigma_x\}^3,$$

$$\kappa = E(\epsilon/\sigma)^4 - 3, \qquad \kappa_x = n^{-1} \sum_{i=1}^{n} \{(x_i - \bar{x})/\sigma_x\}^4 - 3.$$

The bootstrap versions of these expansions are

$$P\{n^{1/2}(\hat{d}^* - \hat{d})\,\sigma_x/\hat{\sigma} \le w \mid \mathcal{X}\}$$
$$= \Phi(w) + n^{-1/2}\,\hat{p}_1(w)\,\phi(w) + n^{-1}\,\hat{p}_2(w)\,\phi(w) + \ldots, \qquad (4.29)$$

$$P\{n^{1/2}(\hat{d}^* - \hat{d})\,\sigma_x/\hat{\sigma}^* \le w \mid \mathcal{X}\}$$
$$= \Phi(w) + n^{-1/2}\,\hat{q}_1(w)\,\phi(w) + n^{-1}\,\hat{q}_2(w)\,\phi(w) + \ldots, \qquad (4.30)$$

where $\hat{p}_j$, $\hat{q}_j$ are obtained from p_j, q_j, respectively, on replacing the population moment $E(\epsilon/\sigma)^k$ by its sample version $n^{-1}\sum(\hat{\epsilon}_i/\hat{\sigma})^k$. In particular,

$$\hat{p}_1(w) = \hat{q}_1(w) = -\tfrac{1}{6}\,\hat{\gamma}\gamma_x(w^2 - 1), \qquad (4.31)$$

$$\hat{p}_2(w) = -w\Big\{\tfrac{1}{24}\,\hat{\kappa}(\kappa_x + 3)(w^2 - 3) + \tfrac{1}{72}\,\hat{\gamma}^2\gamma_x^2(w^4 - 10w^2 + 15)\Big\},$$

$$\hat{q}_2(w) = -w\Big\{2 + \tfrac{1}{24}\,(\hat{\kappa}\kappa_x + 6)(w^2 - 3) + \tfrac{1}{72}\,\hat{\gamma}^2\gamma_x^2(w^4 - 10w^2 + 15)\Big\},$$

where

$$\hat{\gamma} = n^{-1} \sum_{i=1}^{n} (\hat{\epsilon}_i/\hat{\sigma})^3, \qquad \hat{\kappa} = n^{-1} \sum_{i=1}^{n} (\hat{\epsilon}_i/\hat{\sigma})^4 - 3.$$

Inverting the Edgeworth expansions (4.27), (4.29), and (4.30) we obtain Cornish-Fisher expansions of the quantiles v_α, $\hat{u}_\alpha$, $\hat{v}_\alpha$ defined by (4.20) and (4.21),

$$v_\alpha = z_\alpha - n^{-1/2}\,q_1(z_\alpha)$$
$$+ n^{-1}\{q_1(z_\alpha)\,q_1'(z_\alpha) - \tfrac{1}{2}z_\alpha\,q_1(z_\alpha)^2 - q_2(z_\alpha)\} + \ldots, \qquad (4.32)$$

$$\hat{v}_\alpha = z_\alpha - n^{-1/2}\,\hat{q}_1(z_\alpha)$$
$$+ n^{-1}\{\hat{q}_1(z_\alpha)\,\hat{q}_1'(z_\alpha) - \tfrac{1}{2}z_\alpha\,\hat{q}_1(z_\alpha)^2 - \hat{q}_2(z_\alpha)\} + \ldots, \qquad (4.33)$$

$$\hat{u}_\alpha = z_\alpha - n^{-1/2}\,\hat{p}_1(z_\alpha)$$
$$+ n^{-1}\{\hat{p}_1(z_\alpha)\,\hat{p}_1'(z_\alpha) - \tfrac{1}{2}z_\alpha\,\hat{p}_1(z_\alpha)^2 - \hat{p}_2(z_\alpha)\} + \ldots. \qquad (4.34)$$

(Note formulae (3.13)–(3.16).)

Substituting these expressions into formulae for the upper endpoints of the confidence intervals J_1, $\widehat{J}_1$, and $\widehat{I}_1$ defined in (4.22) and (4.24), and remembering that on this occasion, $\hat{p}_1 = \hat{q}_1 = q_1 + O_p(n^{-1/2})$, we deduce that the upper endpoints differ only in terms of $O_p(n^{-1/2} \cdot n^{-1/2} \cdot n^{-1/2}) = O_p(n^{-3/2})$. Therefore $\widehat{I}_1$ and $\widehat{J}_1$ are second-order correct for J_1.

The "other percentile method" confidence interval $\widehat{I}_{12}$, defined in (4.23), is not second-order correct for J_1. As we noted in Section 3.10, this deficiency may be overcome by bias correction. The ordinary bias correction, or BC, method holds no surprises in the regression case. It produces a confidence interval for slope that is not fully corrected, just as it does in the situation considered in Section 3.10. Therefore we pass directly to the one-sided accelerated bias-corrected interval (constructed by the ABC or BC_a method), defined by

$$\widehat{I}_{\mathrm{ABC}} = (-\infty, \, \hat{y}_{\mathrm{ABC}, \alpha}), \qquad (4.35)$$

where

$$\hat{y}_{\mathrm{ABC}, \alpha} = \widehat{H}^{-1}\Big(\Phi\big[\hat{m} + (\hat{m} + z_\alpha)\{1 - \hat{a}(\hat{m} + z_\alpha)\}^{-1}\big]\Big), \quad (4.36)$$

$$\widehat{H}(x) = P(\hat{d}^* \leq x \mid \mathcal{X}),$$

$$\hat{m} = \Phi^{-1}\{\widehat{H}(\hat{d})\}, \qquad (4.37)$$

$$\hat{a} = \text{coefficient of } w^2 \text{ in } n^{-1/2}\{\hat{p}_1(w) + \hat{q}_1(w)\}. \qquad (4.38)$$

(See Section 3.10, particularly (3.92) and (3.93).)   We shall prove that $\widehat{I}_{\mathrm{ABC}}$ is *third*-order correct relative to $\widehat{I}_1$; that is to say, the upper endpoints of the ordinary percentile method interval and the ABC-corrected "other percentile method" interval agree in terms of order $n^{-3/2}$, differing only in terms of order n^{-2} or smaller. Incidentally, the fact that the acceleration constant $\hat{a}$ is generally of size $n^{-1/2}$, and not of smaller order (see (4.38)), indicates that the uncorrected version of the "other percentile method" interval, and the ordinary BC method interval, are not second-order correct.

The main step in verifying our claim that $\widehat{I}_{\mathrm{ABC}}$ is third-order correct for $\widehat{I}_1$ is establishing an expansion of $\hat{y}_{\mathrm{ABC}, \alpha}$. To this end, note that

$$\widehat{H}(\hat{d}) = P\{n^{1/2}(\hat{d}^* - \hat{d})/\hat{\sigma} \leq 0 \mid \mathcal{X}\}$$
$$= \Phi(0) + n^{-1/2}\,\hat{p}_1(0)\,\phi(0) + O_p(n^{-3/2}),$$

using (4.29) and the fact that $\hat{p}_2$ is an odd polynomial. Therefore, by (4.37),

$$\hat{m} = n^{-1/2}\,\hat{p}_1(0)\,\phi(0) + O_p(n^{-3/2})$$
$$= n^{-1/2}\,\tfrac{1}{6}\,\hat{\gamma}\gamma_x + O_p(n^{-3/2}),$$

by (4.31). Furthermore, by (4.31) and (4.38),

$$\hat{a} = -n^{-1/2}\,\tfrac{1}{3}\,\hat{\gamma}\gamma_x. \qquad (4.39)$$

Therefore,

$$\hat{m} + (\hat{m} + z_\alpha)\{1 - \hat{a}(\hat{m} + z_\alpha)\}^{-1}$$

$$= 2\hat{m} + z_\alpha + \hat{a}(z_\alpha^2 + 2\hat{m}\,z_\alpha) + \hat{a}^2 z_\alpha^3 + O_p(n^{-3/2})$$

$$= z_\alpha + n^{-1/2}\,\tfrac{1}{3}\,\hat{\gamma}\gamma_x(1 - z_\alpha^2) + n^{-1}\,\tfrac{1}{9}\,\hat{\gamma}^2\gamma_x^2 z_\alpha(z_\alpha^2 - 1) + O_p(n^{-3/2}),$$

whence, defining

$$\beta = \Phi\big[\hat{m} + (\hat{m} + z_\alpha)\{1 - \hat{a}(\hat{m} + z_\alpha)\}^{-1}\big]$$

and

$$\hat{p}_{21}(w) = \hat{p}_1(w)\hat{p}_1'(w) - \tfrac{1}{2}\,w\hat{p}_1(w)^2 - \hat{p}_2(w),$$

we have by (4.34) and (4.36),

$$\hat{y}_{\text{ABC},\,\alpha}$$

$$= \widehat{H}^{-1}(\beta) = \hat{d} + n^{-1/2}\,\sigma_x^{-1}\,\hat{\sigma}\hat{u}_\beta$$

$$= \hat{d} + n^{-1/2}\,\sigma_x^{-1}\,\hat{\sigma}\Big\{z_\beta - n^{-1/2}\,\hat{p}_1(z_\beta) + n^{-1}\,\hat{p}_{21}(z_\beta) + O_p(n^{-3/2})\Big\}$$

$$= \hat{d} + n^{-1/2}\,\sigma_x^{-1}\,\hat{\sigma}\Big[z_\alpha - n^{-1/2}\,\hat{p}_1(z_\alpha) + n^{-1}\,\hat{p}_{21}(z_\alpha)$$

$$+ n^{-1/2}\,\tfrac{1}{3}\,\hat{\gamma}\gamma_x(1 - z_\alpha^2)$$

$$+ n^{-1}\{\tfrac{1}{9}\,\hat{\gamma}^2\gamma_x^2\,z_\alpha(z_\alpha^2 - 1) - \tfrac{1}{3}\,\hat{\gamma}\gamma_x(1 - z_\alpha^2)\,\hat{p}_1'(z_\alpha)\}$$

$$+ O_p(n^{-3/2})\Big].$$

From this formula, (4.31), (4.34), and the fact that $\hat{p}_{21}$ is an odd polynomial, we deduce that

$$\hat{y}_{\text{ABC},\,\alpha}$$

$$= \hat{d} - n^{-1/2}\,\sigma_x^{-1}\,\hat{\sigma}\Big\{z_{1-\alpha} - n^{-1/2}\,\hat{p}_1(z_{1-\alpha})$$

$$+ n^{-1}\,\hat{p}_{21}(z_{1-\alpha}) + O_p(n^{-3/2})\Big\}$$

$$= \hat{d} - n^{-1/2}\,\sigma_x^{-1}\,\hat{\sigma}\hat{u}_{1-\alpha} + O_p(n^{-2}). \tag{4.40}$$

The upper endpoints of the intervals $\widehat{I}_{\text{ABC}}$ and $\widehat{I}_1$ are, respectively, $\hat{y}_{\text{ABC},\,\alpha}$ and $\hat{d} - n^{-1/2}\,\sigma_x^{-1}\,\hat{\sigma}\hat{u}_{1-\alpha}$, and by (4.40) these points differ only in terms of order n^{-2}, as had to be shown.

Next we consider the issue of coverage accuracy. Recall the definitions of intervals $\widehat{I}_1$, $\widehat{J}_1$, J_1, and $\widehat{I}_{\text{ABC}}$ at (4.22), (4.24), and (4.35). Since the ordinary percentile method interval $\widehat{I}_1$ is, on the present occasion, second-order correct relative to the "ideal" interval J_1, it comes as no surprise to learn that $\widehat{I}_1$ has coverage error $O(n^{-1})$. (A proof of this fact uses an argument similar to that in Section 3.5.) Nevertheless, this degree of coverage accuracy is unusual for percentile method intervals, which generally have coverage error of size $n^{-1/2}$. See subsection 3.5.4.

The accelerated bias-corrected interval $\widehat{I}_{\mathrm{ABC}}$ is third-order equivalent to $\widehat{I}_1$, and so it is to be expected that the coverage error formulae will differ only in terms of order $n^{-3/2}$, implying that $\widehat{I}_{\mathrm{ABC}}$ has coverage error $O(n^{-1})$. Again, this is readily checked using minor modifications of arguments given in Section 3.5, noting particularly the first two paragraphs of subsection 3.5.4. There we showed that a rather general one-sided interval $\mathcal{I}_1$ has coverage error of order n^{-1} if and only if it is second-order correct relative to J_1, and we pointed out that if $\mathcal{I}_1$ and J_1 are jth order equivalent then their coverage probabilities differ only by $O(n^{-j/2})$. The proofs of both these results in the case of regression, in particular for slope estimation, are similar to those in the context of smooth functions of vector means, treated in Chapter 3.

In subsection 3.5.4 we noted that the converse of the result about jth order correctness stated just above fails to be true if $j \geq 3$. The percentile-t interval $\widehat{J}_1$ provides a striking example of this fact. Unusually, the interval $\widehat{J}_1$ has coverage error $O(n^{-3/2})$ in the case of slope estimation for regression, although the error is typically $O(n^{-1})$ in other problems (see subsection 3.5.4), and $\widehat{J}_1$ is not third-order correct for J_1. We now outline a proof that

$$P(d \in \widehat{J}_1) \;=\; \alpha + O(n^{-3/2}). \tag{4.41}$$

Note that

$$\hat{\gamma} \;=\; n^{-1} \sum_{i=1}^{n} (\hat{\epsilon}_i/\hat{\sigma})^3$$

$$=\; \left[n^{-1} \sum_{i=1}^{n} \{\epsilon_i - \bar{\epsilon} - (x_i - \bar{x})(\hat{d} - d)\}^3 \right]$$

$$\times \left[n^{-1} \sum_{i=1}^{n} \{\epsilon_i - \bar{\epsilon} - (x_i - \bar{x})(\hat{d} - d)\}^2 \right]^{-3/2}$$

$$=\; n^{-1} \sum_{i=1}^{n} (\epsilon_i/\sigma)^3 - \tfrac{3}{2}\gamma n^{-1} \sum_{i=1}^{n} \{(\epsilon_i/\sigma)^2 - 1\}$$

$$- 3n^{-1} \sum_{i=1}^{n} (\epsilon_i/\sigma) + O_p(n^{-1})$$

$$=\; \gamma + n^{-1/2} U + O_p(n^{-1}),$$

where

$$U = n^{-1/2} \sum_{i=1}^{n} (\delta_i^3 - \gamma) - \tfrac{3}{2}\gamma n^{-1/2} \sum_{i=1}^{n} (\delta_i^2 - 1) - 3n^{-1/2} \sum_{i=1}^{n} \delta_i$$

and $\delta_i = \epsilon_i/\sigma$. Therefore, by (4.28) and (4.31)–(4.33),

$$
\begin{aligned}
\hat{v}_\alpha &= v_\alpha + n^{-1/2}\{q_1(z_\alpha) - \hat{q}_1(z_\alpha)\} + O_p(n^{-3/2}) \\
&= v_\alpha + n^{-1/2} \tfrac{1}{6} (\hat{\gamma} - \gamma)\gamma_x(z_\alpha^2 - 1) + O_p(n^{-3/2}) \\
&= v_\alpha + n^{-1} \tfrac{1}{6} U\gamma_x(z_\alpha^2 - 1) + O_p(n^{-3/2}).
\end{aligned}
$$

Hence

$$
\begin{aligned}
P(d &\in \hat{J}_1) \\
&= P\big(d \le \hat{d} - n^{-1/2}\sigma_x^{-1}\hat{\sigma}\hat{v}_{1-\alpha}\big) \\
&= P\Big\{n^{1/2}(\hat{d} - d)\sigma_x/\hat{\sigma} \ge v_{1-\alpha} + n^{-1}\tfrac{1}{6}U\gamma_x(z_{1-\alpha}^2 - 1) + O_p(n^{-3/2})\Big\} \\
&= P\big(T + n^{-1}c_1 U \ge v_{1-\alpha}\big) + O(n^{-3/2}),
\end{aligned}
$$

using the delta method, where

$$T = n^{1/2}(\hat{d} - d)\sigma_x/\hat{\sigma} \qquad \text{and} \qquad c_1 = -\tfrac{1}{6}\gamma_x(z_{1-\alpha}^2 - 1).$$

The coverage probability of J_1 is precisely α, that is,

$$\alpha = P(d \in J_1) = P\big(d \le \hat{d} - n^{-1/2}\sigma_x^{-1}\hat{\sigma}v_{1-\alpha}\big) = P(T \ge v_{1-\alpha}).$$

Comparing the last two displayed formulae we see that the desired result will be established if we show that Edgeworth expansions of the distributions of T and $T_1 = T + n^{-1}c_1 U$ differ only in terms of order $n^{-3/2}$.

The terms of orders $n^{-1/2}$ and n^{-1} in Edgeworth expansions of T and T_1 depend only on the first four moments; see Chapter 2. It suffices to prove that the first four moments of T and T_1 differ only in terms of order $n^{-3/2}$,

$$E(T_1^j) = E(T^j) + O(n^{-3/2}), \qquad 1 \le j \le 4. \tag{4.42}$$

Now, $T = S + O_p(n^{-1/2})$, where

$$S = n^{1/2}(\hat{d} - d)\sigma_x/\sigma = n^{-1/2}c_2 \sum_{i=1}^{n} (x_i - \bar{x})\epsilon_i$$

and $c_2 = (\sigma_x\sigma)^{-1}$;

$$
\begin{aligned}
E(T_1^j) &= E(T^j) + n^{-1}jc_1 E(T^{j-1}U) + O(n^{-2}) \\
&= E(T^j) + n^{-1}jc_1 E(S^{j-1}U) + O(n^{-3/2}); \tag{4.43}
\end{aligned}
$$

$E(SU) = 0$, by virtue of the fact that $\sum(x_i - \bar{x}) = 0$ (here we make use of the symmetry conferred by the presence of design points);

$$E(S^3 U) = 3E(S^2) E(SU) + O(n^{-1}) = O(n^{-1});$$

and for all odd j, $E(S^{j-1} U) = O(n^{-1/2})$. The desired formula (4.42) follows on combining the results from (4.43) down.

We close this section by mentioning the analogues of these results for two-sided confidence intervals. Defining $\xi = \frac{1}{2}(1+\alpha)$, we see that two-sided versions of the confidence intervals $\widehat{I}_1$ and $\widehat{J}_1$ are

$$\widehat{I}_2 = \left(\hat{d} - n^{-1/2} \sigma_x^{-1} \hat{\sigma} \, \hat{u}_\xi, \;\; \hat{d} - n^{-1/2} \sigma_x^{-1} \hat{\sigma} \, \hat{u}_{1-\xi} \right),$$

$$\widehat{J}_2 = \left(\hat{d} - n^{-1/2} \sigma_x^{-1} \hat{\sigma} \, \hat{v}_\xi, \;\; \hat{d} - n^{-1/2} \sigma_x^{-1} \hat{\sigma} \, \hat{v}_{1-\xi} \right),$$

respectively. The interval $\widehat{I}_2$ has coverage error of order n^{-1}, as does the two-sided version of $\widehat{I}_{\mathrm{ABC}}$ (which is third-order equivalent to $\widehat{I}_2$). However, the coverage error of $\widehat{J}_2$ is a very low $O(n^{-2})$, as may be proved by a somewhat longer version of the argument just above, noting the parity properties of polynomials in Edgeworth expansions. (We already know from (4.41), and the version of that result with α replaced by $1 - \alpha$, that the coverage error of $\widehat{J}_2$ can be at most $O(n^{-3/2})$.) Short percentile-t bootstrap confidence intervals, defined by

$$\widehat{J}_{\mathrm{SH}} = \left(\hat{d} - n^{-1/2} \sigma_x^{-1} \hat{\sigma} \, \hat{v}_{\alpha+\hat{\beta}}, \;\; \hat{\theta} - n^{-1/2} \sigma_x^{-1} \hat{\sigma} \, \hat{v}_{\hat{\beta}} \right)$$

where $\hat{\beta}$ is chosen to minimize $\hat{v}_{\alpha+\beta} - \hat{v}_\beta$, also have coverage error of order n^{-2}.

4.3.4 Simple Linear Regression: Intercept Parameter and Means

The unusual properties of bootstrap confidence intervals for a slope parameter noted in subsection 4.3.3 do not apply to the case of an intercept parameter or a mean. In this context the relative advantages of different bootstrap methods are just those evidenced in the more classical circumstances treated in Chapter 3. Percentile-t and accelerated bias-corrected intervals are second-order correct and have coverage error $O(n^{-1})$; and both types of percentile method intervals are only first-order correct and have coverage error of size $n^{-1/2}$ in the one-sided case. In the present section we shall briefly describe the details behind these more routine properties.

We continue to assume the simple linear model

$$Y_i = c + x_i d + \epsilon_i,$$

where the design points x_i are fixed and the errors ϵ_i are independent and identically distributed with zero mean and variance σ^2. Estimates $\hat{c}$, $\hat{d}$, $\hat{\sigma}^2$ of c, d, σ^2 were defined at (4.18) and (4.19). Population skewness and kurtosis, γ and κ, respectively, and the sample versions $\hat{\gamma}$ and $\hat{\kappa}$, were introduced in subsection 4.3.3. Our estimate of the mean value

$$y_0 = E(Y \mid x = x_0) = c + x_0 d$$

is

$$\hat{y}_0 = \hat{c} + x_0 \hat{d}.$$

Taking $x_0 = 0$ produces the estimate $\hat{c}$ of the intercept c, and so we may confine attention to estimation of a general y_0.

Define

$$y_i = \sigma_x^{-2}\{(x_0 - \bar{x})(x_i - \bar{x}) + \sigma_x^2\}, \qquad 1 \leq i \leq n,$$

$$\sigma_y^2 = n^{-1} \sum_{i=1}^{n} y_i^2 = 1 + \sigma_x^{-2}(x_0 - \bar{x})^2,$$

$$\gamma_y = n^{-1} \sum_{i=1}^{n} y_i^3, \qquad \kappa_y = n^{-1} \sum_{i=1}^{n} y_i^4 - 3.$$

The variance of $\hat{y}_0$ equals $n^{-1} \sigma^2 \sigma_y^2$, and so $S = n^{1/2}(\hat{y}_0 - y_0)/(\sigma \sigma_y)$ and $T = n^{1/2}(\hat{y}_0 - y_0)/(\hat{\sigma} \sigma_y)$ are (asymptotically) pivotal. The percentile-t method aims to estimate the distribution of T, whereas percentile methods are based on estimating the distribution of $n^{1/2}(\hat{y}_0 - y_0)$.

Recall from subsection 4.3.2 the method of bootstrap resampling in regression models, which produces a resample $\{(x_1, Y_1^*), \ldots, (x_n, Y_n^*)\}$. Let $\hat{c}^*$, $\hat{d}^*$, $\hat{\sigma}^{*2}$ have the same formulae as $\hat{c}$, $\hat{d}$, $\hat{\sigma}^2$, respectively, except that Y_i^* replaces Y_i throughout. The bootstrap versions of $\hat{y}_0$, S, T are

$$\hat{y}_0^* = \hat{c}^* + x_0 \hat{d}^*,$$

$$S^* = n^{1/2}(\hat{y}_0^* - \hat{y}_0)/(\hat{\sigma} \sigma_y), \qquad T^* = n^{1/2}(\hat{y}_0^* - \hat{y}_0)/(\hat{\sigma}^* \sigma_y),$$

respectively.

Edgeworth expansions for the distributions of S and T may be obtained using methods from Chapter 2. They are

$$P(S \leq w) = \Phi(w) + n^{-1/2} p_1(w) \phi(w) + n^{-1} p_2(w) \phi(w) + \ldots,$$

$$P(T \leq w) = \Phi(w) + n^{-1/2} q_1(w) \phi(w) + n^{-1} q_2(w) \phi(w) + \ldots,$$

where, for example,

$$p_1(w) = -\tfrac{1}{6}\gamma\gamma_y(w^2-1), \qquad q_1(w) = -\tfrac{1}{6}\gamma(\gamma_y - 3\sigma_y^{-1})w^2 - \tfrac{1}{6}\gamma\gamma_y,$$

$$p_2(w) = -w\left\{\tfrac{1}{24}\kappa(\kappa_y+3)(w^2-3) + \tfrac{1}{72}\gamma^2\gamma_y^2(w^4 - 10w^2 + 15)\right\},$$

$$q_2(w) = -w\Big[2 + \sigma_y^{-2}\gamma^2 + \tfrac{1}{24}\big\{\kappa\kappa_y + 6 - 8\gamma^2\,\sigma_y^{-1}(\gamma_y - 3\sigma_y^{-1})(w^2-3)\big\}$$

$$+ \tfrac{1}{72}\gamma^2(\gamma_y^2 - 3\sigma_y^{-1})^2(w^4 - 10w^2 + 15)\Big].$$

Note particularly that p_1 and q_1 are quite different polynomials; this is the reason why the percentile method does not perform so well here as in the case of estimating a slope parameter. The bootstrap versions of these Edgeworth expansions are

$$P\big(S^* \le w \mid \mathcal{X}\big) = \Phi(w) + n^{-1/2}\,\hat{p}_1(w)\,\phi(w) + n^{-1}\,\hat{p}_2(w)\,\phi(w) + \dots,$$

$$P\big(T^* \le w \mid \mathcal{X}\big) = \Phi(w) + n^{-1/2}\,\hat{q}_1(w)\,\phi(w) + n^{-1}\,\hat{q}_2(w)\,\phi(w) + \dots,$$

where $\hat{p}_j$ and $\hat{q}_j$ are identical to p_j and q_j, respectively, except that population moments and cumulants in coefficients are replaced by their sample counterparts (e.g., γ, κ are replaced by $\hat{\gamma}$, $\hat{\kappa}$).

These expansions can be employed to determine the (estimate of the) acceleration constant, $\hat{a}$, needed for construction of bias-corrected confidence intervals,

$$\hat{a} = \text{coefficient of } w^2 \text{ in } n^{-1/2}\big\{\hat{p}_1(w) + \hat{q}_1(w)\big\}$$

$$= n^{-1/2}\,\tfrac{1}{6}\,\hat{\gamma}(3\sigma_y^{-1} - 2\gamma_y);$$

see (3.89). The (estimate of the) bias correction, $\hat{m}$, is

$$\hat{m} = \Phi^{-1}\big\{P(\hat{y}_0^* \le \hat{y}_0 \mid \mathcal{X})\big\} = n^{-1/2}\,\tfrac{1}{6}\,\hat{\gamma}\gamma_y + O_p(n^{-3/2});$$

see (3.92). In this notation, and with $\widehat{H}(w) = P(\hat{y}_0^* \le w \mid \mathcal{X})$, a one-sided accelerated bias-corrected confidence interval for y_0, with nominal coverage α, is $(-\infty, \hat{y}_{\text{ABC},\alpha})$, where

$$\hat{y}_{\text{ABC},\alpha} = \widehat{H}^{-1}\big(\Phi[\hat{m} + (\hat{m} + z_\alpha)\{1 - \hat{a}(\hat{m} + z_\alpha)\}^{-1}]\big);$$

compare (3.93). A one-sided percentile-t confidence interval for y_0 is

$$\widehat{J}_1 = \left(-\infty,\; \hat{y}_0 - n^{-1}\hat{\sigma}\sigma_y\hat{v}_{1-\alpha}\right),$$

where $\hat{v}_{1-\alpha}$ is the solution of $P(T^* \le \hat{v}_{1-\alpha} \mid \mathcal{X}) = 1 - \alpha$. The coverage probabilities of either interval may be computed using methods from Section 3.5. For example,

$$P(y_0 \in \widehat{J}_1)$$

$$= \alpha - n^{-1}\,\tfrac{1}{6}\,(\kappa - \tfrac{3}{2}\gamma^2)\,z_\alpha\big\{(\gamma_y - 3\sigma_y^{-1})z_\alpha^2 - \gamma_y\big\}\phi(z_\alpha) + O(n^{-3/2}).$$

Defining $\xi = \frac{1}{2}(1 + \alpha)$ and $\widehat{J}_2 = (\hat{y}_0 - n^{-1/2}\hat{\sigma}\sigma_y\hat{v}_\xi, \ \hat{y}_0 - n^{-1/2}\hat{\sigma}\sigma_y\hat{v}_{1-\xi})$, a two-sided equal-tailed percentile-t interval, we have

$$P(y_0 \in \widehat{J}_2) = \alpha - n^{-1}\tfrac{1}{3}\left(\kappa - \tfrac{3}{2}\gamma^2\right)z_\xi\left\{(\gamma_y - 3\sigma_y^{-1})z_\xi^2 - \gamma_y\right\}\phi(z_\xi) + O(n^{-2}).$$

4.3.5 The Correlation Model

In subsection 4.3.2 we drew the distinction between regression models and correlation models, arguing that the term "regression" should be reserved for settings where inference is conducted conditional on the design points. Our main conclusion was that in regression models there is something special about bootstrap methods for constructing confidence intervals for slope parameters. In particular, the percentile method produces confidence intervals that are second-order correct, despite the fact that they are only first-order correct in most other problems; and the percentile-t method results in confidence intervals whose coverage accuracy is of smaller order than n^{-1}, although it is only of size n^{-1} in other cases. As we explained, this distinction arises because of a particular symmetry of the regression model, on account of the fact that (in the case of simple linear regression)

$$E\left\{n^{-1}\sum_{i=1}^{n}(x_i - \bar{x})\,\epsilon_i^j\right\} = 0 \tag{4.44}$$

for all integers $j \geq 1$. (Here and below, the superscript j denotes an exponent.) For example, see the argument following (4.43). Should the design points x_i be random variables, and should we not be conducting inference conditional on the design, then this symmetry can vanish.

In the case of the usual correlation model we write $x_i = X_i$ and assume that the pairs (X_i, Y_i) are independent and identically distributed. Now,

$$\epsilon_i = Y_i - (c + dX_i),$$

and so

$$
\begin{aligned}
E\{(X_i - \bar{X})\,\epsilon_i^j\} &= (1 - n^{-1})\{E(X_i\,\epsilon_i^j) - E(X_i)\,E(\epsilon_i^j)\} \\
&= (1 - n^{-1})\Big(E[X\{Y - (c + dX)\}^j] \\
&\qquad\qquad - E(X)\,E\{Y - (c + dX)\}^j\Big), \tag{4.45}
\end{aligned}
$$

where (X, Y) denotes a generic pair (X_i, Y_i). Depending on the nature of the dependence between X and Y, the right-hand side of (4.45) can be nonzero for all $j \geq 1$, so the symmetry implicit in (4.44) will not necessarily be in evidence.

However, under the model of correlation with independent errors the pairs (X_i, ϵ_i) are assumed independent, with $E(\epsilon_i) = 0$, and so the quantity in (4.45) vanishes. Therefore the analogue of (4.44) is valid. The conclusion to be drawn is that the unusual properties of bootstrap confidence intervals in regression, described in subsection 4.3.3, are not available under the usual correlation model but do emerge under correlation with independent errors. The same conclusion, and with the same justifying argument, applies in the case of multivariate, multiparameter regression.

4.3.6 Multivariate, Multiparameter Regression

In this subsection we examine the multivariate, multiparameter regression model introduced in subsection 4.3.2,

$$\mathbf{Y}_i = \mathbf{c} + \mathbf{x}_i \mathbf{d} + \boldsymbol{\epsilon}_i,$$

where $\mathbf{c}$, $\mathbf{Y}_i$, $\boldsymbol{\epsilon}_i$ denote $q \times 1$ vectors, $\mathbf{d}$ is a $p \times 1$ vector, and $\mathbf{x}_i$ is a $q \times p$ matrix. We assume that the $\boldsymbol{\epsilon}_i$'s are independent and identically distributed with zero mean and nonsingular variance matrix $\boldsymbol{\Sigma} = E(\boldsymbol{\epsilon}\,\boldsymbol{\epsilon}^{\mathrm{T}})$. Least-squares estimates of $\mathbf{c}$, $\mathbf{d}$, and $\boldsymbol{\Sigma}$ are given by

$$\widehat{\mathbf{d}} = \boldsymbol{\Sigma}_x^{-1} n^{-1} \sum_{i=1}^{n} (\mathbf{x}_i - \bar{\mathbf{x}})^{\mathrm{T}} (\mathbf{Y}_i - \bar{\mathbf{Y}}), \qquad \widehat{\mathbf{c}} = \bar{\mathbf{Y}} - \bar{\mathbf{x}}\widehat{\mathbf{d}},$$

$$\widehat{\boldsymbol{\Sigma}} = n^{-1} \sum_{i=1}^{n} \widehat{\boldsymbol{\epsilon}}_i \widehat{\boldsymbol{\epsilon}}_i^{\mathrm{T}},$$

where $\boldsymbol{\Sigma}_x = n^{-1} \sum (\mathbf{x}_i - \bar{\mathbf{x}})^{\mathrm{T}} (\mathbf{x}_i - \bar{\mathbf{x}})$ and $\widehat{\boldsymbol{\epsilon}}_i = \mathbf{Y}_i - \widehat{\mathbf{c}} - \mathbf{x}_i \widehat{\mathbf{d}}$.

The vector $\mathbf{d}$ is a generalized form of slope and $\mathbf{c}$ is a generalized intercept. Our main conclusions are similar to those reached in the univariate case treated in subsections 4.3.3 and 4.3.4: percentile-t confidence regions are second-order correct for an "ideal" region in general circumstances (e.g., for slope, for intercept, and for the mean $\mathbf{c} + \mathbf{x}_0\,\mathbf{d}$), and percentile regions are second-order correct in the case of slope. The conclusion that percentile regions are second-order correct is unusual, given the more conventional results described in Chapter 3, but is in line with the univariate results for slope described in subsection 4.3.3. It arises because the $n^{-1/2}$ terms in Edgeworth expansions of Studentized and non-Studentized slope estimates are identical. As explained in Section 4.2, this property is sufficient to give the desired result. We therefore confine ourselves to checking the claim about terms in Edgeworth expansions for slope, which is a consequence of the extra symmetry conferred by the presence of design points in regression problems. (In addition, we should remind the reader that, as noted in Section 4.2, the symmetrical construction of many multivariate confidence regions implies that percentile methods can produce second-order correct confidence regions for even intercept and mean.)

Define

$$\mathbf{V} \;=\; n^{-1} \sum_{i=1}^{n} (\mathbf{x}_i - \bar{\mathbf{x}})^{\mathrm{T}} \, \boldsymbol{\Sigma}(\mathbf{x}_i - \bar{\mathbf{x}}),$$

$$\widehat{\mathbf{V}} \;=\; n^{-1} \sum_{i=1}^{n} (\mathbf{x}_i - \bar{\mathbf{x}})^{\mathrm{T}} \, \widehat{\boldsymbol{\Sigma}}(\mathbf{x}_i - \bar{\mathbf{x}}).$$

Then $\widehat{\mathbf{d}}$ has variance matrix $n^{-1}\boldsymbol{\Sigma}_x^{-1}\mathbf{V}\boldsymbol{\Sigma}_x^{-1}$, and so two standardized versions of $\widehat{\mathbf{d}}$ are

$$\mathbf{S} \;=\; n^{1/2}(\boldsymbol{\Sigma}_x^{-1}\mathbf{V}\,\boldsymbol{\Sigma}_x^{-1})^{-1/2}\,(\widehat{\mathbf{d}} - \mathbf{d}),$$
$$\mathbf{T} \;=\; n^{1/2}(\boldsymbol{\Sigma}_x^{-1}\widehat{\mathbf{V}}\,\boldsymbol{\Sigma}_x^{-1})^{-1/2}\,(\widehat{\mathbf{d}} - \mathbf{d}).$$

The latter is Studentized, the former not. Let $\mathbf{R}$ denote either $\mathbf{S}$ or $\mathbf{T}$, let $\boldsymbol{\nu} = (\nu^{(1)},\dots,\nu^{(p)})^{\mathrm{T}}$ be a p-vector of nonnegative integers, let $\mathbf{x} = (x^{(1)},\dots,x^{(p)})^{\mathrm{T}}$ be a general p vector, and define

$$|\boldsymbol{\nu}| = \Sigma |\nu^{(j)}|, \quad \|\mathbf{x}\| = (\Sigma x^{(j)^2})^{1/2}, \quad \boldsymbol{\nu}! = \Pi(\nu^{(j)}!), \quad \mathbf{x}^{\boldsymbol{\nu}} = \Pi\,(x^{(j)})^{\nu^{(j)}},$$

and

$$D^{\boldsymbol{\nu}} \;=\; \prod_{j=1}^{p} \left(\frac{\partial}{\partial x^{(j)}} \right)^{\nu^{(j)}},$$

a differential operator. Moments $\mu_{\boldsymbol{\nu}} = E(\mathbf{R}^{\boldsymbol{\nu}})$ and cumulants $\kappa_{\boldsymbol{\nu}}$, both scalars, are determined by the formal expansions

$$\chi(\mathbf{t}) \;=\; E\{\exp(i\mathbf{t}^{\mathrm{T}}\mathbf{R})\} \;=\; \sum_{\nu \geq 0} \frac{1}{\boldsymbol{\nu}!} \mu_{\boldsymbol{\nu}}(i\mathbf{t})^{\boldsymbol{\nu}}, \qquad (4.46)$$

$$\sum_{\nu \geq 0} \frac{1}{\boldsymbol{\nu}!} \kappa_{\boldsymbol{\nu}}(i\mathbf{t})^{\boldsymbol{\nu}} \;=\; \log \chi(\mathbf{t})$$

$$= \sum_{k \geq 1} (-1)^{k+1} \tfrac{1}{k} \left\{ \sum_{\nu \geq 0} \frac{1}{\boldsymbol{\nu}!} \mu_{\boldsymbol{\nu}}(i\mathbf{t})^{\boldsymbol{\nu}} \right\}^{k}, \qquad (4.47)$$

where the inequality $\boldsymbol{\nu} \geq \mathbf{0}$ is to be interpreted elementwise.

As in the univariate case, cumulants enjoy expansions as power series in n^{-1},

$$\kappa_{\boldsymbol{\nu}} \;=\; n^{-(|\boldsymbol{\nu}|-2)/2}\left(k_{\boldsymbol{\nu},1} + n^{-1}k_{\boldsymbol{\nu},2} + n^{-2}k_{\boldsymbol{\nu},3} + \dots\right);$$

compare (2.20). Since $\mathbf{R}$ has been standardized to have mean of order $n^{-1/2}$ and identity asymptotic variance matrix, then when $|\boldsymbol{\nu}| = 1$, $k_{\boldsymbol{\nu},1} = 0$, and when $|\boldsymbol{\nu}| = 2$, $k_{\boldsymbol{\nu},1} = 1$ if one component of $\boldsymbol{\nu}$ equals 2 and $k_{\boldsymbol{\nu},1} = 0$

otherwise. Thus,

$$
\chi(\mathbf{t}) = \exp\left\{ n^{-1/2} \sum_{|\nu|=1} k_{\nu,2}\,(it)^{\nu} + \sum_{|\nu|=2} k_{\nu,1}\,(it)^{\nu}\,(\nu!)^{-1} \right.
$$
$$
\left. + n^{-1/2} \sum_{|\nu|=3} k_{\nu,1}\,(it)^{\nu}(\nu!)^{-1} + O(n^{-1}) \right\}
$$
$$
= \exp(-\tfrac{1}{2}\,\|\mathbf{t}\|^{2}) \left[1 + n^{-1/2}\Big\{ \sum_{|\nu|=1} k_{\nu,2}\,(it)^{\nu} \right.
$$
$$
\left. + \sum_{|\nu|=3} k_{\nu,1}\,(it)^{\nu}\,(\nu!)^{-1} \Big\} + O(n^{-1}) \right].
$$

Formal inversion of this Fourier transform shows that the density f of $\mathbf{R}$ satisfies

$$
f(\mathbf{x}) = \phi(\mathbf{x})\{1 + n^{-1/2}\,\pi(\mathbf{x}) + O(n^{-1})\},
$$

where ϕ is the p-variate Standard Normal density, the polynomial π is given by

$$
\pi(\mathbf{x}) = \sum_{|\nu|=1} k_{\nu,2}\,H_{\nu}(\mathbf{x}) + \sum_{|\nu|=3} k_{\nu,1}\,H_{\nu}(\mathbf{x})\,(\nu!)^{-1},
$$

and the generalized Hermite function H_{ν} is defined by

$$
H_{\nu}(\mathbf{x})\,\phi(\mathbf{x}) = (-D^{\nu})\,\phi(\mathbf{x})
$$

and has Fourier transform $(it)^{\nu}\,\exp\left(-\tfrac{1}{2}\,\|\mathbf{t}\|^{2}\right).$

In the cases $\mathbf{R} = \mathbf{S}$ and $\mathbf{R} = \mathbf{T}$ we have $\pi = r_1$ and $\pi = s_1$, respectively, where r_1 and s_1 are the polynomials appearing in the expansions (4.4) and (4.3). Our task is to show that $r_1 = s_1$, and for that it suffices to prove that the constants $k_{\nu,2}$ (for $|\nu| = 1$) and $k_{\nu,1}$ (for $|\nu| = 3$) are common to both $\mathbf{S}$ and $\mathbf{T}$. Bearing in mind the relationships (4.46) and (4.47) that express cumulants in terms of moments, we see that it suffices to prove that for each $1 \le i, j, k \le p$,

$$
E(T^{(i)}) = E(S^{(i)}) + O(n^{-1}), \tag{4.48}
$$
$$
E(T^{(i)}\,T^{(j)}\,T^{(k)}) = E(S^{(i)}\,S^{(j)}\,S^{(k)}) + O(n^{-1}). \tag{4.49}
$$

The remainder of the present subsection is dedicated to this purpose.

Defining $\boldsymbol{\xi} = \widehat{\mathbf{d}} - \mathbf{d} = \Sigma_x^{-1}\,n^{-1}\sum(\mathbf{x}_i - \bar{\mathbf{x}})^{\mathrm{T}}\,\epsilon_i$, we see that

$$
\widehat{\epsilon}_i = \epsilon_i - \bar{\epsilon} - (\bar{x}_i - \bar{\mathbf{x}})\,\boldsymbol{\xi},
$$

and so

$$\widehat{\Sigma} = n^{-1} \sum_{i=1}^{n} \widehat{\epsilon}_i \widehat{\epsilon}_i^{\mathrm{T}}$$

$$= n^{-1} \sum_{i=1}^{n} \widehat{\epsilon}_i \widehat{\epsilon}_i^{\mathrm{T}} - \bar{\epsilon}\,\bar{\epsilon}^{\mathrm{T}} - \mathbf{P} - \mathbf{P}^{\mathrm{T}} + \mathbf{Q},$$

where

$$\mathbf{P} = n^{-1} \sum_{i=1}^{n} (\epsilon_i - \bar{\epsilon})\boldsymbol{\xi}^{\mathrm{T}}(\mathbf{x}_i - \bar{\mathbf{x}})^{\mathrm{T}},$$

$$\mathbf{Q} = n^{-1} \sum_{i=1}^{n} (\mathbf{x}_i - \bar{\mathbf{x}})\boldsymbol{\xi}\,\boldsymbol{\xi}^{\mathrm{T}}(\mathbf{x}_i - \bar{\mathbf{x}})^{\mathrm{T}}. \tag{4.50}$$

We shall prove at the end of this subsection that

$$\mathbf{P} = O_p(n^{-1}), \qquad \mathbf{Q} = O_p(n^{-1}) \tag{4.51}$$

as $n \to \infty$. Of course, $\bar{\epsilon}\,\bar{\epsilon}^{\mathrm{T}} = O_p(n^{-1})$, and so

$$\widehat{\Sigma} = n^{-1} \sum_{i=1}^{n} \epsilon_i \epsilon_i^{\mathrm{T}} + O_p(n^{-1}) = \Sigma + n^{-1} \sum_{i=1}^{n} \delta_i + O_p(n^{-1}),$$

where $\delta_i = \epsilon_i \epsilon_i^{\mathrm{T}} - \Sigma$. In consequence,

$$\Sigma_x^{-1} \widehat{\mathbf{V}} \Sigma_x^{-1} = \Sigma_x^{-1} \mathbf{V} \Sigma_x^{-1} \{\mathbf{I} + n^{-1/2}\,\mathbf{U} + O_p(n^{-1})\},$$

where $\mathbf{U} = n^{-1/2} \sum \mathbf{U}_i$ and

$$\mathbf{U}_i = n^{-1} \sum_{j=1}^{n} \Sigma_x \mathbf{V}^{-1}(\mathbf{x}_j - \bar{\mathbf{x}})^{\mathrm{T}} \delta_i(\mathbf{x}_j - \bar{\mathbf{x}}) \Sigma_x^{-1}.$$

Let $\mathbf{A}$ and $\mathbf{AB}$ be symmetric $p \times p$ matrices with $\mathbf{A}$ positive definite, and let η be a scalar. Then

$$\{\mathbf{A}(\mathbf{I} + \eta\mathbf{B})\}^{-1/2} = \mathbf{A}^{-1/2} - \eta\mathbf{C} + O(\eta^2)$$

as $\eta \to 0$, where $\mathbf{C}$ is the unique symmetric matrix satisfying

$$\mathbf{A}^{-1/2}\mathbf{C} + \mathbf{C}\mathbf{A}^{-1/2} = \mathbf{B}\mathbf{A}^{-1}. \tag{4.52}$$

This property entails

$$\left(\Sigma_x^{-1} \widehat{\mathbf{V}} \Sigma_x^{-1}\right)^{-1/2} = \left(\Sigma_x^{-1} \mathbf{V} \Sigma_x^{-1}\right)^{-1/2} - n^{-1/2}\,\mathbf{W} + O_p(n^{-1}),$$

where $\mathbf{W}$ is the unique symmetric matrix satisfying

$$\left(\Sigma_x^{-1} \mathbf{V} \Sigma_x^{-1}\right)^{-1/2}\mathbf{W} + \mathbf{W}\left(\Sigma_x^{-1} \mathbf{V} \Sigma_x^{-1}\right)^{-1/2} = \mathbf{U}\Sigma_x \mathbf{V}^{-1} \Sigma_x. \tag{4.53}$$

Therefore,

$$\mathbf{T} = n^{1/2}\left(\Sigma_x^{-1} \widehat{\mathbf{V}} \Sigma_x^{-1}\right)^{-1/2}\boldsymbol{\xi} = \mathbf{S} - n^{-1/2}\mathbf{S}_1 + O_p(n^{-1}),$$

where

$$\mathbf{S}_1 \;=\; n^{1/2}\,\mathbf{W}\boldsymbol{\xi}. \tag{4.54}$$

Hence (4.48) and (4.49) will follow if we prove that

$$E\big(S_1^{(i)}\big) \;=\; 0, \qquad E\big(S^{(i)}S^{(j)}S_1^{(k)}\big) \;=\; O(n^{-1}) \tag{4.55}$$

whenever $1 \le i,j,k \le p$.

Our first step in establishing (4.55) is to calculate the matrix $\mathbf{W}$. Returning to the general problem of finding the solution $\mathbf{C}$ to equation (4.52), let $\mathbf{v}_1,\dots,\mathbf{v}_p$ be an orthonormal sequence of eigenvectors of $\mathbf{A}^{-1/2}$ with respective positive eigenvalues $\lambda_1,\dots,\lambda_p$. Write

$$\mathbf{B}\mathbf{A}^{-1} \;=\; \sum_{j=1}^{p}\sum_{k=1}^{p} b_{jk}\,\mathbf{v}_j\,\mathbf{v}_k^{\mathrm{T}}, \qquad \mathbf{C} \;=\; \sum_{j=1}^{p}\sum_{k=1}^{p} c_{jk}\,\mathbf{v}_j\,\mathbf{v}_k^{\mathrm{T}}$$

for constants b_{jk}, c_{jk}. Then by (4.52),

$$\mathbf{A}^{-1/2}\mathbf{C} + \mathbf{C}\mathbf{A}^{-1/2} \;=\; \sum_{j=1}^{p}\sum_{k=1}^{p} c_{jk}(\lambda_j + \lambda_k)\,\mathbf{v}_j\,\mathbf{v}_k^{\mathrm{T}} \;=\; \sum_{j=1}^{p}\sum_{k=1}^{p} b_{jk}\,\mathbf{v}_j\,\mathbf{v}_k^{\mathrm{T}},$$

so that $c_{jk} = (\lambda_j + \lambda_k)^{-1}\, b_{jk}$. In our case, equation (4.52) is just (4.53), so that $\mathbf{v}_1,\dots,\mathbf{v}_p$ are eigenvectors and $\lambda_1,\dots,\lambda_p$ are eigenvalues of $\big(\boldsymbol{\Sigma}_x^{-1}\,\mathbf{V}\,\boldsymbol{\Sigma}_x^{-1}\big)^{1/2}$, and b_{jk} is given by

$$\begin{aligned}
b_{jk} &\;=\; \mathbf{v}_j^{\mathrm{T}}\,\mathbf{B}\mathbf{A}^{-1}\,\mathbf{v}_k \;=\; \mathbf{v}_j^{\mathrm{T}}\,\mathbf{U}\boldsymbol{\Sigma}_x\mathbf{V}^{-1}\,\boldsymbol{\Sigma}_x\,\mathbf{v}_k \\
&\;=\; \mathbf{v}_j^{\mathrm{T}}\left(n^{-1/2}\sum_{i=1}^{n}\mathbf{U}_i\,\boldsymbol{\Sigma}_x\mathbf{V}^{-1}\,\boldsymbol{\Sigma}_x\right)\mathbf{v}_k \;=\; n^{-1/2}\sum_{i=1}^{n} b_{ijk},
\end{aligned}$$

where for each $1 \le i \le n$ and $1 \le j,k \le p$,

$$b_{ijk} \;=\; \mathbf{v}_j^{\mathrm{T}}\left\{n^{-1}\sum_{l=1}^{n}\boldsymbol{\Sigma}_x\mathbf{V}^{-1}\,(\mathbf{x}_l - \bar{\mathbf{x}})^{\mathrm{T}}\,\boldsymbol{\delta}_i(\mathbf{x}_l - \bar{\mathbf{x}})\,\mathbf{V}^{-1}\boldsymbol{\Sigma}_x\right\}\mathbf{v}_k.$$

Therefore,

$$\begin{aligned}
\mathbf{W} &\;=\; \sum_{j=1}^{p}\sum_{k=1}^{p} c_{jk}\,\mathbf{v}_j\,\mathbf{v}_k^{\mathrm{T}} \\
&\;=\; \sum_{j=1}^{p}\sum_{k=1}^{p}(\lambda_j + \lambda_k)^{-1}\, b_{jk}\,\mathbf{v}_j\,\mathbf{v}_k^{\mathrm{T}} \\
&\;=\; n^{-1/2}\sum_{i=1}^{n}\sum_{j=1}^{p}\sum_{k=1}^{p}(\lambda_j + \lambda_k)^{-1}\, b_{ijk}\,\mathbf{v}_j\,\mathbf{v}_k^{\mathrm{T}}. \tag{4.56}
\end{aligned}$$

Recall from (4.54) that

$$
\begin{aligned}
\mathbf{S}_1 &= n^{1/2}\,\mathbf{W}\,\xi \\
&= n^{1/2}\,\mathbf{W}\,\boldsymbol{\Sigma}_x^{-1}\,n^{-1}\sum_{i=1}^{n}(\mathbf{x}_i - \bar{\mathbf{x}})^{\mathrm{T}}\,\boldsymbol{\epsilon}_i \\
&= n^{-1}\sum_{i=1}^{n}\sum_{j=1}^{n}\sum_{k=1}^{p}\sum_{l=1}^{p}(\lambda_k + \lambda_l)^{-1}\,b_{ikl}\,\mathbf{v}_k\,\mathbf{v}_l^{\mathrm{T}}\,\boldsymbol{\Sigma}_x^{-1}(\mathbf{x}_j - \bar{\mathbf{x}})^{\mathrm{T}}\,\boldsymbol{\epsilon}_j\,,
\end{aligned}
$$

the last line following from (4.56). More simply,

$$
\mathbf{S} = n^{-1/2}\sum_{i=1}^{n}\left(\boldsymbol{\Sigma}_x^{-1}\mathbf{V}\,\boldsymbol{\Sigma}_x^{-1}\right)^{-1/2}\boldsymbol{\Sigma}_x^{-1}(\mathbf{x}_i - \bar{\mathbf{x}})^{\mathrm{T}}\,\boldsymbol{\epsilon}_i\,.
$$

Defining nonrandom $p \times p$ matrices

$$
\begin{aligned}
\mathbf{A} &= \left(\boldsymbol{\Sigma}_x^{-1}\mathbf{V}\,\boldsymbol{\Sigma}_x^{-1}\right)^{-1/2}\boldsymbol{\Sigma}_x^{-1}\,, \\
\mathbf{A}_{kl} &= (\lambda_k + \lambda_l)^{-1}\,\mathbf{v}_k\,\mathbf{v}_l^{\mathrm{T}}\,\boldsymbol{\Sigma}_x^{-1}\,,
\end{aligned}
$$

we have

$$
\begin{aligned}
\mathbf{S} &= n^{-1/2}\sum_{i=1}^{n}\mathbf{A}(\mathbf{x}_i - \bar{\mathbf{x}})^{\mathrm{T}}\,\boldsymbol{\epsilon}_i\,, \\
\mathbf{S}_1 &= n^{-1}\sum_{i=1}^{n}\sum_{j=1}^{n}\sum_{k=1}^{p}\sum_{l=1}^{p}b_{ikl}\,\mathbf{A}_{kl}\,(\mathbf{x}_j - \bar{\mathbf{x}})^{\mathrm{T}}\,\boldsymbol{\epsilon}_j\,.
\end{aligned}
$$

We are now in a position to verify (4.55). Put

$$
e_{kl} = E\big(\epsilon_i^{(k)}\epsilon_i^{(l)}\big) = (\boldsymbol{\Sigma})^{kl}\,, \qquad e_{klm} = E\big(b_{ikl}\,\epsilon_i^{(m)}\big)\,,
$$

neither depending on i. Then, writing $\mathbf{M}^{ij}$ for the (i,j)th element of a matrix $\mathbf{M}$, we have

$$
\begin{aligned}
E\big(S_1^{(m)}\big) &= n^{-1}\sum_{i=1}^{n}\sum_{j=1}^{n}\sum_{k=1}^{p}\sum_{l=1}^{p}\sum_{r=1}^{q}\big\{\mathbf{A}_{kl}(\mathbf{x}_j - \bar{\mathbf{x}})^{\mathrm{T}}\big\}^{mr}\,E\big(b_{ikl}\,\epsilon_j^{(r)}\big) \\
&= n^{-1}\sum_{j=1}^{n}\sum_{k=1}^{p}\sum_{l=1}^{p}\sum_{r=1}^{q}\big\{\mathbf{A}_{kl}(\mathbf{x}_j - \bar{\mathbf{x}})^{\mathrm{T}}\big\}^{mr}\,e_{klr} \\
&= 0\,,
\end{aligned}
$$

since $\sum (\mathbf{x}_i - \bar{\mathbf{x}}) = \mathbf{0}$; and

$$
n^2 E\big(S^{(m_1)} S^{(m_2)} S_1^{(m_3)}\big)
$$
$$
= E\Bigg(\Bigg[\sum_{i=1}^{n}\sum_{r=1}^{q}\big\{\mathbf{A}(\mathbf{x}_i - \bar{\mathbf{x}})^{\mathrm{T}}\big\}^{m_1 r}\epsilon_i^{(r)}\Bigg]
$$
$$
\times \Bigg[\sum_{i=1}^{n}\sum_{r=1}^{q}\big\{\mathbf{A}(\mathbf{x}_i - \bar{\mathbf{x}})^{\mathrm{T}}\big\}^{m_2 r}\epsilon_i^{(r)}\Bigg]
$$
$$
\times \Bigg[\sum_{i=1}^{n}\sum_{j=1}^{n}\sum_{k=1}^{p}\sum_{l=1}^{p}\sum_{r=1}^{q}\big\{\mathbf{A}_{kl}(\mathbf{x}_i - \bar{\mathbf{x}})^{\mathrm{T}}\big\}^{m_3 r} b_{ikl}\,\epsilon_j^{(r)}\Bigg]\Bigg)
$$
$$
= \sum_{i=1}^{n}\sum_{j=1}^{n}\sum_{r_1=1}^{q}\sum_{r_2=1}^{q}\sum_{r_3=1}^{q}\sum_{k=1}^{p}\sum_{l=1}^{p}\big\{\mathbf{A}(\mathbf{x}_i - \bar{\mathbf{x}})^{\mathrm{T}}\big\}^{m_1 r_1}
$$
$$
\times \big\{\mathbf{A}(\mathbf{x}_i - \bar{\mathbf{x}})^{\mathrm{T}}\big\}^{m_2 r_2}\big\{\mathbf{A}_{kl}(\mathbf{x}_j - \bar{\mathbf{x}})^{\mathrm{T}}\big\}^{m_3 r_3} e_{r_1 r_2}\,e_{klr_3}
$$
$$
+\, 2\sum_{i=1}^{n}\sum_{j=1}^{n}\sum_{r_1=1}^{q}\sum_{r_2=1}^{q}\sum_{r_3=1}^{q}\sum_{k=1}^{p}\sum_{l=1}^{p}\big\{\mathbf{A}(\mathbf{x}_i - \bar{\mathbf{x}})^{\mathrm{T}}\big\}^{m_1 r_1}
$$
$$
\times \big\{\mathbf{A}(\mathbf{x}_j - \bar{\mathbf{x}})^{\mathrm{T}}\big\}^{m_2 r_2}\big\{\mathbf{A}_{kl}(\mathbf{x}_i - \bar{\mathbf{x}})^{\mathrm{T}}\big\}^{m_3 r_3} e_{r_1 r_3}\,e_{klr_2} \,+\, O(n)
$$
$$
= O(n),
$$

since $\sum (\mathbf{x}_j - \bar{\mathbf{x}}) = \mathbf{0}$. (The $O(n)$ term derives from contributions with coefficients $E\big(\epsilon_i^{(r_1)}\epsilon_i^{(r_2)} b_{ikl}\epsilon_i^{(r_3)}\big)$, where all the subscripts i are identical.) This proves (4.55).

It remains only to check (4.51), for which the matrices $\mathbf{P}$, $\mathbf{Q}$ are defined by (4.50). Now, $\boldsymbol{\xi} = O_p(n^{-1/2})$, whence it follows that $\boldsymbol{\xi}^{\mathrm{T}}\boldsymbol{\xi} = O_p(n^{-1})$ and hence that $\mathbf{Q} = O_p(n^{-1})$. To check that $\mathbf{P} = (P^{uv}) = O_p(n^{-1})$, note that

$$
P^{uv} = n^{-2}\sum_{i=1}^{n}\sum_{j=1}^{n}\sum_{k=1}^{q}\sum_{l=1}^{p}\sum_{m=1}^{p}\epsilon_i^{(u)}\epsilon_j^{(k)}(\mathbf{x}_j - \bar{\mathbf{x}})^{kl}\big(\boldsymbol{\Sigma}_x^{-1}\big)^{lm}(\mathbf{x}_i - \bar{\mathbf{x}})^{vm}.
$$

Therefore,

$$
E\big\{(P^{uv})^2\big\}
$$
$$
= n^{-4}\sum_{i_1=1}^{n}\sum_{j_1=1}^{n}\sum_{k_1=1}^{p}\sum_{l_1=1}^{p}\sum_{m_1=1}^{p}\sum_{i_2=1}^{n}\sum_{j_2=1}^{n}\sum_{k_2=1}^{p}\sum_{l_2=1}^{p}\sum_{m_2=1}^{p}\prod_{r=1}^{2}
$$
$$
\big\{(\mathbf{x}_{j_r} - \bar{\mathbf{x}})^{k_r l_r}\big(\boldsymbol{\Sigma}_x^{-1}\big)^{l_r m_r}(\mathbf{x}_{i_r} - \bar{\mathbf{x}})^{v m_r}\big\} E\big(\epsilon_{i_1}^{(u)}\epsilon_{i_2}^{(u)}\epsilon_{j_1}^{(k_1)}\epsilon_{j_2}^{(k_2)}\big). \quad (4.57)
$$

Notice that $E\big(\epsilon_{i_1}^{(u)}\epsilon_{i_2}^{(u)}\epsilon_{j_1}^{(k_1)}\epsilon_{j_2}^{(k_2)}\big) = 0$ unless $i_1 = i_2$ and $i_1 = j_2$, or $i_1 = j_1$ and $j_2 = j_2$, or $i_1 = j_2$ and $i_2 = j_1$. We shall bound the contributions to

$E\{(P^{uv})^2\}$ from the first and second of these cases; the third is similar to the second.

Put

$$c_1 = \sup_j E\big(\epsilon_i^{(j)4}\big), \qquad c_2 = \sup_{i,j} E\big(\Sigma_x^{-1}\big)^{ij}.$$

The contribution to (4.57) from the case $i_1 = i_2$ and $j_1 = j_2$ is in absolute value no greater than

$$n^{-4}c_1c_2^2 \sum_i \sum_j \sum_{k_1} \sum_{l_1} \sum_{m_1} \sum_{k_2} \sum_{l_2}$$
$$\times \sum_{m_2} \left| (\mathbf{x}_j - \bar{\mathbf{x}})^{k_1 l_1} (\mathbf{x}_j - \bar{\mathbf{x}})^{k_2 l_2} (\mathbf{x}_i - \bar{\mathbf{x}})^{vm_1} (\mathbf{x}_i - \bar{\mathbf{x}})^{vm_2} \right|$$

$$= n^{-4}c_1c_2^2 \sum_i \sum_j \Big\{ \sum_k \sum_l |(\mathbf{x}_j - \bar{\mathbf{x}})^{kl}| \Big\}^2 \Big\{ \sum_m |(\mathbf{x}_i - \bar{\mathbf{x}})^{vm}| \Big\}^2$$

$$\leq n^{-4}c_1c_2^2 p^2 q \sum_i \sum_j \sum_k \sum_l \sum_m \{(\mathbf{x}_j - \bar{\mathbf{x}})^{kl}\}^2 \{(\mathbf{x}_i - \bar{\mathbf{x}})^{vm}\}^2$$

$$= O(n^{-2}).$$

The contribution from the case $i_1 = j_1$ and $i_2 = j_2$ is no greater than

$$n^{-4}c_1c_2^2 \sum_i \sum_j \sum_{k_1} \sum_{l_1} \sum_{m_1} \sum_{k_2} \sum_{l_2}$$
$$\times \sum_{m_2} \left| (\mathbf{x}_j - \bar{\mathbf{x}})^{k_1 l_1} (\mathbf{x}_i - \bar{\mathbf{x}})^{k_2 l_2} (\mathbf{x}_j - \bar{\mathbf{x}})^{vm_1} (\mathbf{x}_i - \bar{\mathbf{x}})^{vm_2} \right|$$

$$\leq n^{-4}c_1c_2^2 \left[\sum_i \Big\{ \sum_k \sum_l |(\mathbf{x}_i - \bar{\mathbf{x}})^{kl}| \Big\}^2 \right] \left[\sum_i \Big\{ \sum_m |(\mathbf{x}_i - \bar{\mathbf{x}})^{vm}| \Big\}^2 \right]$$

$$\leq n^{-4}c_1c_2^2 p^2 q \left[\sum_i \sum_k \sum_l \{(\mathbf{x}_i - \bar{\mathbf{x}})^{kl}\}^2 \right] \left[\sum_i \sum_m \{(\mathbf{x}_i - \bar{\mathbf{x}})^{vm}\}^2 \right]$$

$$= O(n^{-2}).$$

Combining the results from (4.57) down we deduce that

$$E(P^{uv})^2 = O(n^{-2}),$$

whence $\mathbf{P} = O_p(n^{-1})$, concluding the proof of (4.51).

4.3.7 Confidence Bands

To introduce the subject of confidence bands we revert to the simple linear regression model,

$$Y_i = c + x_i d + \epsilon_i, \qquad 1 \leq i \leq n, \qquad (4.58)$$

where c, d, x_i, Y_i, and ϵ_i are scalars, and the ϵ_i's are independent and identically distributed random variables with zero mean and variance σ^2. In subsection 4.3.4 we discussed confidence intervals for the mean given $x = x_0$,

$$E(Y \mid x = x_0) = c + x_0 d.$$

A minor modification of the arguments given there allows us to construct simultaneous confidence intervals for any finite number, say m, of means,

$$E\{Y \mid x = x(j)\} = c + x(j)d, \qquad 1 \leq j \leq m.$$

Of more interest is the problem of computing a confidence band for the mean over a set $\mathcal{S}$ containing an infinite number of regressands x,

$$E(Y \mid x) = c + xd, \qquad x \in \mathcal{S}.$$

The set $\mathcal{S}$ would usually be an interval, sometimes the whole real line.

By "confidence band" we mean "simultaneous confidence band." That is, we seek a (random) region $\mathcal{B}$, which depends on the sample and covers the straight line segment $\mathcal{L} = \{c + xd : x \in \mathcal{S}\}$ with given probability α. Typically $\mathcal{B}$ would have the form

$$\mathcal{B} = \{(x, y) : x \in \mathcal{S}, y \in \mathcal{I}(x)\},$$

where $\{\mathcal{I}(x) : x \in \mathcal{S}\}$ is a sequence of intervals whose endpoints form two smooth curves. Figure 4.1 illustrates typical bands. Let the upper curve be $\mathcal{C}_+$ and the lower curve be $\mathcal{C}_-$. We would like the freedom to choose for ourselves the general shape of the envelopes $\mathcal{C}_+$ and $\mathcal{C}_-$, such as linear (for $\widehat{\mathcal{C}}_-$ in Figure 4.1(a)) or "hyperbolic" (for $\widehat{\mathcal{C}}_+$ in the same figure). However, we need a databased rule for selecting the scale of the curves, so as to ensure that the coverage probability is accurate. Naturally we shall suggest bootstrap methods, although the confidence band problem is more than sixty years old and there exists a variety of techniques based on Normal approximation. Bibliographical details will be given at the end of the chapter.

Obviously, confidence bands can be based on confidence intervals. If we construct a sequence of nominal α-level confidence intervals $\mathcal{I}_1(x)$, having the property

$$P\{c + xd \in \mathcal{I}_1(x)\} = \alpha, \qquad \text{all } x \in \mathcal{S},$$

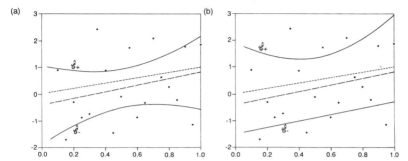

FIGURE 4.1. Bootstrap confidence bands for a regression line. We took $c = 0$, $d = 1$, and $n = 20$, and simulated data with Normal $N(0, 1)$ errors and equally spaced design points on the interval $\mathcal{S} = [0, 1]$. Thus, pairs (x_i, Y_i) were related by the formula $Y_i = x_i + \epsilon_i$, where $x_i = \frac{i}{n}$ and the ϵ_i's were Normal $N(0, 1)$. In each part of the figure, x is graphed on the horizontal axis and Y on the vertical axis. For Figure 4.1(a) the upper template is $f_+ = f_P$ and the lower template, $f_- = f_F \equiv 1$; for Figure 4.1(b), $f_+ = f_- = f_P$. Nominal coverage is $\alpha = 0.95$ in each case, and scale is defined symmetrically; that is, we take $\hat{v}_+ = \hat{v}_- = \hat{v}$, say. This is particularly appropriate in the case of Figure 4.1(b), since the errors are symmetrically distributed and the templates there are identical. Scale is chosen by the percentile-t method, so that $\hat{v}$ is the solution of the equation

$$P\left\{-\hat{v}f_+(x) \leq \Delta^*(x) \leq \hat{v}f_-(x), \text{ all } x \in \mathcal{S} \mid \mathcal{X}\right\} = 0.95.$$

For this definition of $\hat{v}$, the confidence band $\widehat{\mathcal{C}}_+$, $\widehat{\mathcal{C}}_-$ are defined by (4.63). Note that different values of $\hat{v}$ are used for parts (a) and (b) of the figure.

then the naive band

$$\mathcal{B}_1 = \left\{(x, y) : x \in \mathcal{S}, \ y \in \mathcal{I}_1(x)\right\}$$

will undercover the line segment $\mathcal{L}$. However, this difficulty can be remedied by expanding the intervals $\mathcal{I}_1(x)$ in a systematic way. For example, if the intervals $\mathcal{I}_1(x)$ are of the two-sided, equal-tailed percentile-t type, such as the interval $\widehat{J}_2$ described at the end of subsection 4.3.4, then an appropriate widening of the band $\mathcal{B}_1$ produces the "double hyperbola" band illustrated in Figure 4.1(b).

We shall use two "templates" to determine the shapes of the envelopes. The templates are simply known, nonnegative functions f_+ and f_-, for the upper and lower envelopes respectively. There is one sort of template for linear envelopes, another for hyperbolic envelopes, another for V-shaped envelopes, and so on. The templates may depend on the design variables x_i. Two scale factors v_+ and v_- describe the distance of the upper and lower envelopes, respectively, from the least-squares estimate of the line segment $\mathcal{L}$. Specifically, let $\hat{c}$, $\hat{d}$, and $\hat{\sigma}^2$ denote the least-squares estimates of c, d, and σ^2 in the model (4.58); see (4.18) and (4.19). Then the upper and lower envelopes $\mathcal{C}_+$ and $\mathcal{C}_-$ are given by

$$\mathcal{C}_\pm = \left\{ (x,y) : x \in \mathcal{S}, \ y = \hat{c} + x\hat{d} \pm \hat{\sigma} v_\pm f_\pm(x) \right\}. \tag{4.59}$$

The confidence band is

$$\mathcal{B} = \left\{ (x,y) : x \in \mathcal{S}, \ \hat{c} + x\hat{d} - \hat{\sigma} v_- f_-(x) \le y \le \hat{c} + x\hat{d} + \hat{\sigma} v_+ f_+(x) \right\}.$$

The coverage and "skewness" of the band, relative to the line estimate $y = \hat{c} + x\hat{d}$, are governed by the scale factors. For a given pair of templates, and a given coverage level α, we must select v_+, v_- such that

$$P\left\{ \hat{c} + x\hat{d} - \hat{\sigma} v_- f_-(x) \le c + xd \le \hat{c} + x\hat{d} + \hat{\sigma} v_+ f_+(x), \ \text{all } x \in \mathcal{S} \right\} = \alpha,$$

i.e., such that

$$P\left\{ - v_+ f_+(x) \le \Delta(x) \le v_- f_-(x), \quad \text{all } x \in \mathcal{S} \right\} = \alpha, \tag{4.60}$$

where

$$\Delta(x) = (\hat{c} + x\hat{d} - c - xd)/\hat{\sigma}.$$

One candidate for f_+ and f_- is the fixed template $f_F(x) \equiv 1$; another is the root parabolic template

$$f_P(x) = \left[\text{var}\{ n^{1/2}(\hat{c} + x\hat{d})/\sigma \} \right]^{1/2} = \left\{ 1 + \sigma_x^{-2}(x - \bar{x})^2 \right\}^{1/2},$$

where $\sigma_x^2 = n^{-1} \sum (x_i - \bar{x})^2$. This option produces the "hyperbolic" shape present in three of the four envelopes of Figure 4.1. A third candidate is the V-shaped template

$$f_V(x) = 1 + \sigma_x^{-1} |x - \bar{x}|.$$

One-sided bands would be constructed with either f_+ or f_- set at $+\infty$. The band that has $f_+ = f_- = f_F$ is of constant width, since the two envelopes are parallel lines. Of course, such a band is only possible when the set $\mathcal{S}$ is bounded. The band with $f_+ = f_- = f_P$ might be called a "constant probability" band since the pointwise coverage probability

$$P\left\{ \hat{c} + x\hat{d} - \hat{\sigma} v_- f_-(x) \le c + xd \le \hat{c} + x\hat{d} + \hat{\sigma} v_+ f_+(x) \right\}$$

is virtually independent of x; it is approximately equal to

$$P\left\{ - n^{1/2} v_+ < N(0,1) < n^{1/2} v_- \right\},$$

which does not depend on x. The cases $f_+ = f_- = f_P$ and $f_+ = f_- = f_V$ are both feasible when $\mathcal{S}$ denotes the entire real line $\mathbb{R}$.

Having settled on choices of f_+ and f_- we must select scale factors v_+ and v_- such that relation (4.60) holds. Of course, there is an infinite variety of choices of (v_+, v_-), although once one member of the pair is selected the other is determined by (4.60). We suggest two particular choices of (v_+, v_-): the symmetric choice, where $v_+ = v_-$, and the narrowest-width choice, where v_+ and v_- are selected to minimize $v_+ + v_-$ subject to (4.60). In the case of approximately symmetric errors and identical templates the symmetric choice is attractive. However, in other circumstances the narrowest-width choice has the advantage of producing a narrower band with the same level of coverage.

The equal-tailed method is also an option for defining v_+ and v_-. To implement it, let $v_+(\beta)$ and $v_-(\beta)$ be such that

$$P\Big\{ -v_+(\beta)f_+(x) \le \Delta(x), \ \text{ all } x \in \mathcal{S} \Big\}$$

$$= P\Big\{\Delta(x) \le v_-(\beta)f_-(x), \ \text{ all } x \in \mathcal{S} \Big\} = \beta,$$

and define α' to be that value such that

$$P\Big\{ -v_+(\alpha')f_+(x) \le \Delta(x) \le v_-(\alpha')f_-(x), \quad \text{all } x \in \mathcal{S} \Big\} = \alpha.$$

Take $v_+ = v_+(\alpha')$, $v_- = v_-(\alpha')$. For this choice of the scale factors, the confidence band whose boundary curves are defined by (4.59) has the property that the unknown regression line has an equal probability of intersecting either boundary over the set $\mathcal{S}$.

Precise computation of v_+ and v_- requires knowledge of the distribution of the random process Δ, and such information is usually only available in the case of Normal errors. When the error distribution is unknown, the percentile-t bootstrap method may be used to estimate v_+ and v_-. To implement the bootstrap, first define the residuals

$$\hat{\epsilon}_i = Y_i - (\hat{c} + x_i\hat{d}), \quad 1 \le i \le n,$$

and let $\epsilon_1^*, \ldots, \epsilon_n^*$ denote a resample drawn at random, with replacement, from $\{\hat{\epsilon}_1, \ldots, \hat{\epsilon}_n\}$. Put $Y_i^* = \hat{c} + x_i\hat{d} + \epsilon_i^*$, $1 \le i \le n$. Let $\hat{c}^*$, $\hat{d}^*$, $\hat{\sigma}^{*2}$ denote versions of $\hat{c}$, $\hat{d}$, $\hat{\sigma}^2$ computed for the case where the data set $\mathcal{X}$ is replaced by $\mathcal{X}^* = \{(x_i, Y_i^*), 1 \le i \le n\}$. Define

$$\Delta^*(x) = (\hat{c}^* + x\hat{d}^* - \hat{c} - x\hat{d})/\hat{\sigma}^*. \tag{4.61}$$

We estimate solutions v_+, v_- of (4.60) as solutions $\hat{v}_+, \hat{v}_-$ of

$$P\Big\{ -\hat{v}_+ f_+(x) \le \Delta^*(x) \le \hat{v}_- f_-(x), \quad \text{all } x \in \mathcal{S} \mid \mathcal{X} \Big\} = \alpha. \tag{4.62}$$

The symmetric choice has $\hat{v}_- = \hat{v}_+$; the narrowest-width choice has $\hat{v}_+$ and $\hat{v}_-$ taken to minimize $\hat{v}_+ + \hat{v}_-$ subject to (4.62); and there is an obvious bootstrap version of the equal-tailed choice. For any of these rules, the envelopes of a nominal α-level bootstrap confidence band for $y = c + xd$ over the set $\mathcal{S}$ are given by the curves $\widehat{\mathcal{C}}_+$ and $\widehat{\mathcal{C}}_-$, defined by

$$\widehat{\mathcal{C}}_\pm = \left\{ (x,y) : x \in \mathcal{S},\, y = \hat{c} + x\hat{d} \pm \hat{\sigma}\hat{v}_\pm\, f_\pm(x) \right\};\qquad (4.63)$$

compare (4.59).

Towards the end of this section we shall show that the band between the two curves $\mathcal{C}_+$ and $\mathcal{C}_-$ is only $O_p(n^{-1/2})$ wide, and that the curves $\widehat{\mathcal{C}}_+$ and $\mathcal{C}_+$ are $O_p(n^{-3/2})$ apart, uniformly over any finite range. The same applies to the curves $\widehat{\mathcal{C}}_-$ and $\mathcal{C}_-$. In this sense the percentile-t bootstrap confidence bands are second-order accurate, in that they correctly capture second-order features, of size $O_p(n^{-1})$, of the confidence bands. As we shall prove by means of Edgeworth expansion, the bootstrap bands correct for the major departure from Normality of the error distribution, due to skewness.

The case of polynomial regression may be treated as a special case of univariate, multiparameter regression, and so to indicate the generalizations of our argument we shall study the latter. We could also examine multivariate, multiparameter regression, but there the confidence region would be particularly awkward to represent.

First we describe the univariate, multiparameter model. Take the model for the observed data $\mathcal{X} = \left\{ (\mathbf{x}_i, Y_i),\ 1 \le i \le n \right\}$ to be

$$Y_i = c + \mathbf{x}_i\, \mathbf{d} + \epsilon_i,\quad 1 \le i \le n,$$

where $\mathbf{x}_i$ and $\mathbf{d}$ are row and column vectors, respectively, each of length p, and the ϵ_i's are independent and identically distributed random variables with zero mean and variance σ^2. (To facilitate comparison with simple linear regression we separate out an intercept term. The case where the intercept is constrained to be zero, or any other constant, may be treated similarly.) Least-squares estimates of c and $\mathbf{d}$ are

$$\hat{c} = \bar{Y} - \bar{\mathbf{x}}\widehat{\mathbf{d}}\qquad \text{and}\qquad \widehat{\mathbf{d}} = \sigma_x^{-2}\, n^{-1} \sum (\mathbf{x}_i - \bar{\mathbf{x}})(Y_i - \bar{Y}),$$

where $\sigma_x^2 = n^{-1} \sum (\mathbf{x}_i - \bar{\mathbf{x}})(\mathbf{x}_i - \bar{\mathbf{x}})^{\mathsf{T}}$. Our estimate of σ^2 is $\hat{\sigma}^2 = n^{-1} \sum (Y_i - \hat{c} - \mathbf{x}_i\widehat{\mathbf{d}})^2$.

Next we describe confidence bands for the plane segment $\{c + \mathbf{x}\mathbf{d}, \mathbf{x} \in \mathcal{S}\}$ where $\mathcal{S}$ is a given subset of p-dimensional Euclidean space. Again, the percentile-t bootstrap is our method of choice. Only minor modifications of the techniques described for simple linear regression are needed: $c + \mathbf{x}\mathbf{d}$,

$c + \mathbf{x}_i\mathbf{d}$, $c + \bar{\mathbf{x}}\mathbf{d}$ replace $c + xd$, $c + x_id$, $c + \bar{x}d$ at all appearances of the latter; and the templates f_P and f_V are altered to

$$f_P(\mathbf{x}) = \left\{1 + \sigma_x^{-2}(\mathbf{x} - \bar{\mathbf{x}})(\mathbf{x} - \bar{\mathbf{x}})^{\mathrm{T}}\right\}^{1/2},$$

$$f_V(\mathbf{x}) = 1 + \sigma_x^{-1} \sum_{j=1}^{p} |x^{(j)} - \bar{x}^{(j)}|,$$

where $x^{(j)}$, $\bar{x}^{(j)}$ denote the jth elements of the row vectors $\mathbf{x}$, $\bar{\mathbf{x}}$, respectively. In practice it will often be the case that $\mathbf{x}$ is a function of a univariate quantity, for example $x^{(1)}$ (the first component of $\mathbf{x}$). If we were representing polynomial regression of degree p by multiple regression then we would have $\mathbf{x} = (x^{(1)}, \ldots, x^{(1)^p})$. Here our interest would generally be in constructing a confidence region for $c + \mathbf{x}\mathbf{d}$, $x^{(1)} \in \mathcal{S}'$, where $\mathcal{S}'$ is a given interval. We would take $\mathcal{S} = \{(x^{(1)}, \ldots, x^{(1)^p}) : x^{(1)} \in \mathcal{S}'\}$, and recognize that the confidence region over $\mathcal{S}$, say $\{(\mathbf{x}, y) : \mathbf{x} \in \mathcal{S}, h_1(\mathbf{x}) \le y \le h_2(\mathbf{x})\}$, served only as a notational surrogate for the band over $\mathcal{S}'$,

$$\left\{(x^{(1)}, y) : x^{(1)} \in \mathcal{S}', \ h_1(\mathbf{x}) \le y \le h_2(\mathbf{x}) \ \text{where } \mathbf{x} = (x^{(1)}, \ldots, x^{(1)^p})\right\}.$$

In the remainder of this section we develop a theory, based on Edgeworth expansion, to describe the performance of our percentile-t method for constructing confidence bands. We confine attention to the case of simple linear regression, where the model is that in (4.58).

Ideally, the scale factors v_+, v_- would be determined from the equation

$$\beta(v_+, v_-) = \alpha,$$

where

$$\beta(v_+, v_-) = P\left\{-v_+ f_+(x) \le \Delta(x) \le v_- f_-(x), \quad \text{all } x \in \mathcal{S}\right\};$$

see (4.60). Usually the function β would be unknown, and we have suggested that it be replaced by its bootstrap estimate

$$\hat{\beta}(v_+, v_-) = P\left\{-v_+ f_+(x) \le \Delta^*(x) \le v_- f_-(x), \quad \text{all } x \in \mathcal{S} \mid \mathcal{X}\right\}.$$

We shall shortly show that

$$\hat{\beta}(v_+, v_-) - \beta(v_+, v_-) = O_p(n^{-1}) \tag{4.64}$$

uniformly in $v_+, v_- > 0$.

From this fact it is readily proved that the bootstrap estimates $\hat{v}_+, \hat{v}_-$ of solutions v_+, v_- to the equation $\beta(v_+, v_-) = \alpha$, differ from the true v_+ and v_- only in terms of order $n^{-3/2}$. To see how this is done, let us assume for the sake of simplicity and definiteness that we are constructing "symmetric" bands, where $v_+ = v_-$. The cases of narrowest-width or equal-tailed bands may be treated similarly. Now, it may be deduced from the definition of Δ that the process $n^{1/2}\,\Delta(x)$ is asymptotically distributed as $Z_1 + \sigma_x^{-1}(x - \bar{x})\,Z_2$, where Z_1 and Z_2 are independent Standard Normal random variables. Put

$$G_n(w) \;=\; P\Big\{ -wf_+(x) \le n^{1/2}\,\Delta(x) \le wf_-(x), \;\; \text{all } x \in \mathcal{S}\Big\},$$

$$G_\infty(w) \;=\; P\Big\{ -wf_+(x) \le Z_1 + \sigma_x^{-1}(x - \bar{x})\,Z_2 \le wf_-(x), \;\; \text{all } x \in \mathcal{S}\Big\},$$

$$=\; \lim_{n\to\infty} G_n(w)\,.$$

Suppose $\hat{v}(\alpha)$ and $v(\alpha) = \hat{v}(\alpha) - \delta$ are chosen such that

$$\hat{\beta}\{\hat{v}(\alpha), \hat{v}(\alpha)\} \;=\; \beta\{v(\alpha), v(\alpha)\} \;=\; \alpha\,,$$

and let $w_\infty(\alpha)$ be the solution of the equation $G_\infty\{w_\infty(\alpha)\} = \alpha$. Then $n^{1/2}\,\hat{v}(\alpha)$ and $n^{1/2}\,v(\alpha)$ both converge to $w_\infty(\alpha)$. Therefore $\delta = \delta(n) = o(n^{-1/2})$, whence

$$\beta(v - \delta, v - \delta) \;=\; G_n\{n^{1/2}(v - \delta)\}$$
$$=\; G_n(n^{1/2}\,v) - (n^{1/2}\,\delta)\,G_n'(n^{1/2}\,v) + O\{(n^{1/2}\,\delta)^2\}$$
$$=\; \beta(v, v) - (n^{1/2}\,\delta)\,G_\infty'(n^{1/2}\,v) + o(n^{1/2}\,\delta)\,.$$

(These formulae are readily justified by Edgeworth expansions of the type that we shall give shortly.) Hence,

$$0 \;=\; \hat{\beta}\{\hat{v}(\alpha), \hat{v}(\alpha)\} - \beta\{v(\alpha), v(\alpha)\}$$
$$=\; \hat{\beta}\{\hat{v}(\alpha), \hat{v}(\alpha)\} - \beta\{\hat{v}(\alpha), \hat{v}(\alpha)\}$$
$$+\; n^{1/2}\,\delta\,G_\infty'\{n^{1/2}\,v(\alpha)\} + o_p(n^{1/2}\,\delta)\,,$$

whence by (4.64),

$$\delta \;=\; \hat{v}(\alpha) - v(\alpha)$$
$$\sim\; n^{-1/2}\,G_\infty'\{w_\infty(\alpha)\}^{-1}\,\big[\,\beta\{\hat{v}(\alpha), \hat{v}(\alpha)\} - \hat{\beta}\{\hat{v}(\alpha), \hat{v}(\alpha)\}\,\big]$$
$$=\; O_p(n^{-3/2})\,.$$

This establishes the claim made earlier in this paragraph.

The analysis above also demonstrates that the curves $\widehat{\mathcal{C}}_+, \widehat{\mathcal{C}}_-$, defining the envelope of the bootstrap confidence band and given by formula (4.63), are order $n^{-1/2}$ apart, since $\hat{v}(\alpha) \sim n^{-1/2}\,w_\infty(\alpha)$. The fact that

$$\hat{v}(\alpha) - v(\alpha) \;=\; O_p(n^{-3/2})$$

shows that $\widehat{\mathcal{C}}_+$ and $\mathcal{C}_+$ (and also $\widehat{\mathcal{C}}_-$ and $\mathcal{C}_-$) are $O_p(n^{-3/2})$ apart. These conclusions apply also to narrowest-width bands and equal-tailed bands.

It remains to derive (4.64), which we shall do by first developing an Edgeworth expansion of the probability β, then deriving an analogous formula for $\hat{\beta}$, and finally, subtracting these two expressions. Note that the value of β is invariant under changes of the scale σ; this is of course a consequence of the fact that we Studentized when constructing the statistic Δ, so as to make it pivotal. Therefore, we may assume without loss of generality that $\sigma^2 = 1$.

To expand the probability β, observe first that

$$n^{1/2}\,\Delta(x) \;=\; (S_1 + yS_3)\left(1 + n^{-1/2}\,S_2 - n^{-1}\,S_1^2 - n^{-1}\,S_3^2\right)^{-1/2},$$

where $y = \sigma_x^{-1}(x - \bar{x})$,

$$S_1 \;=\; n^{-1/2}\sum_{i=1}^{n} e_i, \qquad S_2 \;=\; n^{-1/2}\sum_{i=1}^{n}(e_i^2 - 1),$$

and

$$S_3 \;=\; n^{-1/2}\sigma_x^{-1}\sum_{i=1}^{n}(x_i - \bar{x})\,e_i.$$

That is, $n^{1/2}\,\Delta(x) = D(S_1, S_2, S_3; y)$, where

$$D(s_1, s_2, s_3; y) \;=\; (s_1 + ys_3)\left(1 + n^{-1/2}s_2 - n^{-1}s_1^2 - n^{-1}s_3^2\right)^{-1/2}. \qquad (4.65)$$

Then

$$\beta(v_+, v_-) \;=\; \int_{\mathcal{Q}} d\mu,$$

where μ is the joint probability measure of (S_1, S_2, S_3) and

$$\mathcal{Q} \;=\; \Big\{(s_1, s_2, s_3) \in \mathbb{R}^3 : -n^{1/2}\,v_+ f_+(x) \;\leq\; D\big[s_1, s_2, s_3; \sigma_x^{-1}(x - \bar{x})\big]$$
$$\leq\; n^{1/2}v_- f_-(x), \quad \text{all } x \in \mathcal{S}\Big\}.$$

For any practical choice of f_+ and f_-, in particular for the versions f_+ and f_- discussed earlier, $\mathcal{Q}$ may be expressed in terms of a finite number of unions and intersections of convex sets, the number of such operations being bounded uniformly in all choices $n \geq 2$ and $v_+, v_- > 0$. Under mild assumptions on the process generating the data (e.g., that the errors have a nonsingular distribution with sufficiently many finite moments, and that the design variables x_i are either regularly spaced in an interval or come from a continuous distribution on an interval), the measure μ admits a short Edgeworth expansion uniformly on convex sets $\mathcal{C} \subseteq \mathbb{R}^3$,

$$\mu(\mathcal{C}) \;=\; \int_{\mathcal{C}}\left\{1 + n^{-1/2}p(\mathbf{s})\right\}\phi(\mathbf{s})\,d\mathbf{s} \;+\; O(n^{-1}),$$

where ϕ is the density of a trivariate Normal distribution having the same mean and covariance structure as $\mathbf{S} = (S_1, S_2, S_3)^{\mathrm{T}}$, p is an odd, third degree polynomial in three variables, and $\mathbf{s} = (s_1, s_2, s_3)^{\mathrm{T}}$. The coefficients depend on moments of the error distribution up to and including the sixth, and on the design variables x_i, but are bounded as $n \to \infty$. This is a version of formula (4.4), truncated after the term of size $n^{-1/2}$. Combining the results of this paragraph we conclude that

$$\beta(v_+, v_-) = \int_Q \left\{1 + n^{-1/2} p(\mathbf{s})\right\} \phi(\mathbf{s}) \, d\mathbf{s} + O(n^{-1}) \qquad (4.66)$$

uniformly in $v_+, v_- > 0$.

The function ϕ appearing in (4.66) is the density of the trivariate Normal distribution with zero mean and variance matrix

$$\Sigma = \begin{bmatrix} 1 & \mu_3 & 0 \\ \mu_3 & \mu_4 - 1 & 0 \\ 0 & 0 & 1 \end{bmatrix}, \qquad (4.67)$$

where $\mu_j = E(e_i/\sigma)^j$. Indicate this fact by writing ϕ as $\phi(\cdot, \Sigma)$.

Next we develop a bootstrap analogue of (4.66). Let $\hat{p}$ be obtained from p on replacing each appearance of $E(e_i^j)$ by the corresponding sample moment $n^{-1} \sum \hat{e}_i^j$, for $j = 1, 2, \dots$. Likewise, define $\hat{\mu}_j = n^{-1} \sum_i (\hat{e}_i/\hat{\sigma})^j$, and let $\hat{\Sigma}$ be the 3×3 matrix obtained on replacing (μ_3, μ_4) by $(\hat{\mu}_3, \hat{\mu}_4)$ in (4.67). Then instead of (4.66), we have

$$\hat{\beta}(v_+, v_-) = \int_Q \left\{1 + n^{-1/2} \hat{p}(\mathbf{s})\right\} \phi(\mathbf{s}, \hat{\Sigma}) \, d\mathbf{s} + O_p(n^{-1}) \qquad (4.68)$$

with probability one as $n \to \infty$. Since the sample moments of the residual sequence $\{\hat{e}_i\}$ are within $O_p(n^{-1/2})$ of the respective population moments of the errors ϵ_i, $\hat{p} - p = O_p(n^{-1/2})$ and $\hat{\Sigma} - \Sigma = O_p(n^{-1/2})$. We may now deduce from (4.68) that

$$\hat{\beta}(v_+, v_-) = \int_Q \phi(\mathbf{s}, \hat{\Sigma}) \, d\mathbf{s} + n^{-1/2} \int_Q p(\mathbf{s}) \, \phi(\mathbf{s}, \Sigma) \, d\mathbf{s} + O_p(n^{-1}).$$

Now, $\phi(\mathbf{s}, \hat{\Sigma})$ and $\phi(\mathbf{s}, \Sigma)$ differ only by $O_p(n^{-1/2})$, and then only in terms involving s_2. Since s_2 enters the formula for $D(s_1, s_2, s_3; y)$ only through the term $n^{-1/2} s_2$ appearing in (4.65), then, noting the definition of Q in terms of $D(s_1, s_2, s_3; y)$, we obtain

$$\int_Q \left\{\phi(\mathbf{s}, \hat{\Sigma}) - \phi(\mathbf{s}, \Sigma)\right\} d\mathbf{s} = O_p(n^{-1/2} n^{-1/2}) = O_p(n^{-1}).$$

Therefore,

$$\hat{\beta}(v_+, v_-) = \int_Q \phi(\mathbf{s}, \Sigma) \, d\mathbf{s} + n^{-1/2} \int_Q p(\mathbf{s}) \, \phi(\mathbf{s}, \Sigma) \, d\mathbf{s} + O_p(n^{-1})$$

$$= \beta(v_+, v_-) + O_p(n^{-1}),$$

the last line coming from (4.66). This establishes the desired result (4.64).

4.4 Nonparametric Density Estimation

4.4.1 Introduction

The purpose of nonparametric density estimation is to construct an estimate of a density function without imposing structural assumptions. Typically the only conditions imposed on the density are that it have at least two bounded derivatives. In this circumstance we may use only local information about the value of the density at any given point. That is, the value of the density of a point x must be calculated from data values that lie in the neighbourhood of x, and to ensure consistency the neighbourhood must shrink as sample size increases. In the case of kernel-type density estimation, to which we shall devote almost all our attention, the radius of the effective neighbourhood is roughly equal to the "bandwidth" or smoothing parameter of the estimator. Under the assumption that the density is univariate with at least two bounded derivatives, and using a nonnegative kernel function, the size of bandwidth that optimizes performance of the estimator in any L^p metric $(1 \le p < \infty)$ is $n^{-1/5}$. (This is the classical case of a "second-order kernel".) The number of "parameters" needed to model the unknown density within a given interval is approximately equal to the number of bandwidths that can be fitted into that interval, and so is roughly of size $n^{1/5}$. Thus, nonparametric density estimation (using a second-order kernel) involves the adaptive fitting of approximately $n^{1/5}$ parameters, this number growing with increasing n.

The argument above indicates that there are going to be marked dissimilarities between our previous experience of the bootstrap and properties of the bootstrap in nonparametric curve estimation. We have not, so far in this monograph, encountered a case where the effective number of parameters grew unboundedly as sample size increased. The dissimilarities are manifested in at least two ways: difficulties with bias and a worsening of the overall convergence rate.

To deal first with bias, note that in many statistical problems the impact of bias increases with the number of unknown parameters. Multiparameter log-linear regression is a well-known example. In the case of nonparametric density estimation, bias is so significant that optimal L^p performance is achieved by constructing the estimator so as to have bias of the same size as the error about the mean. Therefore, far from being negligible, bias has the same importance as the random error component.

The overall convergence rate of a nonparametric density estimator can be quite slow, on account of the fact that only local information is used in its construction. If the density is univariate and the bandwidth of the estimator is h, then the number of data values that lie within one bandwidth of a given point x is of size nh. Effectively, only these observations are used to estimate the density at x, and so the convergence rate is $(nh)^{-1/2}$. If $h \simeq n^{-1/5}$, as suggested three paragraphs above, then this rate is $n^{-2/5}$, which is slower than the $n^{-1/2}$ with which we are familiar. Some of the Edgeworth expansions that we shall use to elucidate properties of nonparametric density estimators are power series in $(nh)^{-1/2}$, rather than $n^{-1/2}$.

There is a wide variety of different bootstrap methods that can be used to construct confidence intervals for $f(x)$. They differ in the manner of resampling, in the ways they account for bias, and in the type of pivoting. Subsection 4.4.2 will motivate and discuss different techniques, and subsections 4.4.3–4.4.6 will develop theory for a particular class of approaches based on the "usual" method for resampling (i.e., resampling directly from the sample), and on bias correction by either undersmoothing or explicit bias estimation. Of course, Edgeworth expansions play a significant role in that theory. It is not our intention to comprehensively treat a wide range of different approaches to constructing confidence intervals; rather, we aim to show that the general principles developed in Chapter 3 for simpler, more traditional problems have straightforward extensions to the present context. This demonstrates that our ideas, particularly the notion of pivoting and the use of Edgeworth expansion methods to elucidate theory, have a broad foundation and a wide range of application. In particular, the concept of second-order correctness emerges in subsection 4.4.5 as an important factor influencing the performance of confidence intervals for densities, albeit with a slightly modified definition. The pivotal percentile-t method produces second-order correct confidence intervals, while the nonpivotal percentile method yields only first-order correct intervals, exactly as noted in Section 3.4.

No matter which method of bootstrap resampling is used to construct confidence intervals, an amount of smoothing is always involved. At best there will be only one smoothing parameter, that used in the construction of the basic density estimator $\hat{f}$ of the unknown density f. At worst there will be several, for determining the amount of smoothing when resampling (if resampling is not done straight from the original sample), for constructing the bias estimator (if bias is estimated explicitly), and of course for constructing the density estimator $\hat{f}$. In the classical theory of density estimation, the appropriate amount of smoothing is usually determined by reference to a "metric," or measure of performance. The most widely accepted criterion is L^2 distance,

$$\left\{ \int E(\hat{f} - f)^2 \right\}^{1/2},$$

although some investigators prefer L^1 or L^∞ distance. What should the criterion be in the case of confidence intervals? We prefer coverage error — that is, we suggest that the smoothing parameter(s) be chosen to minimize, as nearly as possible, the coverage error of the type of confidence interval under construction. This approach is far from straightforward, especially if the bandwidths have to be chosen empirically (as would always be the case in a practical problem). Furthermore, the bandwidths that minimize coverage error are generally of a smaller order of magnitude than those that minimize any L^p distance, for $1 \leq p \leq \infty$. This means that, if our recommendation of minimizing coverage error is followed, then the curve estimator $\hat{f}$ on which the confidence intervals are based will have substantially greater error-about-the-mean than is desirable for a curve estimator. Understandably, this feature does not endear the coverage error criterion to statisticians whose experience is in classical curve estimation problems. However, we would argue that the problems of point (or curve) estimation and interval estimation are distinctly different, with different aims and different ends in mind. There is still a lot to be learned about the interval estimation problem in the context of density estimation, and our theoretical contributions here should be regarded only as a first step towards a deeper, more practically oriented appreciation.

4.4.2 Different Bootstrap Methods in Density Estimation

Our purpose in this subsection is to describe the extensive array of techniques that might be used to construct confidence intervals in nonparametric density estimation. The different methods are based on a variety of approaches to bias correction and pivoting, and on different ways of drawing the resample. For example, we might choose to draw the resample from a smoothed empirical distribution, or we might wish to resample with replacement from the original sample in the usual way. In later subsections we shall focus on specific methods, for the purpose of simplifying our exposition and relating our discussion more closely to the foundations laid in Chapter 3.

Since the principal bootstrap methods differ in their treatment of bias and variance, it is appropriate to preface our account with a brief description of bias and variance properties of density estimators. For the most part we confine attention to kernel-type estimators. Let $\mathcal{X} = \{X_1, \dots, X_n\}$ denote a random sample drawn from a distribution with (univariate) density f, and let K be a bounded function with the property that for some integer $r \geq 1$,

$$\int_{-\infty}^{\infty} y^i K(y) \, dy \quad \begin{cases} = 1 & \text{if } i = 0 , \\ = 0 & \text{if } 1 \leq i \leq r-1 , \\ \neq 0 & \text{if } i = r. \end{cases} \tag{4.69}$$

(If $\int (1 + |y|^r) \, |K(y)| \, dy < \infty$ then the integrals in (4.69) are well defined.) For example, if K is a symmetric density function, such as the Standard Normal density, then (4.69) holds with $r = 2$. This is the case most commonly encountered in practice. A function K satisfying (4.69) is called an rth order kernel and may be used to construct a kernel-type nonparametric density estimator of $f(x)$,

$$\hat{f}(x) \;=\; (nh)^{-1} \sum_{i=1}^{n} K\{(x - X_i)/h\} , \tag{4.70}$$

where h is the "bandwidth" or "window size" of the estimator.

The bandwidth is a very important ingredient of the definition of $\hat{f}$. It determines the sizes of bias and variance, and so plays a major role in governing the performance of $\hat{f}$. To appreciate this influence, define

$$k_1 \;=\; (-1)^r \frac{1}{r!} \int y^r K(y) \, dy , \qquad k_2 \;=\; \int K(y)^2 \, dy$$

and

$$\mu_j(x) \;=\; h^{-1} E\big[K\{(x - X)/h\}^j\big] \;=\; \int K(y)^j \, f(x - hy) \, dy . \tag{4.71}$$

Observe that if $h = h(n) \to 0$ as $n \to \infty$ then bias is given by

$$\begin{aligned} b(x) \;=\; E\{\hat{f}(x)\} - f(x) \;&=\; \mu_1(x) - f(x) \\ &=\; \int K(y) \left\{f(x - hy) - f(x)\right\} dy \\ &=\; \int K(y) \Big\{ - hy \, f'(x) + \tfrac{1}{2}(hy)^2 \, f''(x) - \cdots \\ &\qquad\qquad + (-1)^r \frac{1}{r!} f^{(r)}(x) + \ldots \Big\} \, dy \\ &=\; h^r k_1 \, f^{(r)}(x) + o(h^r) ; \end{aligned} \tag{4.72}$$
$$\tag{4.73}$$

and if additionally $nh \to \infty$,

$$\begin{aligned} \operatorname{var}\{\hat{f}(x)\} \;&=\; (nh)^{-1} \mu_2(x) - n^{-1} \mu_1(x)^2 \\ &=\; (nh)^{-1} \int K^2(y)\{f(x) - hy \, f'(x) + \ldots\} \, dy + O(n^{-1}) \\ &=\; (nh)^{-1} k_2 \, f(x) + o\{(nh)^{-1}\} . \end{aligned} \tag{4.74}$$
$$\tag{4.75}$$

(Rigorous derivation of both these formulae is straightforward if we assume that f has r uniformly bounded, continuous derivatives.) The bias of $\hat{f}$ does not converge to zero unless $h \to 0$, and the variance does not converge to zero unless $nh \to \infty$. Mean squared error is optimized by taking h to be of size $n^{-1/(2r+1)}$,

$$E\{\hat{f}(x) - f(x)\}^2 = (\text{bias})^2 + \text{variance}$$

$$\sim h^{2r}\{k_1 f^{(r)}(x)\}^2 + (nh)^{-1} k_2 f(x),$$

which is minimized by taking $h = k_3(x)\, n^{-1/(2r+1)}$, where

$$k_3(x) = \left[k_2 f(x)(2r)^{-1} \{k_1 f^{(r)}(x)\}^{-2} \right]^{1/(2r+1)}.$$

This is the usual "optimal" prescription for smoothing (see, e.g., Silverman 1986, Section 3.3), and it results in the bias $E\{\hat{f}(x)\} - f(x)$ and error-about-the-mean $\hat{f}(x) - E\{\hat{f}(x)\}$ both being of size $n^{-r/(2r+1)}$. If we mistakenly construct a confidence interval for $E\hat{f}(x)$ instead of $f(x)$, we shall be ignoring bias and committing a centring error that is of the same order as the width of a two-sided interval. This means that even the asymptotic coverage of the interval (when interpreted as an interval for $f(x)$) will be incorrect.

We claim that the "usual" bootstrap algorithm, when applied to the case of nonparametric density estimation, produces a confidence interval for $E\{\hat{f}(x)\}$ and not for $f(x)$, and so suffers from the serious inaccuracies just described. Therefore, some form of bias correction is in order. To appreciate this point, recall that confidence interval methods are traditionally based on the distribution of $\hat{\theta} - \theta_0$ or $(\hat{\theta} - \theta_0)/\hat{\sigma}$. The bootstrap estimates of these distributions are the bootstrap distributions of $\hat{\theta}^* - \hat{\theta}$ and $(\hat{\theta}^* - \hat{\theta})/\hat{\sigma}^*$, respectively. In the case of density estimation, the roles of $\hat{\theta}$ and $\hat{\theta}^*$ are played by $\hat{f}(x)$ and

$$\hat{f}^*(x) = (nh)^{-1} \sum_{i=1}^{n} K\{(x - x_i^*)/h\}, \tag{4.76}$$

respectively, where $\mathcal{X}^* = \{X_1^*, \ldots, X_n^*\}$ denotes a resample drawn randomly, with replacement, from $\mathcal{X}$. Now, the value of $E(\hat{\theta}^* \mid \mathcal{X})$, equal to the bootstrap approximation of $E(\hat{\theta})$, is given by

$$E\{\hat{f}^*(x) \mid \mathcal{X}\} = (nh)^{-1} \sum_{i=1}^{n} E\big[K\{(x - X_i^*)/h\} \mid \mathcal{X}\big]$$

$$= h^{-1} E\big[K\{(x - X_i^*)/h\} \mid \mathcal{X}\big]$$

$$= h^{-1} n^{-1} \sum_{j=1}^{n} K\{(x - X_j)/h\} = \hat{f}(x). \tag{4.77}$$

That is, $E(\hat{\theta}^* \mid \mathcal{X}) = \hat{\theta}$, and so in bootstrapping $\hat{\theta}^* - \hat{\theta}$ (or $(\hat{\theta}^* - \hat{\theta})/\hat{\sigma}^*$) we are actually approximating the distribution of $\hat{\theta} - E\hat{\theta}$ (or $(\hat{\theta} - E\hat{\theta})/\hat{\sigma}$), not the distribution of $\hat{\theta} - \theta_0$ (or $(\hat{\theta} - \theta_0)/\hat{\sigma}$, respectively). In this sense, the bootstrap confuses θ_0 with $E(\hat{\theta})$, and so bias becomes a serious problem.

Put another way, the "usual" bootstrap provides very poor bias estimates in problems of density estimation. The traditional bootstrap estimate of bias, $E(\hat{\theta}^* \mid \mathcal{X}) - \hat{\theta}$ (see Example 1.1 in Section 1.3), is here equal to

$$E\{\hat{f}^*(x) \mid \mathcal{X}\} - \hat{f}(x) = 0,$$

using (4.77). Thus, far from accurately estimating the substantial bias of $\hat{f}(x)$, the bootstrap sets the bias equal to zero.

This is a problem with any estimate that is "linear" in the data, in the sense of

$$\hat{\theta} = \sum_{i=1}^{n} a(X_i) \tag{4.78}$$

for some function a. The bootstrap version of $\hat{\theta}$,

$$\hat{\theta}^* = \sum_{i=1}^{n} a(X_i^*),$$

has the property $E(\hat{\theta}^* \mid \mathcal{X}) = \hat{\theta}$, and so bias is estimated to be zero. Many nonparametric density estimators are either linear (in the sense of (4.78)) or approximately linear. For example, an orthogonal series estimator is given by

$$\hat{f}(x) = \sum_{j=0}^{m} \hat{c}_j \, \psi_j(x), \tag{4.79}$$

where the integer $m \geq 0$ is a smoothing parameter (the analogue of h^{-1}, in a sense); $\psi_0, \psi_1, \ldots$ is a complete orthonormal sequence of functions; and

$$\hat{c}_j = n^{-1} \sum_{i=1}^{n} \psi_j(X_i), \qquad j \geq 0,$$

is an estimate of the generalized Fourier coefficient $c_j = E\{\psi_j(X)\}$. The estimator at (4.79) may be put into the form of (4.78) by defining

$$a(y) = n^{-1} \sum_{j=0}^{m} \psi_j(x) \, \psi_j(y),$$

and of course the kernel estimator at (4.70) is in the form of (4.78).

In view of these difficulties we should pay careful attention to overcoming the problem of bias. There are two obvious options: either estimate bias explicitly, for example by using the asymptotic formula (4.73); or undersmooth, so that bias is rendered negligible. To understand the latter approach, recall from (4.73) and (4.75) that bias is of size h^r, whereas standard deviation is of size $(nh)^{-1/2}$. The so-called "optimal" choice of h, $n^{-1/(2r+1)}$, renders these terms equal in order of magnitude, to $n^{-r/(2r+1)}$. However, by choosing h to be of smaller order than $n^{-1/(2r+1)}$ we may ensure that bias is of smaller order than standard deviation, and therefore is less of a problem than before.

There is a third choice, perhaps obscured by the viewpoint taken in our description of the "usual" bootstrap. We suggested three paragraphs earlier that the "usual" form of the bootstrap involves resampling from the sample. This is reasonable if our aim is to estimate

$$E(\hat{\theta}) = E_F\{\theta(\widehat{F})\},$$

where $\hat{\theta} = \theta(\widehat{F})$ denotes a smooth functional of the empirical distribution function $\widehat{F}$. Here F is the true distribution function and E_F denotes expectation under F. The "usual" bootstrap estimate of this quantity is

$$E(\hat{\theta}^* \mid \mathcal{X}) = E_{\widehat{F}}\{\theta(\widehat{F}^*)\},$$

as argued as far back as Chapter 1; in this formula, $\widehat{F}^*$ is the empirical distribution function of a resample drawn from the sample. However, on the present occasion $\hat{\theta} = \hat{f}$ is not a smooth functional of $\widehat{F}$. It is, instead, a smooth functional of $\hat{f}$ (trivially!). We might therefore interpret our problem as that of estimating

$$E(\hat{\theta}) = E_f\{\theta(\hat{f})\},$$

where $\hat{\theta} = \theta(\hat{f})$ and, in this instance, E_f denotes expectation under the distribution with density f. This interpretation suggests that we estimate $E(\hat{\theta})$ by

$$E_{\hat{f}}\{\theta(\hat{f}^*)\} = \int_{-\infty}^{\infty} K(y)\, \hat{f}(x - hy)\, dy\,.$$

Thus, our estimate of bias would be

$$\hat{b}(x) = E_{\hat{f}}\{\theta(\hat{f}^*)\} - \hat{f}(x)$$

$$= \int_{-\infty}^{\infty} K(y)\, \{\hat{f}(x - hy) - \hat{f}(x)\}\, dy$$

$$= h^r\, k_1\, \hat{f}^{(r)}(x) + o_p(h^r)\,, \tag{4.80}$$

using the argument leading to (4.73).

The bootstrap estimate $\hat{b}(x)$ is identical to that which would result if we simply "plugged in" $\hat{f}$ instead of f at (4.72), in the definition of bias. That is not necessarily a good idea. The performance of $\hat{b}$ as an estimator of b is virtually identical to that of $h^r k_1 \hat{f}^{(r)}$ as an estimator of $h^r k_1 f^{(r)}$ (compare (4.73) and (4.80)), and it is well known that one should use a much larger order of bandwidth when estimating a derivative such as $f^{(r)}$ than when estimating f itself; see for example Härdle, Marron, and Wand (1990). An alternative approach is to estimate bias by using a different smoothed density, say

$$\hat{f}_{(1)}(x) = (nh_1)^{-1} \sum_{i=1}^{n} K\{(x - X_i)/h_1\},$$

constructed using a new, larger bandwidth h_1. The resulting bootstrap bias estimator is then $\hat{b}_1$ where

$$\hat{b}_1(x) = \int_{-\infty}^{\infty} K(y) \{\hat{f}_{(1)}(x - hy) - \hat{f}_{(1)}(x)\} dy.$$

The results produced by this method are very similar to those obtained from estimating $f^{(r)}$ by $\hat{f}_1^{(r)}$ in the asymptotic formula (4.73), and ignoring the remainder, giving the bootstrap estimator $\hat{b}_2$ where

$$\hat{b}_2(x) = h^r k_1 \hat{f}_{(1)}^{(r)}(x).$$

(Here $\hat{f}_{(1)}^{(r)}(x) = (\partial/\partial x)^r \hat{f}_{(1)}(x)$.) This is the explicit bias estimation method, suggested two paragraphs above.

There is yet another viewpoint, which generalizes each of the above approaches. Instead of resampling randomly from $\mathcal{X}$, as suggested earlier for the construction of $\hat{f}^*$, or resampling from the population with density $\hat{f}$, as proposed two paragraphs above, we could resample from another smoothed density, say

$$\hat{f}_{(2)}(x) = (nh_2)^{-1} \sum_{i=1}^{n} K\{(x - X_i)/h_2\},$$

where h_2 is a new, as yet unspecified bandwidth. (We could also use a different kernel at this point.) Let $\mathcal{X}^\dagger = \{X_1^\dagger, \ldots, X_n^\dagger\}$ denote a random sample drawn from the distribution with density $\hat{f}_{(2)}$, conditional on $\mathcal{X}$. Define

$$\hat{f}^\dagger(x) = (nh)^{-1} \sum_{i=1}^{n} K\{(x - X_i^\dagger)/h\},$$

where h is the original bandwidth used to construct $\hat{f}$. Take the conditional distribution of $\hat{f}^\dagger - \hat{f}_2$, given $\mathcal{X}$, as the bootstrap estimate of the distribution of $\hat{f} - f$. Bias corrections can be incorporated if required, and of course we would wish to use this technique in a percentile-t form. If we took $h_2 = 0$ then $\mathcal{X}^\dagger$ would have the same distribution as $\mathcal{X}^*$, given $\mathcal{X}$, and so the "usual" bootstrap approach described earlier (with or without bias correction) is no more than a special case of this more general prescription. Nevertheless, we shall not follow the general route, for reasons that are essentially didactic: the extra smoothing parameter h_2 obscures several of the key issues that we wish to emphasize in our analysis; and the development of methodology based on $\mathcal{X}^*$ rather than $\mathcal{X}^\dagger$ has more points of contact with the work described in Chapter 3.

Thus, the work in this section will take $\hat{f}^*$, defined in (4.76), to be "the" bootstrap form of $\hat{f}$. Recall the definition of $\mu_j(x)$ in (4.71). Calculations similar to those in (4.77) show that

$$\mathrm{var}(\hat{f}^* \mid \mathcal{X}) \; = \; (nh)^{-1} \hat{f}_2 - n^{-1} \hat{f}^2 \,,$$

where

$$\hat{f}_j(x) \; = \; (nh)^{-1} \sum_{i=1}^{n} K\{(x - X_i)/h\}^j$$

is an unbiased estimator of $\mu_j(x)$. Put

$$\hat{f}_j^*(x) \; = \; (nh)^{-1} \sum_{i=1}^{n} K\{(x - X_i^*)/h\}^j \,.$$

(In this notation, $\hat{f}_1 = \hat{f}$ and $\hat{f}_1^* = \hat{f}^*$.)

Since $\hat{f}(x)$ has variance $(nh)^{-1} \mu_2(x) - n^{-1} \mu_1(x)^2$ (see (4.74)), two pivotal versions of $\hat{f}(x)$ are

$$S \; = \; \{\hat{f}(x) - \mu_1(x)\}/\{(nh)^{-1} \mu_2(x) - n^{-1} \mu_1(x)^2\}^{1/2} \,,$$

$$T \; = \; \{\hat{f}(x) - \mu_1(x)\}/\{(nh)^{-1} \hat{f}_2(x) - n^{-1} \hat{f}(x)^2\}^{1/2} \,. \qquad (4.81)$$

The latter is Studentized, the former not. Bootstrap analogues of these quantities are

$$S^* \; = \; \{\hat{f}^*(x) - \hat{f}(x)\}/\{(nh)^{-1} \hat{f}_2(x) - n^{-1} \hat{f}(x)^2\}^{1/2} \,,$$

$$T^* \; = \; \{\hat{f}^*(x) - \hat{f}(x)\}/\{(nh)^{-1} \hat{f}_2^*(x) - n^{-1} \hat{f}^*(x)^2\}^{1/2} \,,$$

respectively. The percentile-t bootstrap takes the form of approximating the distribution of T by the conditional distribution of T^*, given $\mathcal{X}$. In this way we may construct confidence intervals for $E\hat{f}(x)$, which may then be bias-corrected to produce confidence intervals for $f(x)$.

In constructing the pivotal statistics S and T we have employed the full form of asymptotic variance, not the common asymptotic approximation given in (4.75),

$$\mathrm{var}\{\hat{f}(x)\} \; \simeq \; (nh)^{-1} k_2 f(x) \; \simeq \; (nh)^{-1} k_2 \mu_1(x) \,. \qquad (4.82)$$

Our statistic S has variance exactly equal to unity, for all n and h. Alternatively, we could define

$$S_1 = \{\hat{f}(x) - \mu_1(x)\}/\{(nh)^{-1} k_2 \mu_1(x)\}^{1/2},$$

$$T_1 = \{\hat{f}(x) - \mu_1(x)\}/\{(nh)^{-1} k_2 \hat{f}(x)\}^{1/2},$$

$$S_1^* = \{\hat{f}^*(x) - \hat{f}(x)\}/\{(nh)^{-1} k_2 \hat{f}(x)\}^{1/2},$$

$$T_1^* = \{\hat{f}^*(x) - \hat{f}(x)\}/\{(nh)^{-1} k_2 \hat{f}^*(x)\}^{1/2}.$$

Knowing the distribution of either S_1 or T_1, we can readily construct a confidence interval for $\mu_1(x)$, and then bias-correct to obtain a confidence interval for $f(x)$. The conditional distributions of S_1^* and T_1^*, given $\mathcal{X}$, may be used to approximate the unconditional distributions of S_1 and T_1, respectively.

Yet another option is to note that the size of the relative error in the approximate formula (4.82) is reduced from $O(h)$ to $O(h^2)$ (if K is symmetric) if we make instead the approximation

$$\mathrm{var}\{\hat{f}(x)\} \simeq (nh)^{-1} k_2 \mu_1(x) - n^{-1} \mu_1(x)^2.$$

This suggests replacing S_1, T_1, S_1^*, T_1^* by

$$S_2 = \{\hat{f}(x) - \mu_1(x)\}/\{(nh)^{-1} k_2 \mu_1(x) - n^{-1} \mu_1(x)^2\}^{1/2},$$

$$T_2 = \{\hat{f}(x) - \mu_1(x)\}/\{(nh)^{-1} k_2 \hat{f}(x) - n^{-1} \hat{f}(x)^2\}^{1/2},$$

$$S_2^* = \{\hat{f}^*(x) - \hat{f}(x)\}/\{(nh)^{-1} k_2 \hat{f}(x) - n^{-1} \hat{f}(x)^2\}^{1/2},$$

$$T_2^* = \{\hat{f}^*(x) - \hat{f}(x)\}/\{(nh)^{-1} k_2 \hat{f}^*(x) - n^{-1} \hat{f}^*(x)^2\}^{1/2}.$$

A third approach is to observe that since $\mathrm{var}\{\hat{f}(x)\}$ is approximately proportional to $f(x)$ then the square-root transformation is approximately variance stabilizing, and so the asymptotic distribution of

$$S_3 = (nh)^{1/2} \{\hat{f}(x)^{1/2} - \mu_1(x)^{1/2}\}$$

does not depend on unknowns. Of course, the bootstrap estimate of the distribution of S_3 is given by the conditional distribution of

$$S_3^* = (nh)^{1/2} \{\hat{f}^*(x)^{1/2} - \hat{f}(x)^{1/2}\}.$$

An approach to confidence interval construction based on S_3 is attractive, because of its simplicity. And the idea of constructing a confidence interval for $\mu_1(x)$ via a bootstrap estimate of the distribution of S_1 or S_2 is particularly interesting, since this is a rare example of a nonparametric context where the variance of $\hat{\theta}$ is (approximately) a function of the mean of $\hat{\theta}$. Here the debate over percentile and percentile-t is less relevant, although we should bootstrap S_1 or S_2 rather than $\hat{f}(x) - \mu_1(x)$. Nevertheless, we tend to favour the approach described four paragraphs earlier, based on the percentile-t bootstrap estimate of the distribution of T. This avoids errors in approximate formulae for variances, and from the point of view of this monograph is a more direct development of the methods suggested in Chapter 3.

4.4.3 Edgeworth Expansions for Density Estimators

In the classical circumstances described in Chapter 2, an Edgeworth expansion is a power series in $n^{-1/2}$. However, an Edgeworth expansion of the distribution of

$$S = \{\hat{f}(x) - \mu(x)\}/\{(nh)^{-1}\mu_2(x) - n^{-1}\mu_1(x)^2\}^{1/2}$$

is a power series in $(nh)^{-1/2}$. This is readily appreciated if we take the kernel K to be a compactly supported function, for example if K vanishes outside the interval $(-1, 1)$. Then, the number of nonzero terms in the series

$$\hat{f}(x) = (nh)^{-1} \sum_{i=1}^{n} K\{(x - X_i)/h\}$$

is a random variable of size nh, and so nh plays the role of "sample size" — or at least, is roughly proportional to "sample size." More generally, even if K is not compactly supported, the number of terms that make a significant contribution to the series defining $\hat{f}(x)$ is of size nh.

The case of the Studentized version of S,

$$T = \{\hat{f}(x) - \mu_1(x)\}/\{(nh)^{-1}\hat{f}_2(x) - n^{-1}\hat{f}(x)^2\}^{1/2},$$

is more complicated. Here the terms in an Edgeworth expansion are of sizes $(nh)^{-j/2}h^k$ for $j \geq 1$ and $k \geq 0$. This complexity arises from "interactions" between the terms $\hat{f}(x)$ and $\hat{f}_2(x)$ in the definition of T, as may be seen when calculating formulae for the cumulants of T. Cumulants of $\hat{f}(x)$ and $\hat{f}_2(x)$ may be represented individually as power series in $(nh)^{-1}$, but series expansions of cumulants of the vector $(\hat{f}(x), \hat{f}_2(x))$ contain terms of other orders. (See Section 4.2 for a definition of multivariate cumulants.)

Specific Edgeworth expansions are

$P(S \leq y)$

$$= \Phi(y) + (nh)^{-1/2}p_1(y)\phi(y) + (nh)^{-1}p_2(y)\phi(y) + \ldots, \qquad (4.83)$$

where p_j is a polynomial of degree $3j - 1$, is odd or even for even or odd j, respectively, and

$$p_1(y) = -\tfrac{1}{6} \mu_{20}^{-3/2} \mu_{30}(y^2 - 1),\tag{4.84}$$

$$p_2(y) = -\tfrac{1}{24} \mu_{20}^{-2} \mu_{40}\, y(y^2 - 3) - \tfrac{1}{72} \mu_{20}^{-3} \mu_{30}^2\, y(y^4 - 10y^2 + 15),$$

$$\mu_{ij} = h^{-1} E\Big(\big[K\{(x - X)/h\} - EK\{(x - X)/h\}\big]^i$$
$$\times \big[K\{(x - X)/h\}^2 - EK\{(x - X)/h\}^2\big]^j \Big);$$

and

$$P(T \le y) = \Phi(y) + \sum_{j \ge 1} \sum_{k=0}^{j/2} (nh)^{-j/2}\, h^k\, q_{jk}(y)\, \phi(y),$$

where q_{jk} is a polynomial of degree $3j - 1$, is odd or even for even or odd j, respectively, $k \le \tfrac{j}{2}$ means that k does not exceed the integer part of $\tfrac{j}{2}$, and

$$q_{10}(y) = \tfrac{1}{2} \mu_{20}^{-3/2} \mu_{11} - \tfrac{1}{6} \mu_{20}^{-3/2}(\mu_{30} - 3\mu_{11})(y^2 - 1),\tag{4.85}$$

$$q_{20}(y) = -\mu_{20}^{-3} \mu_{30}^2\, y - \big(\tfrac{2}{3} \mu_{20}^{-3} \mu_{30}^2 - \tfrac{1}{12} \mu_{20}^{-2} \mu_{40}\big)\, y(y^2 - 3)$$
$$- \tfrac{1}{18} \mu_{20}^{-3} \mu_{30}^3\, y(y^4 - 10y^2 + 15),$$

$$q_{11}(y) = -\mu_1 \mu_{20}^{-1/2}\, y^2.$$

(It may be shown that for $j \ge 2$, $q_{jk} \equiv 0$ unless $k \le \tfrac{j}{2} - 1$, not just $k \le \tfrac{j}{2}$.) In this notation,

$$P(T \le y) = \Phi(y) + (nh)^{-1/2}\, q_{10}(y)\, \phi(y) + (nh)^{-1}\, q_{20}(y)\, \phi(y)$$
$$+ (h/n)^{1/2}\, q_{11}(y)\, \phi(y) + O\{(nh)^{-3/2} + n^{-1}\}.\tag{4.86}$$

Derivation of (4.83) is relatively straightforward, in that it closely parallels lines developed in Chapter 2. In the remainder of this subsection we shall outline a formal derivation of (4.86), as far as a remainder of order $(nh)^{-3/2} + n^{-1}$. Details of rigour will be given in Section 5.5.

Define $s = (nh)^{1/2}$,

$$\hat{f}_j(x) = (nh)^{-1} \sum_{i=1}^{n} K\{(x - X_i)/h\}^j, \qquad \mu_j(x) = E\{\hat{f}_j(x)\},$$

$$\sigma(x)^2 = \mu_2(x) - h\mu_1(x)^2 = \mu_{20}(x), \qquad \hat{\sigma}(x)^2 = \hat{f}_2(x) - h\hat{f}(x)^2,$$

$$\Delta_j(x) = s\{\hat{f}_j(x) - \mu_j(x)\}/\sigma(x)^j, \qquad \rho(x) = \mu_1(x)/\sigma(x).$$

Dropping the argument x, we have

$$\hat{\sigma}^2 = \sigma^2\{1 + s^{-1} \Delta_2 - 2hs^{-1} \rho\Delta_1 + O_p(n^{-1})\},$$

whence

$$\hat{\sigma}^{-1} = \sigma^{-1}\left\{1 - \tfrac{1}{2}s^{-1}\Delta_2 + \tfrac{3}{8}s^{-2}\Delta_2^2 + hs^{-1}\rho\Delta_1 + O_p(s^{-3} + n^{-1})\right\}.$$

Therefore,

$$
\begin{aligned}
T &= \Delta_1 \sigma \hat{\sigma}^{-1} \\
&= \Delta_1\left(1 - \tfrac{1}{2}s^{-1}\Delta_2 + \tfrac{3}{8}s^{-2}\Delta_2^2 + hs^{-1}\rho\Delta_1\right) + O_p(s^{-3} + n^{-1}).
\end{aligned}
$$

Raising T to a power, taking expectations, and ignoring terms of order $s^{-3} + n^{-1}$ or smaller, we deduce that

$$
\begin{aligned}
E(T) &= -\tfrac{1}{2}s^{-1}E(\Delta_1\Delta_2) + hs^{-1}\rho E(\Delta_1^2), \\
E(T^2) &= E(\Delta_1^2) - s^{-1}E(\Delta_1^2\Delta_2) + s^{-2}E(\Delta_1^2\Delta_2^2), \\
E(T^3) &= E(\Delta_1^3) - \tfrac{3}{2}s^{-1}E(\Delta_1^3\Delta_2) + 3hs^{-1}\rho E(\Delta_1^4), \\
E(T^4) &= E(\Delta_1^4) - 2s^{-1}E(\Delta_1^4\Delta_2) + 3s^{-2}E(\Delta_1^4\Delta_2^2).
\end{aligned}
$$

Now,

$$
\begin{aligned}
E(\Delta_1^2) &= 1, & E(\Delta_1\Delta_2) &= \sigma^{-3}\mu_{11}, \\
E(\Delta_1^3) &= s^{-1}\sigma^{-3}\mu_{30}, & E(\Delta_1^2\Delta_2) &= s^{-1}\sigma^{-4}\mu_{21}, \\
E(\Delta_1^4) &= \sigma^{-4}(3\mu_{20}^2 + s^{-2}\mu_{40}) + O(n^{-1}), \\
E(\Delta_1^3\Delta_2) &= \sigma^{-5}3\mu_{20}\mu_{11} + O(s^{-2}), \\
E(\Delta_1^2\Delta_2^2) &= \sigma^{-6}(2\mu_{11}^2 + \mu_{20}\mu_{02}) + O(s^{-2}), \\
E(\Delta_1^4\Delta_2) &= s^{-1}\sigma^{-6}(4\mu_{11}\mu_{30} + 6\mu_{20}\mu_{21}) + O(s^{-3}), \\
E(\Delta_1^4\Delta_2^2) &= \sigma^{-8}(3\mu_{20}^2\mu_{02} + 12\mu_{11}^2\mu_{20}) + O(s^{-2}).
\end{aligned}
$$

Using these formulae, noting that $\sigma^2 = \mu_{20}$, ignoring terms of order $s^{-3} + n^{-1}$ or smaller, and employing identities expressing cumulants in terms of moments (see (2.6) in Chapter 2), we see that the first four cumulants of T are

$$\kappa_1(T) = -\tfrac{1}{2}s^{-1}\mu_{20}^{-3/2}\mu_{11} + hs^{-1}\rho, \tag{4.87}$$

$$\kappa_2(T) = 1 + s^{-2}\left(\tfrac{7}{4}\mu_{20}^{-3}\mu_{11}^2 + \mu_{20}^{-2}\mu_{02} - \mu_{20}^{-2}\mu_{21}\right),$$

$$\kappa_3(T) = s^{-1}\mu_{20}^{-3/2}(\mu_{30} - 3\mu_{11}) + 6hs^{-1}\rho, \tag{4.88}$$

$$\kappa_4(T) = s^{-2}\left(\mu_{20}^{-2}\mu_{40} - 6\mu_{20}^{-3}\mu_{11}\mu_{30}\right.$$
$$\left. - 6\mu_{20}^{-2}\mu_{21} + 18\mu_{20}^{-3}\mu_{11}^2 + 3\mu_{20}^{-2}\mu_{02}\right).$$

The cumulants $\kappa_j(T)$ for $j \geq 5$ are of order s^{-3} or smaller, as follows from a version of Theorem 2.1 for this setting.

If the nonnegative integers i, j, k, l satisfy $i + 2j = k + 2l$ then $\mu_{ij} - \mu_{kl} = O(h)$. Therefore, ignoring terms of order $s^{-3} + n^{-1}$ or smaller,

$$\kappa_2(T) = 1 + s^{-2}\tfrac{7}{2}\mu_{20}^{-3}\mu_{30}^{2}, \tag{4.89}$$

$$\kappa_4(T) = s^{-2}\left(12\mu_{20}^{-3}\mu_{30}^{2} - 2\mu_{20}^{-2}\mu_{40}\right). \tag{4.90}$$

Combining (4.87)–(4.90), we see that the characteristic function of T satisfies

$$E(e^{itT})$$

$$= e^{-t^2/2}\exp\left[s^{-1}\left\{-\tfrac{1}{2}\mu_{20}^{-3/2}\mu_{11}it + h\rho it\right.\right.$$
$$\left.+ \tfrac{1}{6}\mu_{20}^{-3/2}(\mu_{30} - 3\mu_{11})(it)^3 + h\rho(it)^3\right\}$$
$$\left.+ s^{-2}\left\{\tfrac{7}{8}\mu_{20}^{-3}\mu_{30}^{2}(it)^2 + (\tfrac{1}{2}\mu_{20}^{-3}\mu_{30}^{2} - \tfrac{1}{12}\mu_{20}^{-2}\mu_{40})(it)^4\right\}\right]$$
$$+ O(s^{-3} + n^{-1})$$

$$= e^{-t^2/2}\left[1 + s^{-1}\left\{-\tfrac{1}{2}\mu_{20}^{-3/2}\mu_{11}it\right.\right.$$
$$\left.+ \tfrac{1}{6}\mu_{20}^{-3/2}(\mu_{30} - 3\mu_{11})(it)^3\right\} + hs^{-1}\rho\{it + (it)^3\}$$
$$+ s^{-2}\left\{\mu_{20}^{-3}\mu_{30}^{2}(it)^2 + (\tfrac{2}{3}\mu_{20}^{-3}\mu_{30}^{2} - \tfrac{1}{12}\mu_{20}^{-2}\mu_{40})(it)^4\right.$$
$$\left.\left.+ \tfrac{1}{18}\mu_{20}^{-3}\mu_{30}^{2}(it)^6\right\}\right]$$
$$+ O(s^{-3} + n^{-1}).$$

The desired Edgeworth expansion (4.86) for $P(T \leq y)$ follows, in a formal sense, on inverting the above Fourier-Stieltjes transform.

4.4.4 Bootstrap Versions of Edgeworth Expansions

Recall from subsection 4.4.2 that the distributions of S and T may be approximated by the distributions of S^* and T^* conditional on $\mathcal{X}$ where

$$S^* = \{\hat{f}^*(x) - \hat{f}(x)\}/\{(nh)^{-1}\hat{f}_2(x) - n^{-1}\hat{f}(x)^2\}^{1/2},$$

$$T^* = \{\hat{f}^*(x) - \hat{f}(x)\}/\{(nh)^{-1}\hat{f}_2^*(x) - n^{-1}\hat{f}(x)^2\}^{1/2}.$$

Those approximants also admit Edgeworth expansions, much as described in the more conventional setting of Chapter 3 (Section 3.3). As there, terms in these bootstrap versions of Edgeworth expansions are obtained from their counterparts in the cases of S and T by replacing expectations with respect to F (the distribution of X) by conditional expectations with respect to $\widehat{F}$ (the empirical distribution function of the sample $\mathcal{X}$). In particular, the analogue of μ_{ij} is

$$
\hat{\mu}_{ij} = h^{-1} E\left(\left[K\{(x-X^*)/h\} - E\{K((x-X^*)/h) \mid \mathcal{X}\} \right]^i \right.
$$
$$
\left. \times \left[K\{(x-X^*)/h\}^2 - E\{K((x-X^*)/h)^2 \mid \mathcal{X}\} \right]^j \mid \mathcal{X} \right).
$$

Define $\hat{p}_i$, $\hat{q}_{ij}$ in terms of p_i, q_{ij} respectively, on replacing μ_{kl}, μ_k by $\hat{\mu}_{kl}$, $\hat{\mu}_k$ in the latter. Then, for example,

$$
P(S^* \le y \mid \mathcal{X})
$$
$$
= \Phi(y) + (nh)^{-1/2}\,\hat{p}_1(y)\,\phi(y) + (nh)^{-1}\,\hat{p}_2(y)\,\phi(y)
$$
$$
+ O_p\{(nh)^{-3/2}\}, \tag{4.91}
$$

$$
P(T^* \le y \mid \mathcal{X})
$$
$$
= \Phi(y) + (nh)^{-1/2}\,\hat{q}_{10}(y)\,\phi(y) + (nh)^{-1}\,\hat{q}_{20}(y)\,\phi(y)
$$
$$
+ (h/n)^{1/2}\,\hat{q}_{11}(y)\,\phi(y) + O_p\{(nh)^{-3/2}+n^{-1}\}. \tag{4.92}
$$

The differences $\hat{\mu}_i - \mu_i$, $\hat{\mu}_{ij} - \mu_{ij}$ are of size $(nh)^{-1/2}$, in the sense that both have mean square of order $(nh)^{-1}$; this may be shown using a version of the argument leading to (4.75). It follows that $\hat{p}_i - p_i$, $\hat{q}_{ij} - q_{ij}$ are of size $(nh)^{-1/2}$. Therefore, subtracting expansions (4.83) and (4.91), and also (4.86) and (4.92), we obtain

$$
P(S \le y) - P(S^* \le y \mid \mathcal{X}) = \Phi(y) + O_p\{(nh)^{-1}\},
$$
$$
P(T \le y) - P(T^* \le y \mid \mathcal{X}) = \Phi(y) + O_p\{(nh)^{-1}\}.
$$

These formulae make it clear that in the case of nonparametric density estimation, the bootstrap approximation to the distribution of a pivotal statistic (both S and T are pivotal) is in error by $O_p\{(nh)^{-1}\}$. This rate of convergence is precise, since the convergence rates of $\hat{p}_1 - p_1$ and $\hat{q}_{10} - q_{10}$ to zero are precisely $(nh)^{-1/2}$; indeed, central limit theorems may be proved for each of $(nh)^{1/2}(\hat{p}_1 - p_1)$ and $(nh)^{1/2}(\hat{q}_{10} - q_{10})$. It contrasts with the convergence rate of n^{-1} in simpler problems; see Section 3.1.

On the other hand, the bootstrap approximation to the distribution of the nonpivotal statistic

$$
U = (nh)^{1/2}\{\hat{f}(x) - E\hat{f}(x)\}
$$

is in error by $(nh)^{-1/2}$. To appreciate why, note that $P(U \le y)$ is approximated by $P(U^* \le y \mid \mathcal{X})$, where

$$
U^* = (nh)^{1/2}\{\hat{f}^*(x) - \hat{f}(x)\},
$$

and that by (4.83) and (4.91),

$$P(U \le y) = P(S \le y/\sigma)$$
$$= \Phi(y/\sigma) + (nh)^{-1/2} p_1(y/\sigma) \phi(y/\sigma) + O\{(nh)^{-1}\},$$

$$P(U^* \le y \mid \mathcal{X}) = P(S^* \le y/\hat{\sigma} \mid \mathcal{X})$$
$$= \Phi(y/\hat{\sigma}) + (nh)^{-1/2} \hat{p}_1(y/\hat{\sigma}) \phi(y/\hat{\sigma}) + O_p\{(nh)^{-1}\}.$$

Subtracting, and noting that

$$\hat{p}_1 - p_1 = O_p\{(nh)^{-1/2}\} \qquad \text{and} \qquad \hat{\sigma} - \sigma = O_p\{(nh)^{-1/2}\},$$

we conclude that

$$P(U \le y) - P(U^* \le y \mid \mathcal{X}) = \Phi(y/\sigma) - \Phi(y/\hat{\sigma}) + O_p\{(nh)^{-1}\}.$$

Since $\hat{\sigma} - \sigma$ is of precise order $(nh)^{-1/2}$, in that $(nh)^{1/2}(\hat{\sigma} - \sigma)$ satisfies a central limit theorem, then so is $\Phi(y/\sigma) - \Phi(y/\hat{\sigma})$, and hence also

$$P(U \le y) - P(U^* \le y \mid \mathcal{X}),$$

as had to be shown. (This argument is no more than a version for density estimators of the argument surrounding (3.4).)

4.4.5 Confidence Intervals for $E\hat{f}(x)$

We begin with a little notation. Define $\hat{u}_\alpha$, v_α, $\hat{v}_\alpha$ by

$$P(S^* \le \hat{u}_\alpha \mid \mathcal{X}) = P(T \le v_\alpha) = P(T^* \le \hat{v}_\alpha \mid \mathcal{X}) = \alpha.$$

(Of course, all three quantities depend on x.) Put $\hat{\sigma}^2(x) = \hat{f}_2(x) - h\hat{f}(x)^2$. One-sided percentile and percentile-t confidence intervals for $E\hat{f}(x)$, with nominal coverage α, are given by

$$\widehat{I}_1 = \left(-\infty, \ \hat{f}(x) - (nh)^{-1/2} \hat{\sigma}(x) \hat{u}_{1-\alpha} \right),$$

$$\widehat{J}_1 = \left(-\infty, \ \hat{f}(x) - (nh)^{-1/2} \hat{\sigma}(x) \hat{v}_{1-\alpha} \right),$$

respectively. Each may be regarded as an approximation to the "ideal" interval

$$J_1 = \left(-\infty, \ \hat{f}(x) - (nh)^{-1/2} \hat{\sigma}(x) v_{1-\alpha} \right),$$

whose coverage probability is precisely α.

In view of the properties of Edgeworth expansions noted in subsections 4.4.3 and 4.4.4, we should modify our definition of second-order correctness. Hitherto we have interpreted second-order correctness as requiring agreement between endpoints of confidence intervals up to and including terms of second order in $n^{-1/2}$. However, as pointed out in subsection 4.4.3, the role of n here is played by nh. Therefore we should impose the weaker condition that the endpoints agree in terms of order $\{(nh)^{-1/2}\}^2 = (nh)^{-1}$. With this change, it may be deduced that the percentile-t confidence interval $\widehat{J}_1$ is second-order correct relative to J_1, but that the percentile method interval $\widehat{I}_1$ is only first-order correct.

These results follow easily from Cornish-Fisher expansions, obtained by inverting the Edgeworth expansions (4.86), (4.91), and (4.92),

$$
\begin{aligned}
v_\alpha \;=\;& z_\alpha - (nh)^{-1/2} q_{10}(z_\alpha) \\
& - (nh)^{-1}\Big\{ q_{20}(z_\alpha) + \tfrac{1}{2} z_\alpha q_{10}(z_\alpha) - q_{10}(z_\alpha) q_{10}'(z_\alpha) \Big\} \\
& - (h/n)^{1/2} q_{11}(z_\alpha) + O\{(nh)^{-3/2} + n^{-1}\}, \\[4pt]
\hat{u}_\alpha \;=\;& z_\alpha - (nh)^{-1/2} \hat{p}_1(z_\alpha) \\
& - (nh)^{-1}\Big\{ \hat{p}_2(z_\alpha) + \tfrac{1}{2} z_\alpha \hat{p}_1(z_\alpha) - \hat{p}_1(z_\alpha)\hat{p}_1'(z_\alpha) \Big\} \\
& + O_p\{(nh)^{-3/2}\}, \\[4pt]
\hat{v}_\alpha \;=\;& z_\alpha - (nh)^{-1/2} \hat{q}_{10}(z_\alpha) \\
& - (nh)^{-1}\Big\{ \hat{q}_{20}(z_\alpha) + \tfrac{1}{2} z_\alpha \hat{q}_{10}(z_\alpha) - \hat{q}_{10}(z_\alpha)\hat{q}_{10}'(z_\alpha) \Big\} \\
& - (h/n)^{1/2} \hat{q}_{11}(z_\alpha) + O_p\{(nh)^{-3/2} + n^{-1}\}. \qquad (4.93)
\end{aligned}
$$

Here, the polynomials $\hat{p}_j$, q_{jk}, and $\hat{q}_{jk}$ are as defined in subsections 4.4.3 and 4.4.4. Noting that p_1 and q_{10} have distinctly different forms (compare (4.84) and (4.85)), and that

$$
\hat{p}_i - p_i \;=\; O_p\{(nh)^{-1/2}\} \qquad \text{and} \qquad \hat{q}_{ij} - q_{ij} \;=\; O_p\{(nh)^{-1/2}\},
$$

we see that v_α and $\hat{v}_\alpha$ agree in terms of order $(nh)^{-1/2}$, whereas v_α and $\hat{u}_\alpha$ differ in terms of order $(nh)^{-1/2}$. It follows immediately that $\widehat{J}_1$ is second-order correct relative to J_1, but that $\widehat{I}_1$ is not second-order correct for J_1.

These properties of the endpoints of confidence intervals are reflected in the coverage accuracy of one-sided confidence intervals for $E\hat{f}(x)$, just as they were in the more conventional cases discussed in Section 3.5. In particular, the second-order correct interval $\widehat{J}_1$ has coverage error of order $(nh)^{-1}$, whereas the coverage of the first-order correct interval $\widehat{I}_1$ is in error by $(nh)^{-1/2}$, owing to the fact that it fails to adequately take into account skewness of $\hat{f}(x)$. More concise expansions of coverage are given by

$$P\{E\hat{f}(x) \in \hat{I}_1\} = \alpha - (nh)^{-1/2} \tfrac{1}{2} \gamma_1(x) z_\alpha^2 \phi(z_\alpha)$$
$$+ O\{(nh)^{-1} + (h/n)^{1/2}\}, \qquad (4.94)$$

$$P\{E\hat{f}(x) \in \hat{J}_1\} = \alpha - (nh)^{-1} \tfrac{1}{6} \gamma(x) z_\alpha (2z_\alpha^2 + 1) \phi(z_\alpha)$$
$$+ O\{(nh)^{-3/2} + n^{-1}\}, \qquad (4.95)$$

where

$$\gamma_1(x) = \mu_{20}(x)^{-3f} \mu_{30}(x)$$
$$= f(x)^{-1/2} \left(\int K^2 \right)^{-4f} \int K^3 + o(1),$$

$$\gamma_2(x) = \mu_{20}(x)^{-2} \mu_{40}(x) - \tfrac{3}{2} \mu_{20}(x)^{-3} \mu_{30}(x)^2$$
$$= f(x)^{-1} \left\{ \left(\int K^2 \right)^{-2} \int K^4 - \tfrac{3}{2} \left(\int K^2 \right)^{-3} \left(\int K^3 \right)^2 \right\} + o(1).$$

Related formulae for coverage probabilities of two-sided confidence intervals may be obtained by similar arguments, as we now outline. Define $\xi = \tfrac{1}{2}(1 + \alpha)$ and

$$\hat{I}_2 = \hat{I}_1(\xi) \backslash \hat{I}_1(1 - \xi)$$
$$= \left(\hat{f}(x) - (nh)^{-1/2} \hat{\sigma}(x) \hat{u}_\xi, \ \ \hat{f}(x) - (nh)^{-1/2} \hat{\sigma}(x) \hat{u}_{1-\xi} \right),$$

$$\hat{J}_2 = \hat{J}_1(\xi) \backslash \hat{J}_1(1 - \xi)$$
$$= \left(\hat{f}(x) - (nh)^{-1/2} \hat{\sigma}(x) \hat{v}_\xi, \ \ \hat{f}(x) - (nh)^{-1/2} \hat{\sigma}(x) \hat{v}_{1-\xi} \right).$$

Then by (4.95),

$$P\{E\hat{f}(x) \in \hat{J}_2\} = P\{E\hat{f}(x) \in \hat{J}_1(\xi)\} - P\{E\hat{f}(x) \in \hat{J}_1(1 - \xi)\}$$
$$= \alpha - (nh)^{-1} \tfrac{1}{3} \gamma(x) z_\xi (2z_\xi^2 + 1) \phi(z_\xi)$$
$$+ O\{(nh)^{-3/2} + n^{-1}\}. \qquad (4.96)$$

A more detailed account, passing to additional terms in formula (4.95) and noting the parity properties of polynomials in Edgeworth expansions, shows that the remainder term $O\{(nh)^{-3/2} + n^{-1}\}$ in (4.96) is actually $O\{(nh)^{-2} + n^{-1}\}$.

The coverage probability of the two-sided interval $\hat{I}_2$ may be obtained by a similar argument, first passing to a longer expansion than (4.94), which includes terms of sizes $(nh)^{-1/2}$, $(nh)^{-1}$, $(nh)^{-3/2}$, and $(h/n)^{1/2}$, and a remainder of order $(nh)^{-2} + n^{-1}$. The coefficients of these four terms are even, odd, even, and even polynomials, respectively, multiplied by ϕ. Therefore, the terms of sizes $(nh)^{-1/2}$, $(nh)^{-3/2}$, and $(h/n)^{1/2}$ cancel from the formula

$$P\{E\hat{f}(x) \in \widehat{I_2}\} = P\{E\hat{f}(x) \in \widehat{I_1}(\xi)\} - P\{E\hat{f}(x) \in \widehat{I_1}(1-\xi)\}.$$

It follows that

$$P\{E\hat{f}(x) \in \widehat{I_2}\} = \alpha + O\{(nh)^{-1} + n^{-1}\},$$

which is the same order of magnitude of error derived for the interval $\widehat{J_2}$ in the previous paragraph.

We conclude this subsection by outlining the derivation of (4.95). (A proof of (4.94) is similar but simpler.) Define

$$s = (nh)^{1/2}, \qquad \sigma(x)^2 = \mu_2(x) - h\mu_1(x)^2 = \mu_{20}(x),$$

$$\Delta_j(s) = s\{\hat{f}_j(x) - \mu_j(x)\}/\sigma(x)^j,$$

$$\Delta(x) = \Delta_3(x) - \tfrac{3}{2}\mu_3(x)\,\sigma(x)^{-3}\,\Delta_2(x).$$

Routine Taylor expansion gives

$$\hat{\mu}_{20}^{-3/2}\hat{\mu}_{11} - \mu_{20}^{-3/2}\mu_{11} = s^{-1}\Delta + O_p(s^{-2} + hs^{-1}),$$

$$\hat{\mu}_{20}^{-3/2}\hat{\mu}_{30} - \mu_{20}^{-3/2}\mu_{30} = s^{-1}\Delta + O_p(s^{-2} + hs^{-1}),$$

where here and below we drop the argument x. We may now deduce from (4.85) that

$$\hat{q}_{10}(y) - q_{10}(y) = s^{-1}\tfrac{1}{6}(2y^2 + 1)\Delta + O_p(s^{-2} + hs^{-1}).$$

Therefore, by (4.93),

$$\hat{v}_\alpha - v_\alpha = -s^{-1}\{\hat{q}_{10}(z_\alpha) - q_{10}(z_\alpha)\} + O_p(s^{-3} + n^{-1})$$

$$= -s^{-2}\tfrac{1}{6}(2y^2 + 1)\Delta + O_p(s^{-3} + n^{-1}),$$

whence

$$P(T \le \hat{v}_\alpha) = P\Big\{T + s^{-2}\tfrac{1}{6}(2z_\alpha^2 + 1)\Delta \le v_\alpha + O_p(s^{-3} + n^{-1})\Big\}$$

$$= P(T' \le v_\alpha) + O(s^{-3} + n^{-1}), \qquad (4.97)$$

using the delta method, where $T' = T + s^{-2}\tfrac{1}{6}(2z_\alpha^2 + 1)\Delta$. All but the second of the first four cumulants of T and T' are identical up to (but not including) terms of order $s^{-3} + n^{-1}$, as may be shown using the argument in subsection 3.5.2. Furthermore,

$$\mathrm{var}(T') = \mathrm{var}(T) + s^{-2}\tfrac{1}{3}(2z_\alpha^2 + 1)\gamma_2(x) + O(s^{-4} + n^{-1}).$$

Therefore, comparing Edgeworth expansions for T and T',

$$P(T' \leq y) \ = \ P(T \leq y) \ - \ s^{-2}\tfrac{1}{6}\,(2z_\alpha^2 + 1)\,\gamma_2(x)\,y\phi(y) \ + \ O(s^{-3} + n^{-1}).$$

(This is a direct analogue of (3.36), with essentially the same proof.)

Taking $y = v_\alpha = z_\alpha + O(s^{-1})$, noting that $P(T \leq v_\alpha) = \alpha$, and using (4.97), we deduce that

$$P(T \leq \hat{v}_\alpha) \ = \ \alpha \ - \ s^{-2}\tfrac{1}{6}\,z_\alpha(2z_\alpha^2 + 1)\,\gamma_2(x)\,\phi(z_\alpha) \ + \ O(s^{-3} + n^{-1}).$$

Hence

$$P\{E\hat{f}(x) \in \widehat{J}_1\} \ = \ P(T > \hat{v}_{1-\alpha})$$
$$= \ 1 - P(T \leq \hat{v}_{1-\alpha})$$
$$= \ \alpha \ - \ s^{-2}\tfrac{1}{6}\,z_\alpha(2z_\alpha^2 + 1)\,\gamma_2(x)\,\phi(z_\alpha) \ + \ O(s^{-3} + n^{-1}),$$

which establishes the required result (4.95).

4.4.6 Confidence Intervals for $f(x)$

We take as our starting point the percentile-t confidence intervals $\widehat{J}_1$ and $\widehat{J}_2$ for $E\hat{f}(x)$, introduced in subsection 4.4.5. The idea is to either correct them explicitly for bias, for example by using the asymptotic bias formula (4.73), or to undersmooth them, by appropriate choice of h, to such an extent that they may reasonably be regarded as confidence intervals for $f(x)$ as well as $E\hat{f}(x)$. (As explained in subsection 4.4.2, undersmoothing of $\hat{f}(x)$ reduces the impact of bias.) In the present subsection we briefly outline the effect that each of these approaches to bias correction has on coverage accuracy.

Recall formula (4.73) for bias,

$$b(x) \ = \ E\{\hat{f}(x)\} - f(x) \ = \ h^r\,k_1\,f^{(r)}(x) + o(h^r),$$

as $h \to 0$. We may estimate $f^{(r)}$ by differentiating another kernel estimator, say

$$\tilde{f}(x) \ = \ (nh_1)^{-1} \sum_{i=1}^{n} L\{(x - X_i)/h_1\}, \tag{4.98}$$

where L is an sth order kernel for some $s \geq 1$, and h_1 is a new bandwidth. (Development of theory in this context demands that we assume $r + s$ derivatives of f.) Thus, $\tilde{b}(x) \ = \ h^r\,k_1\,\tilde{f}^{(r)}(x)$ is one form of explicit bias estimator.

Let $\hat{b}(x)$ be a general bias estimator; we allow the possibility $\hat{b}(x) = 0$, which would be appropriate in the case of undersmoothing. In subsection 4.4.5 we regarded $\widehat{J}_1$ and $\widehat{J}_2$ as confidence intervals for $E\hat{f}(x)$. To convert them into confidence intervals for $f(x)$, we simply translate them to the left by an amount $\hat{b}(x)$, obtaining the new intervals

$$\widehat{J}_{b1} = \left(-\infty, \; \hat{f}(x) - \hat{b}(x) - (nh)^{-1/2}\,\hat{\sigma}(x)\,\hat{v}_{1-\alpha} \right),$$

$$\widehat{J}_{b2} = \left(\hat{f}(x) - \hat{b}(x) - (nh)^{-1/2}\,\hat{\sigma}(x)\,\hat{v}_{\xi}, \; \hat{f}(x) - \hat{b}(x) - (nh)^{-1/2}\,\hat{\sigma}(x)\,\hat{v}_{1-\xi} \right)$$

where $\xi = \frac{1}{2}(1+\alpha)$.

The coverage probabilities are given by

$$\beta_1(\alpha) = P\{f(x) \in \widehat{J}_{b1}\} = P(T > \hat{v}_{1-\alpha} - \delta),$$

where

$$T = T(x) = (nh)^{1/2}\{\hat{f}(x) - E\hat{f}(x)\}/\hat{\sigma}(x)$$

and

$$\delta = \delta(x) = (nh)^{1/2}\{\hat{b}(x) - b(x)\}/\hat{\sigma}(x),$$

and by

$$\beta_2(\alpha) = P\{f(x) \in \widehat{J}_{b2}\} = \beta_1(\xi) - \beta_1(1-\xi). \tag{4.99}$$

These coverage probabilities admit Edgeworth expansions, for example,

$$\begin{aligned}
\beta_1(\alpha) &= \alpha + (nh)^{-1}\,w_1(\alpha) + (h/h_1)^{r+1}\,w_2(\alpha) + \eta^2\,w_3(\alpha) \\
&\quad + \eta(nh)^{-1/2}\,w_4(\alpha) + \eta w_5(\alpha) \\
&\quad + o\{(nh)^{-1} + (h/h_1)^{r+1} + \eta^2\},
\end{aligned} \tag{4.100}$$

where $\eta = (nh)^{1/2}\{E\hat{b}(x) - b(x)\}/\sigma(x)$, $\sigma(x)^2 = \mu_2(x) - h\mu_1(x)^2$, and each w_i may be written in the form $r_i(z_\alpha)\,\phi(z_\alpha)$ where r_i is a polynomial. The function r_i is odd for $i = 1, \ldots, 4$ and even for $i = 5$, and the coefficients of r_i depend on x only in the cases $i = 1, 4$. Further details are given in Hall (1991b).

In view of the parity properties of the r_i's we may deduce from (4.99) and (4.100) that

$$\begin{aligned}
\beta_2(\alpha) &= \alpha + 2\Big\{(nh)^{-1}\,w_1(\xi) + (h/h_1)^{r+1}\,w_2(\xi) \\
&\quad\quad\quad\quad + \eta^2\,w_3(\xi) + \eta(nh)^{-1/2}\,w_4(\xi)\Big\} \\
&\quad + o\{(nh)^{-1} + (h/h_1)^{r+1} + \eta^2\}.
\end{aligned} \tag{4.101}$$

It is understood that if the bootstrap estimate $\hat{b}(x)$ is identically zero then we drop the terms involving h_1 in both (4.100) and (4.101). (That is tantamount to taking $h_1 = \infty$ in those terms.)

It will be appreciated from (4.100) and (4.101) that the way in which bias estimation influences coverage accuracy can be highly complex. Let us first treat the simplest case, where $\hat{b}(x) = 0$. Then we should take $h_1 = \infty$ and

$$
\begin{aligned}
\eta &= (nh)^{1/2}\left\{E\hat{b}(x) - b(x)\right\}/\sigma(x) \\
&= -(nh)^{1/2} b(x)/\sigma(x) \\
&\sim (nh)^{1/2} h^r a_1(x),
\end{aligned}
$$

where $a_1(x) = -f^{(r)}(x) f(x)^{-1/2} k_1 k_2^{-1/2}$; note (4.73) and (4.75). Therefore,

$$
\begin{aligned}
\beta_1(\alpha) &= \alpha + (nh)^{-1} w_1(\alpha) + (nh)^{1/2} h^r a_1(x) w_5(\alpha) \\
&\quad + o\left\{(nh)^{-1} + (nh)^{1/2} h^r\right\}, \\
\beta_2(\alpha) &= \alpha + 2\left\{(nh)^{-1} w_1(\xi) + nh^{2r+1} a_1(x)^2 w_3(\xi) + h^r a_1(x) w_4(\xi)\right\} \\
&\quad + o\left\{(nh)^{-1} + nh^{2r+1}\right\}.
\end{aligned}
$$

It follows from these formulae that the values of h that minimize the coverage errors $|\beta_1(\alpha) - \alpha|$ and $|\beta_2(\alpha) - \alpha|$ are of sizes $n^{-3/(2r+3)}$ and $n^{-1/(r+1)}$, respectively. The resulting minimum values of $|\beta_1(\alpha) - \alpha|$ and $|\beta_2(\alpha) - \alpha|$ are of orders $n^{-2r/(2r+3)}$ and $n^{-r/(r+1)}$, respectively, except that on some occasions (depending on the values of the weights w_i, and hence on f, x and the kernel K) it may be possible to choose the constants c_1 and c_2 in the formulae $h \sim c_1 n^{-3/(2r+3)}$ and $h \sim c_2 n^{-1/(r+1)}$ such that $\beta_1(\alpha) - \alpha = o(n^{-2/(2r+3)})$ and $\beta_2(\alpha) - \alpha = o(n^{-r/(r+1)})$, respectively.

We now turn to the case where bias is estimated explicitly, by

$$
\tilde{b}(x) = h^r k_1 \tilde{f}^{(r)}(x),
$$

where $\tilde{f}$ is given by (4.98). Assuming L to be an sth order kernel, and f to have $r + s$ continuous derivatives, we have

$$
E\tilde{f}^{(r)}(x) - f^{(r)}(x) \sim h_1^s l f^{(r+s)}(x),
$$

where $l = (-1)^s (1/s!) \int y^s K(y)\, dy$. (Compare (4.73).) This suggests taking $\eta \sim (nh)^{1/2} h^r h_1^s a_2(x)$, where $a_2(x) = f^{(r+s)}(x) f(x)^{-1/2} k_2^{-1/2} l$. In this event we see from (4.100) and (4.101) that

$$
\begin{aligned}
\beta_1(\alpha) &= \alpha + (nh)^{-1} w_1(\alpha) + (h/h_1)^{r+1} w_2(\alpha) \\
&\quad + (nh)^{1/2} h^r h_1^s a_2(x) w_5(\alpha) \\
&\quad + o\left\{(nh)^{-1} + (h/h_1)^{r+1} + (nh)^{1/2} h^r h_1^s\right\},
\end{aligned}
$$

$$\beta_2(\alpha) = \alpha + 2\Big\{ (nh)^{-1} w_1(\xi) + (h/h_1)^{r+1} w_2(\xi)$$

$$+ nh^{2r+1} h_1^{2s} a_2(x)^2 w_3(\xi) + h^r h_1^s a_2(x) w_4(\xi)\Big\}$$

$$+ o\Big\{ (nh)^{-1} + (h/h_1)^{r+1} + nh^{2r+1} h_1^{2s}\Big\}.$$

The values of h, h_1 minimizing $|\beta_1(\alpha) - \alpha|$ are of size

$$n^{-\{3(r+1)+2s\}/\{(r+1)(2r+3)+2(r+2)s\}},$$

$$n^{-(r+3)/\{(r+1)(2r+3)+2(r+2)s\}}, \tag{4.102}$$

respectively, while the values of h, h_1 minimizing $|\beta_2(\alpha) - \alpha|$ are of size

$$n^{-(r+s+1)/\{(r+1)^2+s(r+2)\}},$$

$$n^{-1/\{(r+1)^2+s(r+2)\}}, \tag{4.103}$$

respectively. In general, the resulting minimum values of $|\beta_1(\alpha) - \alpha|$ and $|\beta_2(\alpha) - \alpha|$ are of orders

$$n^{-2(r+1)(r+s)/\{(r+1)(2r+3)+2(r+2)s\}},$$

$$n^{-(r+1)(r+s)/\{(r+1)^2+s(r+2)\}},$$

respectively, although on occasion judicious choice of the constants in formulae for h and h_1 will allow these orders to be reduced.

The bandwidth h that would optimize the performance of $\hat{f}(x)$ as a point estimate of $f(x)$ is of size $n^{-1/(2r+1)}$ (see, e.g., Rosenblatt 1971). On each occasion in the discussion above, the bandwidth that minimizes coverage error is of smaller order than $n^{-1/(2r+1)}$. (There are four instances, with h of size $n^{-3/(2r+3)}$ or $n^{-1/(r+1)}$ in the case where we take $\hat{b}(x) = 0$, and of a size given by the values in (4.102) or (4.103) in the case of explicit bias estimation.) This makes it clear that, in a variety of different approaches to confidence interval construction, a significantly smaller bandwidth should be used to construct $\hat{f}(x)$ than would be appropriate in the point estimation problem. Bandwidths of smaller order than usual should be employed when bias is estimated by the undersmoothing method, and even explicit bias estimation calls for a degree of undersmoothing.

In assessing which of the two approaches to bias correction — undersmoothing (U) or explicit bias estimation (EBE) — is to be preferred, one should assume a given number of derivatives of the unknown density f. If that number is d then one should take $r = d$ in the case of method U, and

$r + s = d$ in the case of method EBE. Now, the minimum orders of coverage error of one-sided confidence intervals are

$$n^{-2r/(2r+3)} \qquad \text{and} \qquad n^{-2(r+1)(r+s)/\{(r+1)(2r+3)+2(r+2)s\}}$$

under methods U and EBE respectively; and the minimum orders for two-sided intervals are

$$n^{-r/(r+1)} \qquad \text{and} \qquad n^{-(r+1)(r+s)/\{(r+1)^2+s(r+2)\}}$$

under methods U and EBE. Since

$$2(r+s)\big/\{2(r+s)+3\} > 2(r+1)(r+s)\big/\{(r+1)(2r+3)+2(r+2)s\}$$

and

$$(r+s)\big/(r+s+1) > (r+1)(r+s)\big/\{(r+1)^2 + s(r+2)\},$$

then in both the one-sided and two-sided cases, this analysis suggests a preference for the undersmoothing method. However, using either method it would admittedly be difficult to compute the optimal smoothing parameters, given that they depend on a variety of unknown derivatives of f.

4.5 Nonparametric Regression

4.5.1 Introduction

The motivation behind nonparametric regression is similar to that in the problem of nonparametric density estimation, treated in the previous section: we wish to construct an estimate of a smooth curve, this time a regression mean, without making structural assumptions about the curve. The methods used are often similar in both the density and regression cases, and we shall use kernel estimators that are analogues of those employed earlier in our study of density estimation.

A nonparametric regression model for an observed data set (x_i, Y_i), $1 \leq i \leq n$, is

$$Y_i = g(x_i) + \epsilon_i, \qquad 1 \leq i \leq n,$$

where $g(x) = E(Y \mid x)$ is the regression mean, the subject of our interest, and the errors ϵ_i are independent and identically distributed with zero mean and variance σ^2. We follow the principle enunciated in subsection 4.3.2, in that we assume the design variables x_i to be nonrandom. (Of course, this allows them to have a random source but be conditioned upon.) An analysis of the case where the pairs (x_i, Y_i) are regarded as independent and identically distributed random vectors falls within the ambit of a correlation model, not regression.

As in the case of nonparametric density estimation, there is a wide choice of different approaches to constructing bootstrap confidence intervals for $g(x)$, for a specified value of x. Again, bias is an important aspect of the problem, and an "optimal" nonparametric estimator of g is typically constructed so that bias is of the same order as error about the mean. (See for example, Collomb 1981.) The "usual" bootstrap, which involves directly resampling the estimated centred residuals without additional smoothing, fails to adequately take account of bias, for the same reason as in the density estimation case in Section 4.4: the regression curve estimator is linear in the data. Bias should be corrected, either explicitly or by undersmoothing the estimator of g. Other approaches include resampling from a smoothed residual distribution, in which case the extent of bias depends on the possibly different amounts of smoothing used for error estimation and estimation of g; and resampling from a given parametric distribution (such as the Normal) or from a distribution with adjustable skewness, such as the gamma, which can be constructed to have the same first three moments as those estimated for the error distribution. Of course, the issue of percentile versus percentile-t methods also arises.

The discussion in subsection 4.4.2 served to outline the different possible bootstrap methods for curve estimation, and we shall not dwell any further on the options. We shall instead suggest a particular approach, which is little more than a regression adaptation of the method that we used for density estimation.

Let K be an rth order kernel; see (4.69) for the specifications of such a function. Our estimator of g, using kernel K and bandwidth h, is

$$\hat{g}(x) = \frac{\sum_{i=1}^{n} Y_i \, K\{(x - x_i)/h\}}{\sum_{i=1}^{n} K\{(x - x_i)/h\}} \, .$$

This is a standard kernel-type regression estimator; see, e.g., Härdle (1990, Section 3.1) for an introduction to the theory of kernel methods in regression. Choice of bandwidth plays a major role in determining how closely $\hat{g}$ tracks the true curve g. At an extreme, as $h \to 0$ for a given sample, the function $\hat{g}$ converges to a "curve" that takes the value Y_i at x_i and is zero if $x \neq x_i$ for any i. Consistency demands that as $n \to \infty$, $h = h(n) \to 0$ and $nh \to \infty$, just as in the case of density estimation.

The variance of $\hat{g}(x)$ is

$$\tau_x^2 = \mathrm{var}\{\hat{g}(x)\} = \sigma^2 \beta_x^2 \, , \qquad (4.104)$$

where $\sigma^2 = \text{var}(\epsilon_i)$ and

$$\beta_x^2 = \frac{\sum_{i=1}^n K\{(x-x_i)/h\}^2}{[\sum_{i=1}^n K\{(x-x_i)/h\}]^2}.$$

Estimation of τ_x^2 demands that we estimate σ^2, and for that purpose we employ a difference-based method. Let $\{d_j\}$ be a sequence of numbers with the properties

$$\sum d_j = 0, \quad \sum d_j^2 = 1, \quad d_j = 0 \text{ for } j < -m_1 \text{ or } j > m_2, \quad d_{-m_1} d_{m_2} \neq 0,$$

where $m_1, m_2 \geq 0$. Put $m = m_1 + m_2$. Assume that the sample

$$\mathcal{X} = \{(x_i, Y_i),\ 1 \leq i \leq n\}$$

has been ordered such that $x_1 < \ldots < x_n$. Then our variance estimate is

$$\hat{\sigma}^2 = (n-m)^{-1} \sum_{j=m_1+1}^{n-m_2} \left(\sum_j d_j Y_{i+j} \right)^2.$$

Our estimate of τ_x^2 is obtained by replacing σ^2 by $\hat{\sigma}^2$ in formula (4.104), $\hat{\tau}_x^2 = \hat{\sigma}^2 \beta_x^2$.

Under mild regularity conditions on g and the design points x_i, $\hat{\sigma}^2$ is $\sqrt{n}$-consistent for σ^2. For example, if g has a bounded derivative on the interval $[0, 1]$, and if the x_i's represent ordered values of observations derived by sampling randomly from a distribution whose density is bounded away from zero on $[0, 1]$, then $n^{1/2}(\hat{\sigma}^2 - \sigma^2)$ is asymptotically Normal $N(0, a^2)$, where

$$a^2 = \kappa \sigma^4 + 2\sigma^4 \sum_j \left(\sum_k d_k d_{j+k} \right)^2$$

and $\kappa = E(\epsilon^4)\sigma^{-4} - 3$ denotes the standardized kurtosis of the error distribution. References to places where this result and related work on difference-based variance estimates may be found are given at the end of the chapter.

The fact that we may estimate σ^2 $\sqrt{n}$-consistently sets the regression problem apart from its density estimation counterpart, where the variance estimate was itself a curve estimate with a slow rate of convergence. There are significant and interesting differences between our conclusions in these two cases, which derive from the fact that the Studentized regression estimate is a ratio of two quantities which of themselves have quite different Edgeworth expansions. To delineate these differences it is helpful to recall not just the conclusions about bootstrap density estimation drawn in Section 4.4, but also our earlier work on the classical case in Chapter 3, as follows.

(a) In the classical case, the bootstrap estimates the distribution of a pivotal statistic with accuracy n^{-1} in probability (i.e., the error equals $O_p(n^{-1})$), and of a nonpivotal statistic with accuracy $n^{-1/2}$. Furthermore, use of pivotal methods to construct confidence intervals results in coverage errors of size n^{-1} for both one- and two-sided intervals, whereas the coverage errors are $n^{-1/2}$ in the case of nonpivotal methods and one-sided intervals. These results were derived in Chapter 3.

(b) In the case of density estimation, the bootstrap estimates the distribution of a pivotal statistic (in this case T, defined in (4.81)) with accuracy $(nh)^{-1}$, and of a nonpivotal statistic (i.e., $(nh)^{1/2} \{\hat{f}(x) - E\hat{f}(x)\}$) with accuracy $(nh)^{-1/2}$. One-sided confidence intervals for $E\hat{f}(x)$ constructed by pivotal and nonpivotal methods have coverage errors of sizes $(nh)^{-1}$ and $(nh)^{-1/2}$, respectively. See Section 4.4.

(c) In the case of nonparametric regression, the bootstrap estimates the distribution of a pivotal statistic (the statistic T, defined in (4.106) below) with accuracy $n^{-1} h^{-1/2}$, and of a nonpivotal statistic (i.e., $(nh)^{1/2} \{\hat{g}(x) - E\hat{g}(x)\}$) with accuracy $n^{-1/2}$. One-sided confidence intervals for $E\hat{g}(x)$ constructed by pivotal methods and nonpivotal methods have coverage errors of sizes n^{-1} and $n^{-1} + (h/n)^{1/2}$, respectively. These results will be established later in this section.

The striking aspect of the conclusions in (c) is that pivotal methods perform so well, achieving coverage accuracy of order n^{-1} even in the infinite-parameter problem of nonparametric regression. Furthermore, it transpires that Edgeworth expansions of distributions of Studentized and non-Studentized versions of the regression mean (i.e., T and S, defined in (4.106) and (4.105)) agree to first order, which here means to order $(nh)^{-1/2}$. This property fails in most classical problems, as pointed out in Chapter 3. It also fails in the case of estimating a density; note the differences between the polynomials p_1 and q_{10} defined in (4.84) and (4.85) respectively. The property does hold when estimating slope in parametric regression, as pointed out in Section 4.3, but there the reason was the extra symmetry conferred by design, and that has no bearing on the present case. Indeed, we showed in subsection 4.3.4 that in the problem of estimating a mean in parametric regression, the first-order terms in Edgeworth expansions of distributions of pivotal and nonpivotal statistics are quite different.

Subsection 4.5.2 will develop Edgeworth expansions for nonparametric regression estimators, and subsection 4.5.3 will describe bootstrap versions of those expansions. The differences between bootstrap expansions of pivotal and nonpivotal statistics, and their implications, will be discussed. Subsection 4.5.4 will apply these results to the problem of bootstrap confidence intervals for $E\hat{g}(x)$, dealing with issues such as coverage accuracy. For the sake of simplicity, we do not concern ourselves, on this occasion, with issues of bias correction. Our main purpose is to point out how fundamental issues and techniques, such as the notion of pivoting and the use of Edgeworth expansions, extend in a natural way to the case of nonparametric regression.

4.5.2 Edgeworth Expansions for Nonparametric Regression Estimators

First we introduce the polynomials that appear in Edgeworth expansions. Define

$$a_1 = a_1(x) = (nh)^{1/2} \frac{\sum_{i=1}^n K\{(x-x_i)/h\}^3}{(\sum_{i=1}^n K\{(x-x_i)/h\}^2)^{3/2}}$$

$$a_2 = a_2(x) = (nh)^2 \frac{\sum_{i=1}^n K\{(x-x_i)/h\}^4}{[\sum_{i=1}^n K\{(x-x_i)/h\}^2]^2}$$

$$b = b(x) = -\frac{1}{2}(nh)^{-1/2} \frac{\sum_{i=1}^n K\{(x-x_i)/h\}}{(\sum_{i=1}^n K\{(x-x_i)/h\}^2)^{1/2}}$$

$$p_1(y) = -\frac{1}{6} a_1(y^2 - 1), \qquad p_2(y) = -\frac{1}{24} a_2 y(y^2 - 3),$$

$$p_3(y) = -\frac{1}{72} a_1^2 y(y^4 - 10y^2 + 15), \qquad p_4(y) = -by^2.$$

Note that the polynomials $p_1, \ldots, p_4$ depend on the data distribution only through the design sequence $\{x_i\}$; they do not depend on the error distribution or on the regression mean g, and neither are they influenced by the difference sequence $\{d_j\}$ used to construct $\hat{\sigma}^2$. Furthermore, the coefficients a_1, a_2, and b are bounded as $n \to \infty$. For example, if the design points $\{x_1, \ldots, x_n\}$ are confined to the interval $I = [0,1]$, and if they are either regularly spaced on I or represent the first n random observations of a distribution whose continuous density f is bounded away from zero in a neighbourhood of x, then

$$a_1 \sim f(x)^{-1/2} \left(\int K^2 \right)^{-3/2} \int K^3, \qquad a_2 \sim f(x)^{-1} \left(\int K^2 \right)^{-2} \int K^4,$$

$$b \sim -\frac{1}{2} f(x)^{1/2} \left(\int K^2 \right)^{-1/2},$$

where we take $f(x) = 1$ in the case of regularly spaced design.

The "ordinary" and "Studentized" versions of $\hat{g} - E\hat{g}$, both standardized for scale, are

$$S = \{\hat{g}(x) - E\hat{g}(x)\}/\tau_x = \lambda_x^{-1}\sigma^{-1}\sum_{i=1}^{n}\epsilon_i K\{(x - x_i)/h\}, \quad (4.105)$$

$$T = \{\hat{g}(x) - E\hat{g}(x)\}/\hat{\tau}_x = \lambda_x^{-1}\hat{\sigma}^{-1}\sum_{i=1}^{n}\epsilon_i K\{(x - x_i)/h\}, \quad (4.106)$$

respectively, where

$$\lambda_x = \left[\sum_{i=1}^{n} K\{(x - x_i)/h\}^2\right]^{1/2}.$$

Our conclusions about pivotal and nonpivotal forms of the bootstrap hinge on differences between the distributions of S and T. To describe these differences, write $\gamma = E\{(\epsilon_i/\sigma)^3\}$ and $\kappa = E\{(\epsilon_i/\sigma)^4\} - 3$ for the standardized skewness and kurtosis of the error distribution. Then

$$\begin{aligned} P(S \le y) = \ & \Phi(y) + (nh)^{-1/2}\gamma p_1(y)\,\phi(y) \\ & + (nh)^{-1}\{\kappa p_2(y) + \gamma^2 p_3(y)\}\,\phi(y) \\ & + O\{n^{-1} + (nh)^{-3/2}\}, \end{aligned} \quad (4.107)$$

$$\begin{aligned} P(T \le y) = \ & \Phi(y) + (nh)^{-1/2}\gamma p_1(y)\,\phi(y) \\ & + (nh)^{-1}\{\kappa p_2(y) + \gamma^2 p_3(y)\}\,\phi(y) \\ & + (h/n)^{1/2}\gamma p_4(y)\,\phi(y) + O\{n^{-1} + (nh)^{-3/2}\}. \end{aligned} \quad (4.108)$$

The most striking consequence of (4.107) and (4.108) is that the distributions of S and T agree in terms of sizes $(nh)^{-1/2}$ and $(nh)^{-1}$, which on the present occasion amount to first and third orders in the expansions. Second-order terms are of order $(h/n)^{1/2}$, and are included in the expansion for T but not that for S. (Admittedly this discussion of "order" is a little imprecise since the relative sizes of $(nh)^{-1/2}, (h/n)^{1/2}$, and $(nh)^{-1}$ depend on the size of h. However, it would typically be the case in practice that h is of larger order than $n^{-1/3}$, and in that circumstance $(nh)^{-1/2} \gg (h/n)^{1/2} \gg (nh)^{-1}$.)

4.5.3 Bootstrap Versions of Edgeworth Expansions

We begin by developing a bootstrap algorithm for the case of nonparametric regression. Observe from (4.105) and (4.106) that the mean function g does not influence the distribution of S, and enters the distribution of T only through $\hat{\sigma}$. It turns out that the effect of g on $\hat{\sigma}$ is relatively minor; in particular, g enters only the remainder term on the right-hand side of (4.108). Therefore, when using the bootstrap to approximate these distributions we shall, in effect, make the fictitious assumption that $g \equiv 0$.

To estimate the error distribution, first compute the sample residuals

$$\tilde{\epsilon}_i = Y_i - \hat{g}(x_i), \qquad i \in \mathcal{J},$$

where $\mathcal{J}$ is an appropriate set of indices (e.g., the set of i's such that x_i is not too close to the boundary of the interval on which inference is being conducted).[1] Define n' to equal the number of elements of $\mathcal{J}$, let $\sum_i'$ denote summation over $i \in \mathcal{J}$, and put

$$\bar{\tilde{\epsilon}} = n'^{-1} \sum_i' \tilde{\epsilon}_i, \quad \hat{\epsilon}_i = \tilde{\epsilon}_i - \bar{\tilde{\epsilon}},$$

the latter being centred residuals. (Centring was not necessary in the case of linear regression, since in that circumstance $\sum \hat{\epsilon}_i = 0$.) Conditional on the sample $\mathcal{X} = \{(x_i, Y_i), \; 1 \le i \le n\}$, draw a resample $\{\epsilon_1^*, \dots, \epsilon_n^*\}$ at random, with replacement, from $\{\hat{\epsilon}_i, \; i \in \mathcal{J}\}$. Define $m = m_1 + m_2$,

$$\check{\sigma}^2 = n'^{-1} \sum_i' \hat{\epsilon}_i^2, \qquad \hat{\sigma}^{*2} = (n-m)^{-1} \sum_{i=m_1+1}^{n-m_2} \left(\sum_j d_j \, \epsilon_{i+j}^* \right)^2,$$

$$S^* = \tau_x^{-1} \check{\sigma}^{-1} \sum_{i=1}^n \epsilon_i^* \, K\{(x - x_i)/h\},$$

$$T^* = \tau_x^{-1} \hat{\sigma}^{*-1} \sum_{i=1}^n \epsilon_i^* \, K\{(x - x_i)/h\}.$$

The conditional distributions of S^* and T^*, given $\mathcal{X}$, are good approximations to the distributions of S and T respectively. Indeed, if we define

$$\hat{\mu}_j = E\{(\epsilon_1^*/\check{\sigma})^j \mid \mathcal{X}\} = \check{\sigma}^{-j} n'^{-1} \sum_i' \hat{\epsilon}_i^j,$$

and write $\hat{\gamma} = \hat{\mu}_3$ and $\hat{\kappa} = \hat{\mu}_4 - 3$ for estimates of (standardized) skewness and kurtosis, then we have the following bootstrap analogues of (4.107) and (4.108):

$$
\begin{aligned}
P(S^* \le u \mid \mathcal{X}) = \; & \Phi(u) + (nh)^{-1/2} \, \hat{\gamma} \, p_1(u) \, \phi(u) \\
& + (nh)^{-1} \left\{ \hat{\kappa} \, p_2(u) + \hat{\gamma}^2 \, p_3(u) \right\} \phi(u) \\
& + O_p\{ n^{-1} + (nh)^{-3/2} \},
\end{aligned} \qquad (4.109)
$$

$$
\begin{aligned}
P(T^* \le u \mid \mathcal{X}) = \; & \Phi(u) + (nh)^{-1/2} \, \hat{\gamma} \, p_1(u) \, \phi(u) \\
& + (nh)^{-1} \left\{ \hat{\kappa} \, p_2(u) + \hat{\gamma}^2 \, p_3(u) \right\} \phi(u) \\
& + (h/n)^{1/2} \, \hat{\gamma} \, p_4(u) \, \phi(u) \\
& + O_p\{ n^{-1} + (nh)^{-3/2} \}.
\end{aligned} \qquad (4.110)
$$

---

[1] The x_i's must be kept away from the boundary of the interval so as to ensure that edge effects do not play a significant role in construction of the residuals $\hat{\epsilon}_i$. If the kernel K vanishes outside $(-c, c)$ then it is sufficient to ask that for each $i \in \mathcal{J}$, x_i is no closer to the boundary than ch. See Härdle and Bowman (1988).

The polynomials $p_1, \ldots, p_4$ in (4.109) and (4.110) are exactly as they were in (4.107) and (4.108).

In virtually all cases of interest, $(nh)^{-3/2}$ is of smaller order than n^{-1}; consider for example the common circumstance where K is a second-order kernel and h is of size $n^{-1/5}$ (Härdle 1990, Section 4.1). Therefore, the remainders $O\{n^{-1} + (nh)^{-3/2}\}$ in (4.107)–(4.110) are, generally, $O(n^{-1})$. Furthermore, it is generally the case that $\hat{\gamma} - \gamma = O_p(n^{-1/2})$ and $\hat{\kappa} - \kappa = O_p(n^{-1/2})$. We may therefore deduce from (4.107)–(4.110) that

$$P(S^* \leq u \mid \mathcal{X}) - P(S \leq u) = O_p(n^{-1} h^{-1/2}),$$

$$P(T^* \leq u \mid \mathcal{X}) - P(T \leq u) = O_p(n^{-1} h^{-1/2}).$$

These are precise rates of convergence, not simply upper bounds, because the convergence rate of $\hat{\gamma}$ to γ is precisely $n^{-1/2}$. Thus, the exact rate of convergence of bootstrap approximations to the distributions of S and T is $n^{-1} h^{-1/2}$, which is slightly poorer than the rate n^{-1} found in classical finite-parameter problems, but slightly better than the corresponding rate $(nh)^{-1}$ found in applications of the bootstrap to density estimation.

We are now in a position to explain why bootstrap methods for estimating the distributions of pivotal and nonpivotal statistics admit the properties described in point (c) of subsection 4.5.1. The pivotal percentile-t method approximates the distribution of T by that of T^*, and as we have just shown the resulting error is of size $n^{-1} h^{-1/2}$. On the other hand, the nonpivotal percentile method approximates the distribution of

$$U = \tau_x^{-1}\{\hat{g}(x) - E\hat{g}(x)\} = \lambda_x^{-1} \sum_{i=1}^{n} \epsilon_i K\{(x - x_i)/h\}$$

by the conditional distribution of

$$U^* = \lambda_x^{-1} \sum_{i=1}^{n} \epsilon_i^* K\{(x - x_i)/h\}.$$

To appreciate the significant errors that arise in this approximation, observe from (4.107) and (4.109) that

$$P(U^* \leq u \mid \mathcal{X}) - P(U \leq u)$$
$$= P(S^* \leq u/\check{\sigma} \mid \mathcal{X}) - P(S \leq u/\sigma)$$
$$= \Phi(u/\check{\sigma}) - \Phi(u/\sigma) + O_p(n^{-1} h^{-1/2}). \quad (4.111)$$

(Here we have used the fact that $\check{\sigma} - \sigma = O_p(n^{-1/2})$.) Since $\check{\sigma} - \sigma$ is of precise size $n^{-1/2}$, $\Phi(u/\check{\sigma}) - \Phi(u/\sigma)$ is of precise size $n^{-1/2}$. It therefore follows from (4.111) that the bootstrap approximation to the distribution of U will have errors of at least $n^{-1/2}$, as stated in subsection 4.5.1.

4.5.4 Confidence Intervals for $E\,\hat{g}(x)$

We begin by defining the confidence intervals. Let u_α, $\hat{u}_\alpha$, v_α, $\hat{v}_\alpha$ be given by the equations

$$P(S \le u_\alpha) \;=\; P(S^* \le \hat{u}_\alpha \mid \mathcal{X}) \;=\; P(T \le v_\alpha) \;=\; P(T^* \le \hat{v}_\alpha \mid \mathcal{X}) \;=\; \alpha.$$

One-sided percentile and percentile-t confidence intervals for $E\hat{g}(x)$ are given by

$$\widehat{I}_1 \;=\; \widehat{I}_1(\alpha) \;=\; \left(-\infty,\; \hat{g}(x) - \tau_x^{-1}\,\hat{\sigma}\hat{u}_{1-\alpha}\right),$$

$$\widehat{J}_1 \;=\; \widehat{J}_1(\alpha) \;=\; \left(-\infty,\; \hat{g}(x) - \tau_x^{-1}\,\hat{\sigma}\,\hat{v}_{1-\alpha}\right),$$

respectively, and may be viewed as approximations to the "ideal" interval

$$J_1 \;=\; \left(-\infty,\; \hat{g}(x) - \tau_x^{-1}\,\hat{\sigma}v_{1-\alpha}\right).$$

Each has nominal coverage α, and J_1 has perfect coverage accuracy in the sense that $P\{E\hat{g}(x) \in J_1\} = \alpha$.

To compare properties of these intervals, let us assume that h is of larger order than $n^{-1/2}$, which would typically be the case in practice. Then the upper endpoint of $\widehat{I}_1$ differs from that of J_1 in terms of size n^{-1}, whereas the upper endpoints of $\widehat{J}_1$ and J_1 differ in terms of order $n^{-3/2}\,h^{-1}$, which is smaller than n^{-1}. In this sense, $\widehat{J}_1$ is more accurate than $\widehat{I}_1$ as an approximation to J_1. To verify our claims about the sizes of the endpoints of $\widehat{I}_1$ and $\widehat{J}_1$, we follow the usual route of deriving Cornish-Fisher expansions by inverting the Edgeworth expansions (4.107)–(4.110) and then subtracting appropriate pairs of expansions. Arguing in this way, and using longer expansions than (4.107)–(4.110) to ascertain the order of the remainder, we may conclude that

$$\hat{u}_{1-\alpha} - u_{1-\alpha} \;=\; (nh)^{-1/2}\,(\hat{\gamma} - \gamma)\,p_1(z_\alpha) + O_p(n^{-3/2}\,h^{-1}), \qquad (4.112)$$

$$\hat{v}_{1-\alpha} - v_{1-\alpha} \;=\; (nh)^{-1/2}\,(\hat{\gamma} - \gamma)\,p_1(z_\alpha) + O_p(n^{-3/2}\,h^{-1}), \qquad (4.113)$$

$$u_{1-\alpha} - v_{1-\alpha} \;=\; (h/n)^{1/2}\,\gamma\,p_4(z_\alpha) + O(n^{-1}). \qquad (4.114)$$

Noting that τ_x is of size $(nh)^{-1/2}$ and $\hat{\gamma}-\gamma$ of size $n^{-1/2}$, put $\rho_x = (nh)^{1/2}\,\tau_x$ and $\Delta = n^{1/2}\,(\hat{\gamma} - \gamma)$. Then, assuming h is of larger order than $n^{-1/2}$,

$$\tau_x^{-1}\,\hat{\sigma}\,\hat{u}_{1-\alpha} - \tau_x^{-1}\,\hat{\sigma}\,v_{1-\alpha} \;=\; \tau_x^{-1}\,\hat{\sigma}(\hat{u}_{1-\alpha} - u_{1-\alpha} + u_{1-\alpha} - v_{1-\alpha})$$

$$= \rho_x^{-1}\,\hat{\sigma}\{n^{-3/2}\,h^{-1}\,\Delta\,p_1(z_\alpha) + n^{-1}\,\gamma\,p_4(z_\alpha)\}$$
$$+ O_p(n^{-3/2}\,h^{-1/2}),$$

$$\tau_x^{-1}\,\hat{\sigma}\,\hat{v}_{1-\alpha} - \tau_x^{-1}\,\hat{\sigma}\,v_{1-\alpha} \;=\; \rho_x^{-1}\,\hat{\sigma}n^{-3/2}\,h^{-1}\,\Delta\,p_1(z_\alpha)$$
$$+ O_p(n^{-3/2}\,h^{-1/2}).$$

This verifies our claims.

The coverage probabilities of $\widehat{I}_1$ and $\widehat{J}_1$ are given by

$$\beta_{1,I} = P\{E\hat{g}(x) \in \widehat{I}_1\} = P(T > \hat{u}_{1-\alpha}),$$
$$\beta_{1,J} = P\{E\hat{g}(x) \in \widehat{J}_1\} = P(T > \hat{v}_{1-\alpha}),$$

respectively. We claim that $\beta_{1,J} - \alpha = O(n^{-1})$, whereas

$$\beta_{1,I} - \alpha = -(h/n)^{1/2} \gamma p_4(z_\alpha) \phi(z_\alpha) + O(n^{-1}).$$

That is, the coverage error of $\widehat{J}_1$ equals $O(n^{-1})$, whereas that of $\widehat{I}_1$ is of size $n^{-1/2} h^{1/2}$. Since h is necessarily of larger order than n^{-1} (since $nh \to \infty$), then once again $\widehat{J}_1$ emerges as the more accurate interval.

To check our claims about the sizes of coverage error, note first that it may be shown by the usual methods for Edgeworth expansion that the distribution functions of T and

$$T_1(y) = T - (nh)^{-1/2}(\hat{\gamma} - \gamma) p_1(y)$$

differ only in terms of order n^{-1}. Therefore, using the delta method and formulae (4.112)–(4.114),

$$\begin{aligned}
\beta_{1,I}(\alpha) &= P(T > \hat{u}_{1-\alpha}) \\
&= P\left\{T_1(z_\alpha) > v_{1-\alpha} + (h/n)^{1/2} \gamma p_4(z_\alpha) + O_p(n^{-1})\right\} \\
&= P\left\{T > v_{1-\alpha} + (h/n)^{1/2} \gamma p_4(z_\alpha)\right\} + O(n^{-1}) \\
&= 1 - \left\{1 - \alpha + (h/n)^{1/2} \gamma p_4(z_\alpha) \phi(z_\alpha)\right\} + O(n^{-1}) \\
&= \alpha - (h/n)^{1/2} \gamma p_4(z_\alpha) \phi(z_\alpha) + O(n^{-1}), \tag{4.115}
\end{aligned}$$

$$\begin{aligned}
\beta_{1,J}(\alpha) &= P(T > \hat{v}_{1-\alpha}) \\
&= P\left\{T_1(z_\alpha) > v_{1-\alpha} + O_p(n^{-1})\right\} \\
&= P\{T_1(z_\alpha) > v_{1-\alpha}\} + O(n^{-1}) \\
&= \alpha + O(n^{-1}). \tag{4.116}
\end{aligned}$$

This verifies the claims.

To construct two-sided versions of the intervals $\widehat{I}_1$ and $\widehat{J}_1$, put

$\xi = \frac{1}{2}(1 + \alpha)$ and

$$
\begin{aligned}
\widehat{I}_2 &= \widehat{I}_1(\xi) \backslash \widehat{I}_1(1 - \xi) \\
&= \left(\hat{g}(x) - \tau_x^{-1} \hat{\sigma} \, \hat{u}_\xi, \ \ \hat{g}(x) - \tau_x^{-1} \hat{\sigma} \, \hat{u}_{1-\xi} \right), \\
\widehat{J}_2 &= \widehat{J}_1(\xi) \backslash \widehat{J}_1(1 - \xi) \\
&= \left(\hat{g}(x) - \tau_x^{-1} \hat{\sigma} \, \hat{v}_\xi, \ \ \hat{g}(x) - \tau_x^{-1} \hat{\sigma} \, \hat{v}_{1-\xi} \right).
\end{aligned}
$$

Both of these intervals have coverage error $O(n^{-1})$, when viewed as confidence intervals for $E\hat{f}(x)$. That is clear from (4.115) and (4.116); for example,

$$
\begin{aligned}
\beta_{2,I}(\alpha) &= P\{E\hat{g}(x) \in \widehat{I}_2\} \\
&= \beta_{1,I}(\xi) - \beta_{1,I}(1 - \xi) \\
&= \xi - (h/n)^{1/2} p_4(z_\xi) \phi(z_\xi) + O(n^{-1}) \\
&\quad - \left\{ 1 - \xi - (h/n)^{1/2} p_4(z_{1-\xi}) \phi(z_{1-\xi}) + O(n^{-1}) \right\} \\
&= \alpha + O(n^{-1}),
\end{aligned}
$$

since $z_{1-\xi} = -z_\xi$ and p_4 is an even polynomial.

4.6 Bibliographical Notes

Our theory for multivariate confidence regions, described in Section 4.2, involves straightforward application of multivariate Edgeworth expansions, of which detailed accounts may be found in Bhattacharya and Rao (1976), Bhattacharya and Ghosh (1978), and Barndorff-Nielsen and Cox (1989, Section 6.3). Beran and Millar (1985) propose a general asymptotic theory for optimal multivariate confidence sets, and (1986) discuss bootstrap confidence bands for multivariate distribution functions. Hall (1987b) introduces multivariate likelihood-based confidence regions. The method of empirical likelihood, mentioned in Section 4.2 and due to Owen (1988, 1990), is surveyed by Hall and La Scala (1990).

Bickel and Freedman (1981) and Freedman (1981) treat the problem of consistency and asymptotic Normality of bootstrap methods for regression. Shorack (1982) introduces bootstrap methods in robust regression, Bickel and Freedman (1983) study multiparameter bootstrap regression, Freedman and Peters (1984) and Peters and Freedman (1984a, 1984b) discuss empirical results for bootstrap regression, Morey and Schenck (1984) report the results of small-sample simulation studies of bootstrap and jack-knife regression, Weber (1984) establishes a general result about asymptotic Normality in bootstrap regression and (1986) discusses bootstrap and jackknife methods in regression, Stine (1985) studies bootstrap prediction intervals in regression, de Wet and van Wyk (1986) treat bootstrap methods in regression with dependent errors, Wu (1986) describes (among other topics) bootstrap methods for regression with heteroscedastic errors, and Shao (1988) analyses resampling methods for variance and bias estimation in linear models. The work in our Section 4.3 is based on theory for bootstrap regression developed by Robinson (1987), Hall (1988d), and Hall and Pittelkow (1990).

Bootstrap methods for nonparametric curve estimation are in comparative infancy, compared with the evolution of bootstrap technology in more conventional problems such as those described in Chapter 3. The bootstrap has a role to play in several areas of curve estimation, for example in estimating mean squared error and determining an empirical version of the "optimal" smoothing parameter (e.g., Taylor 1989, Faraway and Jhun 1990, Hall 1990b) or in constructing confidence intervals and confidence bands (as described in our Sections 4.4 and 4.5). Härdle and Bowman (1988), Härdle (1990), and Härdle and Marron (1991) study nonparametric regression, paying particular attention to the problem of bias. In each case those researchers use nonpivotal percentile methods and concern themselves with first-order asymptotic theory — i.e., with ensuring that coverage error converges to zero. The techniques discussed in Sections 4.4 and 4.5 are principally pivotal percentile-t, and our theoretical arguments focus on second-order features such as the size of coverage error. Our analysis is based on unpublished material, including Hall (1990f) in the case of Section 4.4 and Hall (1990g) for Section 4.5. Hall (1990h) develops Edgeworth expansions for the coverage error of confidence bands, as distinct from confidence intervals, in the case of nonparametric density estimation. The reader is referred to Tapia and Thompson (1978), Prakasa Rao (1983, Chapters 2 and 3), Devroye and Györfi (1985), and Silverman (1986) for comprehensive accounts of the technology of density estimation. Nonparametric regression is treated in detail by Prakasa Rao (1982, Section 4.2) and Härdle (1990). Substantial contributions to bootstrap methods in curve estimation are obtained by Mammen (1991) in his Habilitationsschrift.

Work on difference-based estimation of error variance in nonparametric regression includes contributions by Rice (1984), Gasser, Sroka, and Jennen-Steinmetz (1986), Müller and Stadtmüller (1987, 1988), Müller (1988, p. 99ff), and Hall, Kay, and Titterington (1990). The latter paper treats the particular estimate $\hat{\sigma}^2$ considered in Section 4.5.

5

Details of Mathematical Rigour

5.1 Introduction

In previous chapters we have been somewhat cavalier about technical details, from at least three points of view. Firstly, we have been imprecise about the regularity conditions under which our results hold; secondly, we have not been specific about properties of remainder terms in expansions, preferring to write the remainders as simply "...'; and thirdly, our proofs have been more heuristic than rigorous. Our aim in the present chapter is to redress these omissions, by developing examples of concise theory under explicit assumptions and with rigorous proofs.

The majority of the asymptotic expansions described in this monograph can be deduced rather easily from Edgeworth expansions for distributions of sums of independent random variables or random vectors. Often those vectors are identically distributed, as in the case of Theorem 2.2 (which is one of relatively few results that we *have* stated under explicit regularity conditions). Identical distribution is a consequence of the fact that we are dealing with repeated sampling or resampling from the same population. However, in some instances the summands have nonidentical distributions, as in the regression case treated in Section 4.3; there, weights based on the design points must be incorporated into the sums.

Sections 5.2 and 5.3 give a rigorous treatment of the classical case considered in Chapter 3. In Section 5.2 we provide proofs of Edgeworth and Cornish-Fisher expansions for bootstrap statistics, of the type described in Sections 3.3 and 3.4, and in Section 5.3 we derive an Edgeworth expansion of coverage probability, making rigorous the results in Section 3.5. Sections 5.4 and 5.5 address simple linear regression and nonparametric density estimation, respectively, encountered earlier in Chapter 4. Many of the techniques needed there are similar to those employed to treat the more conventional problems from Chapter 3, and so we make our discussion in Sections 5.4 and 5.5 as brief as possible, referring the reader to the more detailed accounts in Sections 5.2 and 5.3 for descriptions of the methods.

Our emphasis is on the bootstrap case, and so we do not rederive standard results about Edgeworth expansions of "ordinary" (as distinct from "bootstrap") distributions. Those results are stated as lemmas on the occasions where they are needed.

5.2 Edgeworth Expansions of Bootstrap Distributions

5.2.1 Introduction

In this section we give rigorous proofs of results in Section 3.3, such as Edgeworth expansions (3.17)–(3.19) and Cornish-Fisher expansions (3.20)–(3.22). The context is that of the "smooth function model" introduced in Section 2.4, and so we pause here to describe its main features.

Let $\mathcal{X} = \{\mathbf{X}_1, \dots, \mathbf{X}_n\}$ be a random sample from a d-variate population, and write $\overline{\mathbf{X}} = n^{-1} \sum \mathbf{X}_i$ for the sample mean. We assume that the parameter of interest has the form $\theta_0 = g(\boldsymbol{\mu})$, where $g : \mathbb{R}^d \to \mathbb{R}$ is a smooth function and $\boldsymbol{\mu} = E(\mathbf{X})$ denotes the population mean. The parameter estimate is $\hat{\theta} = g(\overline{\mathbf{X}})$, which is assumed to have asymptotic variance $n^{-1} \sigma^2 = n^{-1} h(\boldsymbol{\mu})^2$, where h is a known, smooth function. It is supposed that $\sigma^2 \neq 0$. The sample estimate of σ^2 is $\hat{\sigma}^2 = h(\overline{\mathbf{X}})^2$.

We wish to estimate the distribution of either $(\hat{\theta} - \theta_0)/\sigma$ or $(\hat{\theta} - \theta_0)/\hat{\sigma}$. Note that both these statistics may be written in the form $A(\overline{\mathbf{X}})$, where

$$A(\mathbf{x}) = \{g(\mathbf{x}) - g(\boldsymbol{\mu})\}/h(\boldsymbol{\mu}) \tag{5.1}$$

in the former case and

$$A(\mathbf{x}) = \{g(\mathbf{x}) - g(\boldsymbol{\mu})\}/h(\mathbf{x}) \tag{5.2}$$

in the latter. To construct the bootstrap distribution estimates, let

$$\mathcal{X}^* = \{\mathbf{X}_1^*, \dots, \mathbf{X}_n^*\}$$

denote a sample drawn randomly, with replacement, from $\mathcal{X}$, and write $\overline{\mathbf{X}}^* = n^{-1} \sum \mathbf{X}_i^*$ for the resample mean. Define

$$\hat{\theta}^* = g(\overline{\mathbf{X}}^*) \quad \text{and} \quad \hat{\sigma}^* = h(\overline{\mathbf{X}}^*).$$

Both $(\hat{\theta}^* - \hat{\theta})/\hat{\sigma}$ and $(\hat{\theta}^* - \hat{\theta})/\hat{\sigma}^*$ may be expressed as $\widehat{A}(\overline{\mathbf{X}}^*)$, where

$$\widehat{A}(\mathbf{x}) = \{g(\mathbf{x}) - g(\overline{\mathbf{X}})\}/h(\overline{\mathbf{X}}), \tag{5.3}$$

$$\widehat{A}(\mathbf{x}) = \{g(\mathbf{x}) - g(\overline{\mathbf{X}})\}/h(\mathbf{x}) \tag{5.4}$$

in the respective cases. The bootstrap estimate of the distribution of $A(\overline{\mathbf{X}})$ is given by the distribution of $\widehat{A}(\overline{\mathbf{X}}^*)$, conditional on $\mathcal{X}$.

Subsection 5.2.2 describes bootstrap versions of Edgeworth expansions and Cornish-Fisher expansions, and compares them with their nonbootstrap counterparts. The main proof is given in subsection 5.2.3.

5.2.2 Main Results

Theorem 2.2 in Section 2 gave an Edgeworth expansion of the distribution of $A(\overline{\mathbf{X}})$, which may be stated as follows. Assume that for an integer $\nu \geq 1$, the function A has $\nu + 2$ continuous derivatives in a neighbourhood of $\boldsymbol{\mu}$, that $A(\boldsymbol{\mu}) = 0$, that $E(\|\mathbf{X}\|^{\nu+2}) < \infty$, that the characteristic function χ of $\mathbf{X}$ satisfies

$$\limsup_{\|\mathbf{t}\| \to \infty} |\chi(\mathbf{t})| < 1, \tag{5.5}$$

and that the asymptotic variance of $n^{1/2} A(\overline{\mathbf{X}})$ equals 1. (This is certainly the case if A is given by (5.1) or (5.2).) Then

$$P\{n^{1/2} A(\overline{\mathbf{X}}) \leq x\} = \Phi(x) + \sum_{j=1}^{\nu} n^{-j/2} \pi_j(x) \phi(x) + o(n^{-\nu/2}) \tag{5.6}$$

uniformly in x, where π_j is a polynomial of degree of $3j - 1$, odd for even j and even for odd j, with coefficients depending on moments of $\mathbf{X}$ up to order $j + 2$. The form of π_j depends very much on the form of A. We denote π_j by p_j if A is given by (5.1), and by q_j if A is given by (5.2).

Now we describe the bootstrap version of this result. Let $\widehat{A}$ be defined by either (5.3) or (5.4), put $\pi_j = p_j$ in the former case and $\pi_j = q_j$ in the latter, and let $\hat{\pi}_j$ be the polynomial obtained from π_j on replacing population moments by the corresponding sample moments. (The sample moment corresponding to the population moment

$$E\left(X^{(i_1)} \ldots X^{(i_j)}\right)$$

is

$$n^{-1} \sum_{k=1}^{n} X_k^{(i_1)} \ldots X_k^{(i_j)} . \ )$$

THEOREM 5.1. *Let $\lambda > 0$ be given, and let $l = l(\lambda)$ denote a sufficiently large positive number (whose value we do not specify). Assume that g and h each have $\nu + 3$ bounded derivatives in a neighbourhood of $\boldsymbol{\mu}$, that $E(\|\mathbf{X}\|^l) < \infty$, and that Cramér's condition (5.5) holds. Then there exists a constant $C > 0$ such that*

$$P\left[\sup_{-\infty < x < \infty} \left| P\{n^{1/2} \widehat{A}(\overline{\mathbf{X}}^*) \leq x \mid \mathcal{X}\} - \Phi(x) \right. \right.$$
$$\left. \left. - \sum_{j=1}^{\nu} n^{-j/2} \hat{\pi}_j(x) \phi(x) \right| > Cn^{-(\nu+1)/2} \right] = O(n^{-\lambda}) \tag{5.7}$$

and

$$P\left\{ \max_{1 \leq j \leq \nu} \sup_{-\infty < x < \infty} (1 + |x|)^{-(3j-1)} |\hat{\pi}_j(x)| > C \right\} = O(n^{-\lambda}). \tag{5.8}$$

A proof of Theorem 5.1 will be given in subsection 5.2.3.

If we take $\lambda > 1$ in this theorem and apply the Borel-Cantelli lemma, we may deduce from (5.7) that

$$\sup_{-\infty < x < \infty} \left| P\{n^{1/2}\, \widehat{A}(\overline{\mathbf{X}}^*) \leq x \mid \mathcal{X}\} - \Phi(x) - \sum_{j=1}^{\nu} n^{-j/2}\, \hat{\pi}_j(x)\, \phi(x) \right|$$

$$= O(n^{-(\nu+1)/2})$$

with probability one. This is a strong form of the Edgeworth expansions (3.17)–(3.19), all of which are corollaries. Formula (5.8) declares that the coefficients of the polynomials $\hat{\pi}_j$ are virtually bounded — the chance that they exceed a constant value is no more than $O(n^{-\lambda})$.

The value of $l(\lambda)$ is not specified in Theorem 5.1 since it is most unlikely that the upper bound to l produced by our proof is anywhere near the best possible.

As explained in Section 2.5, an Edgeworth expansion may be inverted to give a Cornish-Fisher expansion for quantiles of the distribution of $n^{1/2}\, A(\overline{\mathbf{X}})$. To describe that expansion, define polynomials π_{j1}, $j \geq 1$, by the analogue of formula (2.49)

$$\Phi\left\{ z_\alpha + \sum_{j \geq 1} n^{-i/2}\, \pi_{j1}(z_\alpha) \right\} + \sum_{i \geq 1} n^{-i/2}\, p_i\left\{ z_\alpha + \sum_{j \geq 1} n^{-j/2}\, \pi_{j1}(z_\alpha) \right\}$$

$$\times\, \phi\left\{ z_\alpha + \sum_{j \geq 1} n^{-j/2}\, \pi_{j1}(z_\alpha) \right\} = \alpha, \qquad 0 < \alpha < 1,$$

where $z_\alpha = \Phi^{-1}(\alpha)$. Then π_{j1} is of degree $j+1$, odd for even j and even for odd j. Writing

$$w_\alpha = \inf\left\{ t : P[n^{1/2}\, A(\overline{\mathbf{X}}) < t] \geq \alpha \right\},$$

we may deduce directly from (5.6) that for any $0 < \epsilon < \frac{1}{2}$,

$$\sup_{\epsilon \leq \alpha \leq 1-\epsilon} \left| w_\alpha - z_\alpha - \sum_{j=1}^{\nu} n^{-j/2}\, \pi_{j1}(z_\alpha) \right| = o(n^{-\nu/2}); \qquad (5.9)$$

see Section 2.5.

The bootstrap version of this formula is an expansion of the quantile

$$\hat{w}_\alpha = \inf\left\{ t : P[n^{1/2}\, \widehat{A}(\overline{\mathbf{X}}^*) \leq t \mid \mathcal{X}] \geq \alpha \right\},$$

where $\widehat{A}$ is defined by either (5.3) or (5.4). Indeed, if we derive the bootstrap Cornish-Fisher expansion from the bootstrap Edgeworth expansion (5.7), then we may replace the range $\epsilon \leq \alpha \leq 1-\epsilon$ in (5.9) by $n^{-\epsilon} \leq \alpha \leq 1-n^{-\epsilon}$, for some $\epsilon > 0$, owing to the fact that the error bound given by (5.7) is $O(n^{-(\nu+1)/2})$ rather than simply $o(n^{-\nu/2})$. The following result may be derived directly from Theorem 5.1, using arguments that involve only algebraic methods.

THEOREM 5.2. *Under the conditions of Theorem 5.1, and for each* $\delta, \lambda > 0$, *there exist constants* $C, \epsilon > 0$ *such that*

$$P\left\{ \sup_{n^{-\epsilon} \leq \alpha \leq 1 - n^{-\epsilon}} \left| \hat{w}_\alpha - z_\alpha - \sum_{j=1}^{\nu} n^{-j/2} \hat{\pi}_{j1}(z_\alpha) \right| > C n^{-(1/2)(\nu+1)+\delta} \right\}$$

$$= O(n^{-\lambda}).$$

A proof of Theorem 5.2 is given at the end of this subsection.

Taking $\lambda > 1$ in Theorem 5.2 we may deduce from the Borel-Cantelli lemma that

$$\sup_{n^{-\epsilon} \leq \alpha \leq 1 - n^{-\epsilon}} \left| \hat{w}_\alpha - z_\alpha - \sum_{j=1}^{\nu} n^{-j/2} \hat{\pi}_{j1}(z_\alpha) \right| = O(n^{-(1/2)(\nu+1)+\delta})$$

with probability one. This is a strong form of the Cornish-Fisher expansions (3.20) and (3.21). In the case where $\hat{A}$ is given by (5.3) we should take $(\pi, \hat{w}_\alpha) = (p, \hat{u}_\alpha)$, which is the context of (3.20); and when $\hat{A}$ is given by (5.4), $(\pi, \hat{w}_\alpha) = (q, \hat{v}_\alpha)$, which is the context of (3.21).

Similar methods may be used to derive bootstrap versions of the other kind of Cornish-Fisher expansion, where z_α is expressed in terms of $\hat{w}_\alpha$ by a power series in $n^{-1/2}$. However, since results of this type have not been used in earlier chapters, we do not consider them here.

We close this subsection with a proof of Theorem 5.2. By (5.8), with probability $1 - O(n^{-\lambda})$ the coefficients of $\hat{\pi}_j$ (and hence those of $\hat{\pi}_{1j}$) are dominated by a fixed constant C_1 uniformly in $1 \leq j \leq \nu$. It follows that with probability $1 - O(n^{-\lambda})$, the function

$$\hat{\Psi}(x) = \Phi(x) + \sum_{j=1}^{\nu} n^{-j/2} \hat{\pi}_j(x) \phi(x)$$

is strictly increasing in $|x| \leq \log n$. Therefore, given $\delta > 0$ we may choose $\epsilon \in (0, \frac{1}{4})$ so small that the function $\hat{\Psi}^{-1}(\alpha)$ is well defined within the interval $n^{-2\epsilon} \leq \alpha \leq 1 - n^{-2\epsilon}$ and satisfies

$$P\left\{ 0 \leq \left| (d/d\alpha) \hat{\Psi}^{-1}(\alpha) \right| \leq C_3 n^\delta, \text{ all } n^{-2\epsilon} \leq \alpha \leq 1 - n^{-2\epsilon} \right\} = 1 - O(n^{-\lambda}),$$

$$P\left\{ \sup_{n^{-2\epsilon} \leq \alpha \leq 1 - n^{-2\epsilon}} \left| \hat{\Psi}^{-1}(\alpha) - z_\alpha - \sum_{j=1}^{\nu} n^{-j/2} \hat{\pi}_{j1}(z_\alpha) \right| > C_4 n^{-(1/2)(\nu+1)+\delta} \right\}$$

$$= O(n^{-\lambda}) \quad (5.10)$$

for constants $C_3, C_4 > 0$. (Note that for all $c > 0$, $z_{1-n^{-c}} \sim (2c \log n)^{1/2}$ as $n \to \infty$.) By (5.7), for the value of C in that formula,

$$P\left[\sup_{0<\alpha<1} \left|P\{n^{1/2}\,\widehat{A}(\overline{\mathbf{X}}^*) \le \hat{w}_\alpha \mid \mathcal{X}\} - \alpha\right| > 2C\,n^{-(\nu+1)/2}\right] = O(n^{-\lambda}),$$

whence (again using (5.7)),

$$P\left\{\sup_{0<\alpha<1} \left|\alpha - \widehat{\Psi}(\hat{w}_\alpha)\right| > 3C\,n^{-(\nu+1)/2}\right\} = O(n^{-\lambda}).$$

If both the events

$$0 \le (d/d\alpha)\,\widehat{\Psi}^{-1}(\alpha) \le C_3\,n^\delta, \quad \text{for all } n^{-2\epsilon} \le \alpha \le 1 - n^{-2\epsilon},$$

$$\sup_{0<\alpha<1} \left|\alpha - \widehat{\Psi}(\hat{w}_\alpha)\right| \le 3C\,n^{-(\nu+1)/2}$$

hold, then

$$\widehat{\Psi}^{-1}(\alpha) \le \widehat{\Psi}^{-1}\left\{\widehat{\Psi}(\hat{w}_\alpha) + 3C\,n^{-(\nu+1)/2}\right\}$$

$$\le \hat{w}_\alpha + 3CC_3\,n^{-(1/2)(\nu+1)+\delta}$$

uniformly in $n^{-\epsilon} \le \alpha \le 1 - n^{-\epsilon}$. Similarly,

$$\widehat{\Psi}^{-1}(\alpha) \ge \hat{w}_\alpha - 3CC_3\,n^{-(1/2)(\nu+1)+\delta}.$$

Therefore

$$P\left\{\sup_{n^{-\epsilon} \le \alpha \le 1-n^{-\epsilon}} \left|\widehat{\Psi}^{-1}(\alpha) - \hat{w}_\alpha\right| > 3CC_3\,n^{-(1/2)(\nu+1)+\delta}\right\} = O(n^{-\lambda}).$$

Theorem 5.2 follows from this result and (5.10).

5.2.3 Proof of Theorem 5.1

Our proof is divided into four parts: definitions, error bounds for the bootstrap Edgeworth expansion, identification of the expansion, and derivation of a lemma from the second part of the proof.

(i) Notation

For real-valued d-vectors $\mathbf{v} = (v^{(1)}, \ldots, v^{(d)})^{\mathrm{T}}$, we define

$$\|\mathbf{v}\| = \left\{\sum_{i=1}^{d} (v^{(i)})^2\right\}^{1/2}.$$

In the case of vectors whose components are all integers, say

$$\mathbf{r} = (r^{(1)}, \ldots, r^{(d)})^{\mathrm{T}},$$

we write

$$|\mathbf{r}| = \sum_{i=1}^{d} |r^{(i)}|.$$

Let $\mathcal{C} = \mathbb{R}^n$ denote the class of all possible samples $\mathcal{X}$. If $\mathcal{E}$ is a subset of $\mathcal{C}$, we write $\tilde{\mathcal{E}} = \mathcal{E}^{\sim}$ for the complement of $\mathcal{E}$, i.e., $\mathcal{C}\backslash\mathcal{E}$. Given a function F of bounded variation and an absolutely integrable function f, both from $\mathbb{R}^d$ to $\mathbb{R}$, we define the Fourier-Stieltjes transform of F to be

$$\int_{\mathbb{R}^d} \exp(it^{\mathrm{T}}\mathbf{x})\, dF(\mathbf{x})$$

and the Fourier transform of f to be

$$\int_{\mathbb{R}^d} \exp(it^{\mathrm{T}}\mathbf{x})\, f(\mathbf{x})\, d\mathbf{x}.$$

Let $\mathcal{B}_d$ denote the class of all Borel subsets of $\mathbb{R}^d$. If $\mathbf{X}$ is a d-variate random vector then the probability measure Π on $(\mathbb{R}^d, \mathcal{B}_d)$ induced by $\mathbf{X}$ is defined to be

$$\Pi(B) = P(\mathbf{X} \in B), \qquad B \in \mathcal{B}_d.$$

The symbols $C_1, C_2, \ldots$ denote fixed constants, possibly depending on the distribution of $\mathbf{X}$ but not depending on n. The Kronecker delta is denoted by δ_{ij}.

If all moments of a random d-vector $\mathbf{X} = (X^{(1)}, \ldots, X^{(d)})^{\mathrm{T}}$ are finite, we may define

$$\mu_{\mathbf{r}} = E\big((X^{(1)})^{r_1}, \ldots, (X^{(d)})^{r_d}\big)$$

for nonnegative integer vectors $\mathbf{r} = (r^{(1)}, \ldots, r^{(d)})^{\mathrm{T}}$. In this case the characteristic function χ of $\mathbf{X}$ may be expressed as a formal power series,

$$\chi(\mathbf{t}) = E\{\exp(it^{\mathrm{T}}\mathbf{X})\} = \sum_{\mathbf{r} \geq 0} \tfrac{1}{\mathbf{r}!}\, \mu_{\mathbf{r}}(it)^{\mathbf{r}},$$

where the inequality $\mathbf{r} \geq \mathbf{0}$ is to be interpreted elementwise and

$$(it)^{\mathbf{r}} = \prod_{j=1}^{d} (it^{(j)})^{r^{(j)}}, \qquad \mathbf{r}! = \prod_{j=1}^{d} (r^{(j)}!).$$

The (multivariate) cumulants of $\mathbf{X}$ form a sequence $\{\kappa_{\mathbf{r}}, \ \mathbf{r} \geq \mathbf{0}\}$, defined formally by the relationship

$$\sum_{\mathbf{r} > 0} \tfrac{1}{\mathbf{r}!}\, \kappa_{\mathbf{r}}(it)^r = \log \chi(\mathbf{t})$$

$$= \sum_{k \geq 1} (-1)^{k+1} \tfrac{1}{k} \Big\{ \sum_{\mathbf{r} > 0} \tfrac{1}{\mathbf{r}!}\, \mu_{\mathbf{r}}(it)^{\mathbf{r}} \Big\}^k.$$

This formula expresses $\kappa_{\mathbf{r}}$ as a polynomial in $\mu_{\mathbf{s}}$ for $\mathbf{s} \leq \mathbf{r}$. (We described the univariate case in Section 2.2, and there (2.6) defines κ_r in terms of μ_s, $s \leq r$.)

Let $\check{P}_j(\mathbf{t}, \{\kappa_{\mathbf{r}}\})$, $j \geq 1$, denote polynomials defined by the formal expansions

$$1 + \sum_{j \geq 1} \check{P}_j(\mathbf{t}, \{\kappa_{\mathbf{r}}\})\, u^j \;=\; \exp\left\{\sum_{k \geq 1}\Big(\sum_{\mathbf{r}:|\mathbf{r}|=k+2} \tfrac{1}{\mathbf{r}!}\, \mathbf{t}^{\mathbf{r}}\Big) u^k\right\}$$

$$= 1 + \sum_{l \geq 1} \tfrac{1}{l!}\left\{\sum_{k \geq 1}\Big(\sum_{\mathbf{r}:|\mathbf{r}|=k+2} \tfrac{1}{\mathbf{r}!}\, \kappa_{\mathbf{r}}\, \mathbf{t}^{\mathbf{r}}\Big) u^k\right\}^l,$$

where u is a real number. Thus, $\check{P}_j(\mathbf{t} : \{\kappa_{\mathbf{r}}\})$ is a polynomial of degree $3j$, odd or even according to whether j is odd or even, respectively, and depends on the sequence $\{\kappa_{\mathbf{r}}\}$ only through those $\kappa_{\mathbf{r}}$'s with $|\mathbf{r}| \leq j + 2$. We may define $\check{P}_j(\mathbf{t} : \{\kappa_{\mathbf{r}}\})$ whenever the distribution of $\mathbf{X}$ satisfies $E(\|\mathbf{X}\|^{j+2}) < \infty$; we do not need all moments to be finite.

Given a d-variate cumulant sequence $\{\kappa_{\mathbf{r}}\}$ (which need only be defined for $|\mathbf{r}| \leq j + 2$) and a $d \times d$ variance matrix $\mathbf{V}$, let $P_j(-\phi_{0,\mathbf{V}} : \{\kappa_{\mathbf{r}}\})$ denote the function whose Fourier transform equals $\check{P}_j(it : \{\kappa_{\mathbf{r}}\}) \exp(-\tfrac{1}{2}\,\mathbf{t}^{\mathrm{T}}\mathbf{V}\mathbf{t})$,

$$\int_{\mathbb{R}^d} P_j(-\phi_{0,\mathbf{V}} : \{\kappa_{\mathbf{r}}\})(\mathbf{x})\, \exp(it^{\mathrm{T}}\mathbf{x})\, d\mathbf{x} \;=\; \check{P}_j(it : \{\kappa_{\mathbf{r}}\}) \exp\Big(-\tfrac{1}{2}\,\mathbf{t}^{\mathrm{T}}\mathbf{V}\mathbf{t}\Big).$$

It is readily shown by integration by parts that

$$P_j(-\phi_{0,\mathbf{V}} : \{\kappa_{\mathbf{r}}\})(\mathbf{x}) \;=\; \check{P}_j(-\mathbf{D} : \{\kappa_{\mathbf{r}}\})\, \phi_{0,\mathbf{V}}(\mathbf{x}),$$

where $\phi_{0,\mathbf{V}}$ denotes the density of the d-variate Normal $N(\mathbf{0}, \mathbf{V})$ distribution, $\mathbf{D} = (\partial/\partial x^{(1)}, \dots, \partial/\partial x^{(d)})^{\mathrm{T}}$ is the vector differential operator, and $\check{P}_j(-\mathbf{D} : \{\kappa_{\mathbf{r}}\})$ is the polynomial differential operator defined in terms of a polynomial in $-\mathbf{D}$. Let $P_j(-\Phi_{0,\mathbf{V}} : \{\kappa_{\mathbf{r}}\})$ be the signed measure whose density is $P_j(-\phi_{0,\mathbf{V}} : \{\kappa_{\mathbf{r}}\})$, i.e.,

$$P_j(-\Phi_{0,\mathbf{V}} : \{\kappa_{\mathbf{r}}\})(\mathcal{R}) \;=\; \int_{\mathcal{R}} P_j(-\phi_{0,\mathbf{V}} : \{\kappa_{\mathbf{r}}\})(\mathbf{x})\, d\mathbf{x}, \qquad \mathcal{R} \subseteq \mathbb{R}^d.$$

A detailed account of the properties of $\check{P}_j(\mathbf{t} : \{\kappa_{\mathbf{r}}\})$, $P_j(-\phi_{0,\mathbf{V}} : \{\kappa_{\mathbf{r}}\})$, and $P_j(-\Phi_{0,\mathbf{V}} : \{\kappa_{\mathbf{r}}\})$ is given by Bhattacharya and Rao (1976, pp. 51–57).

(ii) Bootstrap Edgeworth Expansion

We begin by defining four subsets $\mathcal{E}_1, \dots, \mathcal{E}_4$ of $\mathcal{C}$.

The random coefficients of polynomials $\hat{q}_i$, $1 \leq i \leq \nu$, are continuous functions of bootstrap population moments, i.e., of sample moments, of order $3\nu - 1$ or less. A typical sample moment of order r has the form

$$M = n^{-1} \sum_{i=1}^{n} \prod_{j=1}^{d} (X_i^{(j)})^{l_j},$$

where $l_1, \ldots, l_d$ are nonnegative integers satisfying $l_1 + \ldots + l_d = r$. The corresponding population moment is $E(M)$ and does not depend on n. Under sufficiently stringent moment conditions on $\mathbf{X}$, it follows by Markov's inequality that each such sample moment appearing in the coefficients satisfies

$$P(|M - EM| > \epsilon) = O(n^{-\lambda})$$

for all $\epsilon > 0$. Therefore, we may choose $C_1 > 0$ and $\mathcal{E}_1 \subseteq \mathcal{C}$ such that (i) $P(\tilde{\mathcal{E}}_1) = O(n^{-\lambda})$, and (ii) whenever $\mathcal{X} \in \mathcal{E}_1$, the absolute values of each coefficient of $\hat{\pi}_j$, $1 \le j \le \nu$, and the absolute values of the first $\nu + 3$ derivatives of A evaluated at $\overline{\mathbf{X}}$, are less than or equal to C_1. This proves (5.8). The remainder of our proof is directed at establishing (5.7).

Let $\mathbf{X}^*$ denote a generic $\mathbf{X}_i^*$, and observe that for any $r > 2$,

$$E(\|\mathbf{X}^* - \overline{\mathbf{X}}\|^r \mid \mathcal{X}) = n^{-1} \sum_{i=1}^{n} \left\{ \sum_{j=1}^{d} (X_i^{(j)} - \overline{X}^{(j)})^2 \right\}^{r/2}$$

$$\le (2d)^r n^{-1} \sum_{i=1}^{n} \sum_{j=1}^{d} (|X_i^{(j)}|^r + |\overline{X}^{(j)}|^r).$$

Therefore, by Markov's inequality,

$$P\left\{ E(\|\mathbf{X}^* - \overline{\mathbf{X}}\|^r \mid \mathcal{X}) > (2d)^r \sum_{j=1}^{d} (E|X^{(j)}|^r + |EX^{(j)}|^r) + (2d)^{r+1} \right\}$$

$$\le P\left\{ \sum_{j=1}^{d} \left| \sum_{i=1}^{n} (|X_i^{(j)}|^r - E|X^{(j)}|^r) \right| + \sum_{j=1}^{d} ||\overline{X}^{(j)}|^r - |E\overline{X}^{(j)}|^r| > 2dn \right\}$$

$$\le \sum_{j=1}^{d} n^{-2\lambda} E\left\{ \left| \sum_{i=1}^{n} (|X_i^{(j)}|^r - E|X^{(j)}|^r) \right|^\lambda \right\} + \sum_{j=1}^{d} E|\overline{X}^{(j)} - E\overline{X}^{(j)}|^{r\lambda}$$

$$= O(n^{-\lambda}),$$

the last identity following by Rosenthal's inequality (Hall and Heyde 1980, p. 23) under sufficiently stringent moment conditions on $\mathbf{X}$. Hence we may choose $C_2 > 1$ such that the set

$$\mathcal{E}_2 = \left\{ \mathcal{X} : E(\|\mathbf{X}^* - \overline{\mathbf{X}}\|^{\max(2\nu+3,\, d+1)} \mid \mathcal{X}) \le C_2 \right\}$$

has $P(\tilde{\mathcal{E}}_2) = O(n^{-\lambda})$.

Let $\mathbf{V} = (v_{ij})$ denote the variance matrix of $\mathbf{X}$. Cramér's condition (5.5) ensures that $\mathbf{V}$ is positive definite. The sample estimate of $\mathbf{V}$ is $\hat{\mathbf{V}} = (\hat{v}_{ij})$, where

$$\hat{v}_{ij} \;=\; n^{-1} \sum_{k=1}^{n} (\mathbf{X}_k - \overline{\mathbf{X}})^{(i)} \, (\mathbf{X}_k - \overline{\mathbf{X}})^{(j)} \,.$$

Under sufficiently stringent moment conditions on $\mathbf{X}$, and for each $\epsilon > 0$,

$$P\big(\max_{i,j} |\hat{v}_{ij} - v_{ij}| > \epsilon \big) \;=\; O(n^{-\lambda}) \,.$$

Hence for some $C_3 > 1$, the set

$$\mathcal{E}_3 \;=\; \big\{ \mathcal{X} : \text{all eigenvalues of } \hat{\mathbf{V}} \text{ lie within } (C_3^{-1}, C_3) \big\}$$

has $P(\tilde{\mathcal{E}}_3) = O(n^{-\lambda})$.

Given a resample $\mathcal{X}^* = \{\mathbf{X}_1^*, \dots, \mathbf{X}_n^*\}$, define

$$\mathbf{Y}_i \;=\; \begin{cases} \hat{\mathbf{V}}^{-1/2}(\mathbf{X}_i^* - \overline{\mathbf{X}}) & \text{if } \|\hat{\mathbf{V}}^{-1/2}(\mathbf{X}_i^* - \overline{\mathbf{X}})\| \leq n^{1/2} \,, \\ 0 & \text{otherwise} \,, \end{cases}$$

and $\mathbf{X}_i^\dagger = \mathbf{Y}_i - E(\mathbf{Y}_i \mid \mathcal{X})$. Let $\mathbf{Y}$, $\mathbf{X}^\dagger$ denote generic versions of $\mathbf{Y}_i$, $\mathbf{X}_i^\dagger$, respectively. Write Q and $Q^\dagger$ for the probability measures on $\mathbb{R}^d$ generated by the random variables $n^{-1/2} \sum (\mathbf{X}_i^* - \overline{\mathbf{X}})$ and $n^{-1/2} \sum \mathbf{X}_i^\dagger$ respectively, both conditional on $\mathcal{X}$. Let $\hat{\mathbf{V}}^\dagger = (\hat{v}_{ij}^\dagger)$ be the variance matrix and $\{\hat{\kappa}_r^\dagger\}$ the cumulant sequence of $\mathbf{X}^\dagger$, again conditional on $\mathcal{X}$. Write $\{\hat{\kappa}_r\}$ for the conditional cumulant sequence of $\hat{\mathbf{V}}^{-1/2}(\mathbf{X}^* - \overline{\mathbf{X}})$. Observe that $\hat{\mathbf{V}}^\dagger \to \mathbf{I} = (\delta_{ij})$ with probability one as $n \to \infty$. If $\mathbf{X}$ has sufficiently many finite moments then for each $\epsilon > 0$,

$$P\big(\max_{i,j} |\hat{v}_{ij}^\dagger - \delta_{ij}| > \epsilon \big) \;=\; O(n^{-\lambda}) \,.$$

It follows that the set

$$\mathcal{E}_4 \;=\; \big\{ \mathcal{X} : \text{all eigenvalues of } \hat{\mathbf{V}}^\dagger \text{ lie within } (\tfrac{1}{2}, 2) \big\}$$

satisfies $P(\tilde{\mathcal{E}}_4) = O(n^{-\lambda})$.

Throughout the remainder of our proof we consider only samples

$$\mathcal{X} \in \mathcal{E} \;=\; \mathcal{E}_1 \cap \mathcal{E}_2 \cap \mathcal{E}_3 \cap \mathcal{E}_4.$$

Note that

$$P(\tilde{\mathcal{E}}) \;=\; O(n^{-\lambda}). \tag{5.11}$$

Let $\mathcal{R}$ denote a Borel subset of $\mathbb{R}^d$, possibly random but measurable in the σ-field generated by $\mathcal{X}$. Indeed, we shall need only sets of the form

$$\mathcal{R} \;=\; \mathcal{R}(t) \;=\; \big\{ \mathbf{x} \in \mathbb{R}^d : n^{1/2} \, \hat{A}(\overline{\mathbf{X}} + n^{-1/2} \mathbf{x}) \leq t \big\}, \tag{5.12}$$

$$-\infty < t < \infty.$$

Put

$$\mathcal{R}^{\dagger} = \hat{\mathbf{V}}^{-1/2} \mathcal{R} - n^{1/2} E(\mathbf{Y} \mid \mathcal{X})$$

$$= \{\hat{\mathbf{V}}^{-1/2} \mathbf{x} - n^{1/2} E(\mathbf{Y} \mid \mathcal{X}) : \mathbf{x} \in \mathcal{R}\}$$

and

$$R = Q^{\dagger} - \sum_{j=0}^{\nu+3} n^{-j/2} P_j(-\Phi_{0,\hat{\mathbf{V}}^{\dagger}} : \{\hat{\kappa}_{\mathbf{r}}^{\dagger}\}).$$

Since the coefficients of $P_j(-\Phi_{0,\hat{\mathbf{V}}^{\dagger}} : \{\kappa_{\mathbf{r}}^{\dagger}\})$ are polynomials in the conditional moments of $\hat{\mathbf{V}}^{\dagger-1/2} \mathbf{X}^{\dagger}$, up to order $j+2$, and since for $j \geq \nu+1$ and $\mathcal{X} \in \mathcal{E}$,

$$E\big(\|\hat{\mathbf{V}}^{\dagger-1/2} \mathbf{X}^{\dagger}\|^{j+2} \mid \mathcal{X}\big) \leq 2^{(j+2)/2} E\big(\|\mathbf{X}^{\dagger}\|^{j+2} \mid \mathcal{X}\big)$$

$$\leq 2^{3(j+2)/2} E\big(\|\mathbf{Y}\|^{j+2} \mid \mathcal{X}\big)$$

$$\leq 2^{3(j+2)/2} n^{(j-\nu-1)/2} E\big(\|\mathbf{Y}\|^{\nu+3} \mid \mathcal{X}\big)$$

$$\leq 2^{3(j+2)/2} C_2 C_3^{(\nu+3)/2} n^{(j-\nu-1)/2},$$

then for a constant $C_4 > 0$,

$$\int_{\mathbb{R}^d} \left| d\left[Q^{\dagger} - \sum_{j=0}^{\nu} n^{-j/2} P_j\big(-\Phi_{0,\hat{\mathbf{V}}^{\dagger}} : \{\kappa_{\mathbf{r}}^{\dagger}\}\big) - R\right]\right|$$

$$\leq C_4 \sum_{j=\nu+1}^{\nu+d} n^{-j/2} n^{(j-\nu-1)/2} = C_4 d n^{-(\nu+1)/2}. \qquad (5.13)$$

Next we show that for a constant $C_5 > 0$,

$$\sup_{\mathcal{R} \subseteq \mathbb{R}^d} \left| \int_{\mathcal{R}} d\left[Q - \sum_{j=0}^{\nu} n^{-j/2} P_j\big(-\Phi_{0,\hat{\mathbf{v}}} : \{\hat{\kappa}_{\mathbf{r}}\}\big)\right] \right.$$

$$\left. - \int_{\mathcal{R}^{\dagger}} d\left[Q^{\dagger} - \sum_{j=0}^{\nu} n^{-j/2} P_j\big(-\Phi_{0,\hat{\mathbf{V}}^{\dagger}} : \{\kappa_{\mathbf{r}}^{\dagger}\}\big)\right]\right|$$

$$\leq C_5 n^{-(\nu+1)/2} \qquad (5.14)$$

uniformly in $\mathcal{X} \in \mathcal{E}$. If $\mathcal{X} \in \mathcal{E}_3$ then

$$\|\hat{\mathbf{V}}^{-1/2}(\mathbf{X}^* - \bar{\mathbf{X}})\| \leq C_3^{1/2} \|\mathbf{X}^* - \bar{\mathbf{X}}\|,$$

and so by Markov's inequality

$$P\left\{ \|\widehat{\mathbf{V}}^{-1/2}(\mathbf{X}^* - \overline{\mathbf{X}})\| > n^{1/2} \mid \mathcal{X}\right\}$$

$$\le\ C_3^{(\nu+3)/2}\, n^{-(\nu+3)/2}\, E(\,\|\mathbf{X}^* - \overline{\mathbf{X}}\|^{\nu+3}\mid \mathcal{X})$$

$$\le C_2\, C_3^{(\nu+3)/2}\, n^{-(\nu+3)/2}\,.$$

Therefore, with $C_6 = C_2\, C_3^{(\nu+3)/2}$,

$$P\left\{ \max_{1\le i\le n} \|\widehat{\mathbf{V}}^{-1/2}(\mathbf{X}_i^* - \overline{\mathbf{X}})\| > n^{1/2} \mid \mathcal{X}\right\}\ \le\ C_6\, n^{-(\nu+1)/2}\,,$$

whence

$$\sup_{\mathcal{R}\subseteq\mathbb{R}}\ |Q(\mathcal{R}) - Q^\dagger(\mathcal{R}^\dagger)|\ \le\ C_6\, n^{-(\nu+1)/2}\,. \tag{5.15}$$

Moreover, if $j \le \nu + 2$ then any conditional jth moment of $\widehat{\mathbf{V}}^{-1/2}(\mathbf{X}^* - \overline{\mathbf{X}})$ differs from its counterpart for $\mathbf{X}^\dagger$ by $O(n^{-(\nu+1)/2})$, as follows from the inequalities

$$E\left\{ \|\mathbf{X}^\dagger - \widehat{\mathbf{V}}^{-1/2}(\mathbf{X}^* - \overline{\mathbf{X}})\|^j \mid \mathcal{X}\right\}$$

$$\le\ 2^j\, E\left[\|\widehat{\mathbf{V}}^{-1/2}(\mathbf{X}^* - \overline{\mathbf{X}})\|^j\, I\{\|\widehat{\mathbf{V}}^{-1/2}(\mathbf{X}^* - \overline{\mathbf{X}})\| > n^{1/2}\} \mid \mathcal{X}\right]$$

$$\le\ 2^j\, n^{\{j-(2\nu+3)\}/2}\, E\left\{ \|\widehat{\mathbf{V}}^{-1/2}(\mathbf{X}^* - \overline{\mathbf{X}})\|^{2\nu+3} \mid \mathcal{X}\right\}$$

$$\le 2^{\nu+2}\, C_2\, C_3^{(2\nu+3)/2}\, n^{-(\nu+1)/2}\,.$$

Therefore, for a constant $C_7 > 0$ not depending on n,

$$\sup_{\mathcal{R}\subseteq\mathbb{R}}\ \left| \int_{\mathcal{R}} d\Big[\sum_{j=0}^{\nu} n^{-j/2}\, P_j(-\Phi_{\mathbf{0},\hat{\mathbf{v}}} : \{\hat{\kappa}_\mathbf{r}\})\Big]\right.$$

$$\left. - \int_{\mathcal{R}^\dagger} d\Big[\sum_{j=0}^{\nu} n^{-j/2}\, P_j(-\Phi_{\mathbf{0},\hat{\mathbf{v}}^\dagger} : \{\hat{\kappa}_\mathbf{r}^\dagger\})\Big]\right|$$

$$\le\ C_7\, n^{-(\nu+1)/2}$$

uniformly in $\mathcal{X} \in \mathcal{E}$. The claimed result (5.14) follows from this inequality and (5.15).

Combining (5.13) and (5.14) and taking $C_8 = C_4\, d + C_5$, we deduce that

$$\sup_{\mathcal{R}\subseteq\mathbb{R}^d}\ \left| \int_{\mathcal{R}} d\Big[Q - \sum_{j=0}^{\nu} n^{-j/2}\, P_j(-\Phi_{\mathbf{0},\hat{\mathbf{v}}} : \{\hat{\kappa}_\mathbf{r}\})\Big] - \int_{\mathcal{R}^\dagger} dR\right|$$

$$\le\ C_8\, n^{-(\nu+1)/2}\,. \tag{5.16}$$

Next we develop an upper bound to $\left|\int_{\mathcal{R}\dagger} dR\right|$, by smoothing the measure R. Write $B(\mathbf{x}, r)$ for the closed sphere in $\mathrm{I\!R}^d$ centred at $\mathbf{x}$ and with radius r. Let L be a probability density on $\mathrm{I\!R}^d$ with support confined to $B(\mathbf{0}, 1)$ and Fourier transform (i.e., characteristic function)

$$\check{L}(\mathbf{t}) \; = \; \int_{\mathrm{I\!R}^d} \exp(i\mathbf{t}^{\mathrm{T}}\mathbf{x}) \, L(\mathbf{x}) \, dx$$

satisfying

$$|D^{\mathbf{r}} \, \check{L}(\mathbf{t})| \; \leq \; C_9 \, \exp(-\|\mathbf{t}\|^{1/2}) \, ,$$

for all $\mathbf{t} \in \mathrm{I\!R}^d$ and all d-vectors $\mathbf{r} = (r^{(1)}, \dots, r^{(d)})^{\mathrm{T}}$ with nonnegative integer components such that

$$|\mathbf{r}| \; = \; \sum_{j=1}^{d} r^{(j)} \; \leq \; d+1 \, .$$

Here $D^{\mathbf{r}}$ denotes the differential operator

$$D^{\mathbf{r}} \; = \; \prod_{j=1}^{d} \left(\frac{\partial}{\partial \, t^{(j)}} \right)^{r^{(j)}} \, .$$

Define $L_\delta(\mathbf{x}) = \delta^{-d} L(\delta^{-1}\mathbf{x})$ for $0 < \delta \leq 1$, and let $\check{L}_\delta(\mathbf{t}) = \check{L}(\delta\mathbf{t})$ be the corresponding Fourier transform. There exists a constant $C_{10} > 0$ such that for any function $g : \mathrm{I\!R}^d \to \mathrm{I\!R}$ with Fourier transform $\check{g}$,

$$\int_{\mathrm{I\!R}^d} |g(\mathbf{x})| \, dx \; \leq \; C_{10} \max_{\mathbf{r}:|\mathbf{r}|=0, d+1} \int_{\mathrm{I\!R}^d} |D^{\mathbf{r}} \, \check{g}(\mathbf{t})| \, dt \, . \tag{5.17}$$

Take g to be the density of the measure G defined by

$$G(\mathcal{R}) \; = \; \int_{\mathrm{I\!R}^d} L_\delta(\mathbf{y}) \, dy \int_{\mathcal{R}-\mathbf{y}} dR \, .$$

Then $\check{g} = \check{L}_\delta \, \rho$, where ρ is the Fourier-Stieltjes transform of the distribution function corresponding to the measure R, and so

$$\int_{\mathrm{I\!R}^d} |g(\mathbf{x})| \, dx \; \leq \; C_{10} \max_{|\mathbf{r}|=0, \; d+1} \int_{\mathrm{I\!R}^d} \left| D^{\mathbf{r}}\{\check{L}_\delta(\mathbf{t}) \, \rho(\mathbf{t})\} \right| \, dt$$

$$\leq \; C_{11} \max_{\substack{0 \leq \mathbf{r} \leq \mathbf{s} \\ |\mathbf{s}| \leq d+1}} \int_{\mathrm{I\!R}^d} \left| D^{\mathbf{r}} \, \check{L}_\delta(\mathbf{t}) \right| \left| D^{\mathbf{s}-\mathbf{r}} \, \rho(\mathbf{t}) \right| \, dt \, , \tag{5.18}$$

where the inequality $\mathbf{0} \leq \mathbf{r} \leq \mathbf{s}$ is to be interpreted elementwise. Write $\partial\mathcal{R}$ for the boundary of $\mathcal{R}$, and $(\partial\mathcal{R})^\epsilon$ for the set of all points in $\mathrm{I\!R}^d$ distant no more than ϵ from $\partial\mathcal{R}$. Then, since L vanishes outside $B(\mathbf{0}, 1)$, we have for any $\mathcal{R} \subseteq \mathbb{R}^d$,

$$\left| \int_{\mathcal{R}} dR \right| \leq \int_{\mathbb{R}^d} |g(\mathbf{x})| \, d\mathbf{x}$$

$$+ \sum_{j=0}^{\nu+d} n^{-j/2} \int_{(\partial \mathcal{R})^{2\delta}} \left| P_j\bigl(-\phi_{\mathbf{0},\hat{\mathbf{V}}^\dagger} : \{\hat{\kappa}_\nu^\dagger\}\bigr)(\mathbf{x}) \right| d\mathbf{x}, \quad (5.19)$$

where g is as defined just above. Replacing $\mathcal{R}$ by $\mathcal{R}^\dagger$ and combining (5.18) and (5.19) we deduce that

$$\left| \int_{\mathcal{R}^\dagger} dR \right| \leq C_{11} \max_{\substack{0 \leq \mathbf{r} \leq \mathbf{s} \\ |\mathbf{s}| \leq d+1}} \int_{\mathbb{R}^d} \left| D^{\mathbf{r}} \check{L}_\delta(\mathbf{t}) \right| \left| D^{\mathbf{s}-\mathbf{r}} \rho(\mathbf{t}) \right| d\mathbf{t}$$

$$+ \sum_{j=0}^{\nu+d} n^{-j/2} \int_{(\partial \mathcal{R}^\dagger)^{2\delta}} \left| P_j\bigl(-\phi_{\mathbf{0},\hat{\mathbf{V}}^\dagger} : \{\hat{\kappa}_\nu^\dagger\}\bigr)(\mathbf{x}) \right| d\mathbf{x}. \quad (5.20)$$

To simplify the first term on the right-hand side of (5.20) we need a standard result on Taylor expansion of characteristic functions, which we state here without proof. The lemma is a special case of Theorem 9.9 of Bhattacharya and Rao (1976, p. 77) and of Lemma 8 of Sazanov (1981, p. 55). It applies to an expansion of the characteristic function of a general random variable, which for simplicity we write as $\mathbf{X}$.

LEMMA 5.1. *Let χ denote the characteristic function of a d-variate random variable $\mathbf{X}$ with zero mean, nonsingular variance matrix $\mathbf{V}$, and cumulant sequence $\{\kappa_\nu\}$. Assume that $E\|\mathbf{X}\|^m < \infty$ for some integer $m \geq 3$. Let $C > 0$ be a constant such that $\mathbf{t}^T \mathbf{V}^2 \mathbf{t} \leq C$ implies $|\chi(\mathbf{t}) - 1| \leq \frac{1}{2}$. There exist absolute constants C_{12} and C_{13}, depending only on d and m, such that for all vectors $\mathbf{r} = (r^{(1)}, \ldots, r^{(d)})^T$ with nonnegative integer components satisfying $\sum |r^{(j)}| \leq m$,*

$$\left| D^{\mathbf{r}} \Bigl[\chi(n^{-1/2} \mathbf{V}^{-1/2} \mathbf{t})^n - \exp\bigl(-\tfrac{1}{2} \|\mathbf{t}\|^2 \bigr) \sum_{j=0}^{m-3} n^{-j/2} \check{P}_j\bigl(i\mathbf{V}^{-1/2} : \{\kappa_\nu\}\bigr) \Bigr] \right|$$

$$\leq C_{12}\, E\|\mathbf{V}^{-1/2} \mathbf{X}\|^m\, n^{-(m-2)/2}$$

$$\times \bigl(\|\mathbf{t}\|^{m-|\mathbf{r}|} + \|\mathbf{t}\|^{3(m-2)+|\mathbf{r}|} \bigr) \exp\bigl(-\tfrac{1}{4} \|\mathbf{t}\|^2 \bigr) \quad (5.21)$$

uniformly in $\|\mathbf{t}\| \leq n^{1/2} \min\bigl\{ C^{1/2}, C_{13}(E\|\mathbf{V}^{-1/2} \mathbf{X}\|^m)^{-1/(m-2)} \bigr\}$.

In the lemma, replace $\mathbf{X}$ by $\mathbf{X}^\dagger$ (conditional on $\mathcal{X}$) and take

$$m = \max(\nu + 3, d+1).$$

Then $\mathbf{V}$ changes to $\hat{\mathbf{V}}^\dagger$ and $\chi(\mathbf{t})$ is replaced by

$$\chi^\dagger(\mathbf{t}) = E\bigl\{ \exp(i\mathbf{t}^T \mathbf{X}^\dagger) \mid \mathcal{X} \bigr\}.$$

If $\mathcal{X} \in \mathcal{E}_4$ then

$$|1 - \chi^\dagger(\mathbf{t})| \leq \mathbf{t}^T \widehat{\mathbf{V}}^\dagger \mathbf{t} \leq 2 \mathbf{t}^T \mathbf{t} \qquad \text{and} \qquad \mathbf{t}^T \widehat{\mathbf{V}}^{\dagger^2} \mathbf{t} \geq \tfrac{1}{4} \mathbf{t}^T \mathbf{t},$$

whence $\mathbf{t}^T \widehat{\mathbf{V}}^{\dagger^2} \mathbf{t} \leq \frac{1}{16}$ implies $|1 - \chi^\dagger(\mathbf{t})| \leq \frac{1}{2}$. Therefore, we may take $C = \frac{1}{16}$ in the lemma. Furthermore, if $\mathcal{X} \in \mathcal{E}_2 \cap \mathcal{E}_3 \cap \mathcal{E}_4$,

$$
\begin{aligned}
E(\|\widehat{\mathbf{V}}^{\dagger -1/2} \mathbf{X}^\dagger\|^m \mid \mathcal{X}) &\leq 2^{m/2} E(\|\mathbf{X}^\dagger\|^m \mid \mathcal{X}) \\
&\leq 2^{3m/2} E(\|\mathbf{Y}\|^m \mid \mathcal{X}) \\
&\leq 2^{3m/2} C_3^{m/2} E(\|\mathbf{X}^* - \overline{\mathbf{X}}\|^m \mid \mathcal{X}) \\
&\leq 2^{3m/2} C_3^{m/2} C_2 = C_{14},
\end{aligned}
$$

say. Defining $C_{15} = \min\left(\frac{1}{4}, C_{13} C_{14}^{-1/(m-2)}\right)$, we deduce from (5.21) that if $\mathcal{X} \in \mathcal{E}$ and $|\mathbf{r}| \leq m$,

$$|D^{\mathbf{r}} \rho(\mathbf{t})| \leq C_{12} C_{14} n^{-(m-2)/2} \left(\|\mathbf{t}\|^{m-|\mathbf{r}|} + \|\mathbf{t}\|^{3(m-2)+|\mathbf{r}|}\right) \exp\left(-\tfrac{1}{4} \|\mathbf{t}\|^2\right)$$

uniformly in $\|\mathbf{t}\| \leq C_{15} n^{1/2}$. Since $|D^{\mathbf{r}} L_\delta(\mathbf{t})| \leq C_9$ then

$$\max_{\substack{0 \leq \mathbf{r} \leq \mathbf{s} \\ |\mathbf{s}| \leq d+1}} \int_{\|\mathbf{t}\| \leq C_{15} n^{1/2}} |D^{\mathbf{r}} \check{L}_\delta(\mathbf{t})| |D^{\mathbf{s}-\mathbf{r}} \rho(\mathbf{t})| \, d\mathbf{t} \leq C_{16} n^{-(\nu+1)/2}, \quad (5.22)$$

where

$$C_{16} = C_9 C_{12} C_{14} \max_{0 \leq r \leq m} \int_{\mathbb{R}^d} \left(\|\mathbf{t}\|^{m-r} + \|\mathbf{t}\|^{3(m-2)+r}\right) \exp\left(-\tfrac{1}{4} \|\mathbf{t}\|^2\right) d\mathbf{t}.$$

Next we develop a formula analogous to (5.22), in which the integral over $\|\mathbf{t}\| \leq C_{15} n^{1/2}$ on the left-hand side is replaced by an integral over $\|\mathbf{t}\| > C_{15} n^{1/2}$. Let ψ denote the characteristic function corresponding to the probability measure $Q^\dagger$, and observe that

$$\rho(\mathbf{t}) = \psi(\mathbf{t}) - \exp\left(-\tfrac{1}{2} \|\mathbf{t}\|^2\right) \sum_{j=0}^{\nu+3} n^{-j/2} \check{P}_j\left(i\widehat{\mathbf{V}}^{\dagger -1/2} \mathbf{t} : \{\hat{\kappa}_\nu^\dagger\}\right).$$

If $\mathcal{X} \in \mathcal{E}$ then for a constant $C_{17} > 0$,

$$\max_{|\mathbf{r}| \leq d+1} \left| D^{\mathbf{r}} \left[\exp\left(-\tfrac{1}{2} \|\mathbf{t}\|^2\right) \sum_{j=0}^{\nu+3} n^{-j/2} \check{P}_j\left(i\widehat{\mathbf{V}}^{\dagger -1/2} \mathbf{t} : \{\hat{\kappa}_\nu^\dagger\}\right)\right]\right|$$

$$\leq C_{17} \exp\left(-\tfrac{1}{4} \|\mathbf{t}\|^2\right)$$

uniformly in $\mathbf{t}$, and so

$$\max_{0 \leq \mathbf{r} \leq \mathbf{s}, |\mathbf{s}| \leq d+1} \int_{\|\mathbf{t}\| > C_{15} n^{1/2}} |D^{\mathbf{r}} \check{L}_\delta(\mathbf{t})| |D^{\mathbf{s}-\mathbf{r}} \{\rho(\mathbf{t}) - \psi(\mathbf{t})\}| \, d\mathbf{t}$$

$$\leq C_{18} n^{-(\nu+1)/2}. \qquad (5.23)$$

The lemma below provides an upper bound to

$$a_\delta = \max_{\substack{0 \le r \le s \\ |s| \le d+1}} \int_{\|\mathbf{t}\| > C_{15}\, n^{1/2}} \left| D^{\mathbf{r}}\, \check{L}_\delta(\mathbf{t}) \right| \left| D^{\mathbf{s}-\mathbf{r}}\, \psi(\mathbf{t}) \right| d\mathbf{t}\,.$$

Let $\chi^*(\mathbf{t}) = E\{\exp(i\mathbf{t}^T \mathbf{X}^*) \mid \mathcal{X}\}$ denote the empirical characteristic function of the sample, and note that $\chi = E(\chi^*)$.

LEMMA 5.2. *Assume that the distribution of* $\mathbf{X}$ *satisfies Cramér's condition. Then there exist positive constants* $C_{19} \in (0,1)$, C_{20}, C_{21}, *and* C_{22} *such that whenever* $\mathcal{X} \in \mathcal{E}$, $0 < \delta < 1$, $n > 2(d+1)$, *and* $m \in [1,\, n-(d+1)]$,

$$a_\delta \le C_{20}\, n^{d+1} \int \left[C_{19}^n + \{ C_{21}\, |\chi^*(n^{-1/2}\, \mathbf{t}) - \chi(n^{-1/2}\, \mathbf{t})| \}^m \right]$$

$$\times \exp\left(-C_{22}\, \|\delta \mathbf{t}\|^{1/2} \right) d\mathbf{t}\,. \tag{5.24}$$

A proof will be given in step (iv). In the meantime, take $\delta = n^{-u}$, where u is yet to be chosen. Since

$$E\{ |\chi^*(\mathbf{t}) - \chi(\mathbf{t})|^m \} \le C_{23}(m)\, n^{-m/2}$$

uniformly in $\mathbf{t}$, then by Markov's inequality, the set

$$\mathcal{E}_5 = \left\{ \mathcal{X} : \int \left| \chi^*(n^{-(1/2)+u}\, \mathbf{t}) - \chi(n^{-(1/2)+u}\, \mathbf{t}) \right|^m \exp(-C_{22}\, \|\mathbf{t}\|^{1/2})\, d\mathbf{t} \right.$$
$$\left. \le n^{-d(u+1)-(\nu+3)/2} \right\}$$

satisfies $P(\tilde{\mathcal{E}}_5) = O(n^{-\lambda})$, provided m (fixed, not depending on n) is chosen sufficiently large. Henceforth we shall assume that $\mathcal{X} \in \mathcal{E}' = \mathcal{E} \cap \mathcal{E}_5$; note that by (5.11),

$$P(\tilde{\mathcal{E}}) = O(n^{-\lambda})\,.$$

For such an $\mathcal{X}$ we may deduce from (5.24) that

$$a_\delta \le C_{20}\, n^{d+1} \left(C_{19}^n\, n^{du} + C_{21}^m\, n^{-d-(\nu+3)/2} \right) \le C_{24}\, n^{-(\nu+1)/2}\,.$$

Combining this result with (5.20) and (5.22) we conclude that with $\delta = n^{-u}$,

$$\left| \int_{\mathcal{R}^\dagger} dR \right| \le C_{11}(C_{16} + C_{24})\, n^{-(\nu+1)/2} + b_\delta\,, \tag{5.25}$$

where

$$b_\delta = \sum_{j=0}^{\nu+d} n^{-j/2} \int_{(\partial \mathcal{R}^\dagger)^{2\delta}} \left| P_j \big(-\phi_{\mathbf{0},\hat{\mathbf{V}}^\dagger} : \{\hat{\kappa}_\nu^\dagger\} \big)(\mathbf{x}) \right| d\mathbf{x}$$

uniformly in $\mathcal{X} \in \mathcal{E}'$ and $\mathcal{R} \subseteq \mathbb{R}^d$.

The remainder of our proof is directed at obtaining an upper bound for b_δ, uniformly in sets $\mathcal{R}$ of the form (5.12). Note that the function

$$P_j\left(-\phi_{0,\widehat{\mathbf{V}}^\dagger} : \{\hat{\kappa}_\nu^\dagger\}\right)$$

equals a polynomial of degree $3j$ multiplied by $\exp\left(-\tfrac{1}{2}\mathbf{x}^\mathrm{T}\,\widehat{\mathbf{V}}^{\dagger^{-1}}\mathbf{x}\right)$. The coefficients in the polynomial, albeit random, are bounded uniformly in $\mathcal{X} \in \mathcal{E}$, and so

$$b_\delta \leq C_{25} \int_{(\partial\mathcal{R}^\dagger)^{2\delta}} \left(1 + \|\mathbf{x}\|^{3(\nu+d)}\right) \exp\left(-\tfrac{1}{2}\mathbf{x}^\mathrm{T}\,\widehat{\mathbf{V}}^{\dagger^{-1}}\mathbf{x}\right) d\mathbf{x}.$$

Define $\mathbf{a} = -n^{1/2}\,\widehat{\mathbf{V}}^{1/2}\,E(\mathbf{Y} \mid \mathcal{X})$, recall that $\tilde{\mathcal{S}}$ denotes the complement of a set $\mathcal{S}$, and observe that if $\mathcal{X} \in \mathcal{E}$,

$$\left\{\widehat{\mathbf{V}}^{1/2}\mathbf{x} : \mathbf{x} \in (\partial\mathcal{R}^\dagger)^{2\delta}\right\}$$

$$= \left\{\widehat{\mathbf{V}}^{1/2}\mathbf{x} : \mathbf{x} \in \mathcal{R}^\dagger \text{ and } B(\mathbf{x},2\delta) \cap (\mathcal{R}^\dagger)^{\sim} \neq \emptyset,\right.$$
$$\left. \text{or } \mathbf{x} \in (\mathcal{R}^\dagger)^{\sim} \text{ and } B(\mathbf{x},2\delta) \cap \mathcal{R}^\dagger \neq \emptyset\right\}$$

$$= \left\{\mathbf{x} : \mathbf{x} \in \mathcal{R} + \mathbf{a} \text{ and } [\widehat{\mathbf{V}}^{1/2}B(\widehat{\mathbf{V}}^{-1/2}\mathbf{x},2\delta)] \cap (\mathcal{R}+\mathbf{a})^{\sim} \neq \emptyset,\right.$$
$$\left. \text{or } \mathbf{x} \in (\mathcal{R}+\mathbf{a})^{\sim} \text{ and } [\widehat{\mathbf{V}}^{1/2}B(\widehat{\mathbf{V}}^{-1/2}\mathbf{x},2\delta)] \cap (\mathcal{R}+\mathbf{a}) \neq \emptyset\right\}$$

$$\subseteq \left\{\partial(\mathcal{R}+\mathbf{a})\right\}^\eta,$$

where $\eta = C_{26}\,\delta$. Furthermore,

$$\|\mathbf{a}\| = n^{1/2}\,\left\|E\left[(\mathbf{X}^* - \overline{\mathbf{X}})\,I\{\|\widehat{\mathbf{V}}^{-1/2}(\mathbf{X}^* - \overline{\mathbf{X}})\| > n^{1/2}\} \mid \mathcal{X}\right]\right\|$$

$$\leq n^{1/2}\,C_3^{1/2}\,E\left[\|\widehat{\mathbf{V}}^{-1/2}(\mathbf{X}^* - \overline{\mathbf{X}})\|\,I\{\|\widehat{\mathbf{V}}^{-1/2}(\mathbf{X}^* - \overline{\mathbf{X}})\| > n^{1/2}\} \mid \mathcal{X}\right]$$

$$\leq C_3^{1/2}\,E\{\|\widehat{\mathbf{V}}^{-1/2}(\mathbf{X}^* - \overline{\mathbf{X}})\|^2 \mid \mathcal{X}\} = C_3^{1/2}.$$

Therefore,

$$b_\delta \leq C_{27} \int_{(\partial\mathcal{R})^\eta} \exp\left(-C_{28}\|\mathbf{x}\|^2\right) d\mathbf{x}. \qquad (5.26)$$

Assume $\mathcal{R} = \mathcal{R}(t)$ has the form (5.12), and note that we may write

$$n^{1/2}\,A(\overline{\mathbf{X}} + n^{-1/2}\mathbf{x}) = A_1(\mathbf{x}) + A_2(\mathbf{x}), \qquad (5.27)$$

where

$$A_1(\mathbf{x}) = \sum_i A_{(i)}(\overline{\mathbf{X}})\,x^{(i)} + \tfrac{1}{2}n^{-1/2}\sum_{i_1}\sum_{i_2} A_{(i_1,i_2)}(\overline{\mathbf{X}})\,x^{(i_1)}\,x^{(i_2)} + \dots$$

$$+ \frac{1}{(\nu+2)!}\,n^{-(\nu+1)/2}\sum_{i_1}\cdots\sum_{i_{\nu+2}} A_{(i_1,\dots,i_{\nu+2})}(\overline{\mathbf{X}})\,x^{(i_1)}\dots x^{(i_{\nu+2})},$$

$$A_{(i_1,\dots,i_j)}(\mathbf{x}) \;=\; \left(\partial^j / \partial x^{(i_1)} \dots \partial x^{(i_j)}\right) A(\mathbf{x})$$

and

$$|A_2(\mathbf{x})| \;\leq\; C_{29}\, n^{-(\nu+2)/2}\, (\log n)^{\nu+3}\,,$$

the latter uniformly in $\mathcal{X} \in \mathcal{E}$ and $\|\mathbf{x}\| \leq \log n$. All the derivatives $A_{(i_1,\dots,i_j)}(\overline{\mathbf{X}})$, $1 \leq j \leq \nu+3$, are bounded uniformly in $\mathcal{X} \in \mathcal{E}_1$. Take $u = \frac{1}{2}(\nu+2)$, and remember that $\eta = C_{26}\, n^{-u}$. In consequence of (5.27), and provided $\mathcal{X} \in \mathcal{E}$,

$$\{\partial\mathcal{R}(t)\}^\eta \;=\; \Big\{\mathbf{x}_1 \in \mathbb{R}^d: \text{ for some } \mathbf{x}_2 \in \mathbb{R}^d,\ n^{1/2}\, A(\overline{\mathbf{X}} + n^{-1/2}\, \mathbf{x}_2) = t\,,$$
$$\text{and } \|\mathbf{x}_1 - \mathbf{x}_2\| \leq \eta\Big\}$$

$$\subseteq\; \mathcal{R}_1(t) \cup \{B(\mathbf{0}, \log n)\}^\sim\,,$$

where

$$\mathcal{R}_1(t)$$
$$=\; \Big\{\mathbf{x} \in \mathbb{R}^d: \|\mathbf{x}\| \leq \log n \text{ and } A_1(\mathbf{x}) \in \Big[t - C_{29}\, n^{-(\nu+2)/2}(\log n)^{\nu+3}\,,$$
$$t + C_{29}\, n^{-(\nu+2)/2}(\log n)^{\nu+3}\Big]\Big\}.$$

It follows from this relation and the lemma below that for $\mathcal{X} \in \mathcal{E}$,

$$\int_{\{\partial\mathcal{R}(t)\}^\eta} \exp\left(-C_{28}\, \|\mathbf{x}\|^2\right) d\mathbf{x} \;\leq\; C_{30}\, n^{-(\nu+1)/2}\,. \tag{5.28}$$

LEMMA 5.3. *Let* $\mathbf{U}$ *have the* d-*variate Normal* $N(\mathbf{0}, \mathbf{I})$ *distribution, let* $\beta, b > 0$, *and let*

$$r(\mathbf{x}) = \sum_i c_i\, x^{(i)} + n^{-1/2} \sum_{i_1} \sum_{i_2} c_{i_1 i_2}\, x^{(i_1)}\, x^{(i_2)} + \dots$$
$$+ n^{-(\nu+1)/2} \sum_{i_1} \cdots \sum_{i_{\nu+2}} c_{i_1\dots i_{\nu+2}}\, x^{(i_1)} \dots x^{(i_{\nu+2})}$$

be a polynomial in $\mathbf{x}$ *and* $n^{-1/2}$ *whose coefficients* $c_i, c_{i_1 i_2}, \dots, c_{i_1\dots i_{\nu+2}}$ *satisfy*

$$\max_{i_1,\dots,i_{\nu+2}} \left(|c_{i_1}|\,|c_{i_1 i_2}|, \dots, |c_{i_1\dots i_{\nu+2}}|\right) \;\leq\; b\,,$$

and for some i $(1 \leq i \leq d)$, $|c_i| > \frac{1}{b}$. *There exists* $C > 0$, *depending only on* β, b, *and* d, *such that*

$$\sup_{-\infty < t < \infty} P\Big\{\|\mathbf{U}\| \leq b \log n \text{ and } r(\mathbf{U}) \in [t - n^{-\beta}, t + n^{-\beta}]\Big\} \;\leq\; C\, n^{-\beta}\,.$$

Lemma 5.3 is trivial if $d = 1$, and may be proved by induction for general d.

Results (5.26) and (5.28) together imply that

$$b_\delta \leq C_{27}\, C_{30}\, n^{-(\nu+1)/2}$$

uniformly in sets $\mathcal{R} = \mathcal{R}(t)$ of the form (5.12) and samples $\mathcal{X} \in \mathcal{E}$. Combining this with (5.16) and (5.25) we conclude that for $\mathcal{X} \in \mathcal{E}'$,

$$\sup_{-\infty<t<\infty} \left| \int_{\mathcal{R}(t)} d\Big[Q - \sum_{j=0}^{\nu} n^{-j/2}\, P_j(-\Phi_{0,\hat{\mathbf{v}}} : \{\hat{\kappa}_{\mathbf{r}}\}) \Big] \right| \leq C_{31}\, n^{-(\nu+1)/2},$$

where $C_{31} = C_8 + C_{11}(C_{16} + C_{24}) + C_{27}\,C_{30}$. Equivalently,

$$\sup_{-\infty<t<\infty} \left| P\{n^{1/2}\, A(\overline{\mathbf{X}}^*) \leq t \mid \mathcal{X}\} - \sum_{j=0}^{\nu} n^{-j/2}\, P_j\big(-\Phi_{0,\hat{\mathbf{v}}} : \{\hat{\kappa}_{\mathbf{r}}\}\big)\, \{\mathcal{R}(t)\} \right|$$

$$\leq C_{31}\, n^{-(\nu+1)/2} \tag{5.29}$$

uniformly in $\mathcal{X} \in \mathcal{E}'$.

Bearing in mind the definition of P_j given in part (i) of this proof, it is straightforward to Taylor expand the series in P_j in (5.29), obtaining

$$\sup_{-\infty<x<\infty} \left| P\{n^{1/2}\, \widehat{A}(\overline{\mathbf{X}}^*) \leq x \mid \mathcal{X}\} - \Phi(t) - \sum_{j=0}^{\nu} n^{-j/2}\, \hat{r}_j(x)\, \phi(x) \right|$$

$$\leq C_{32}\, n^{-(\nu+1)/2} \tag{5.30}$$

uniformly in $\mathcal{X} \in \mathcal{E}$, where $\hat{r}_j$ is a polynomial of degree $3j - 1$, odd or even for even or odd j, and with coefficients depending on the cumulant sequence $\{\hat{\kappa}_{\mathbf{r}}\}$ up to order $j+2$. In order to establish result (5.7) it remains only to prove that $\hat{r}_j = \hat{p}_j$ or $\hat{q}_j$, depending on the definition taken for A.

(iii) Identifying the Expansion

Let the function A be defined by (5.1) or (5.2), and let r_j denote the version of $\hat{r}_j$ when the cumulant sequence $\{\hat{\kappa}_{\mathbf{r}}\}$ is replaced by $\{\kappa_{\mathbf{r}}\}$. A substantially simpler version of the argument leading to (5.30) shows that under identical regularity conditions,

$$\sup_{-\infty<x<\infty} \left| P\{n^{1/2}\, A(\overline{\mathbf{X}}) \leq x\} - \Phi(x) - \sum_{j=1}^{\nu} n^{-j/2}\, r_j(x)\, \phi(x) \right|$$

$$= O(n^{-(\nu+1)/2}). \tag{5.31}$$

Therefore, it suffices to prove that $r_j = p_j$ or q_j, depending on the circumstance. This is simply a matter of equating the coefficients in r_j with those in p_j or q_j, which depend only on a finite number of population moments or cumulants. All possible sets of values of these quantities are assumed in the case where the distribution of $\mathbf{X}$ is essentially bounded (i.e., for some $C > 0$, $P(\|\mathbf{X}\| \leq C) = 1$), and so we may suppose without loss of generality that $\mathbf{X}$ enjoys this property.

Let

$$\pi_j = p_j \quad \text{if} \quad A(\mathbf{x}) = \{g(\mathbf{x}) - g(\boldsymbol{\mu})\}/h(\boldsymbol{\mu})$$

and

$$\pi_j = q_j \quad \text{if} \quad A(\mathbf{x}) = \{g(\mathbf{x}) - g(\boldsymbol{\mu})\}/h(\mathbf{x}).$$

We must prove that $r_j = \pi_j$. If the distribution of $\mathbf{X}$ is essentially bounded then we may Taylor expand $S_n = n^{1/2} A(\bar{\mathbf{X}})$, as at (2.28) and (2.29), obtaining

$$S_n = S_{n,\nu+1} + R_n$$

where

$$S_{n,\nu+1} = \sum_{i=1}^d a_i Z^{(i)} + n^{-1/2} \tfrac{1}{2} \sum_{i_1=1}^d \sum_{i_2=1}^d a_{i_1 i_2} Z^{(i_1)} Z^{(i_2)} + \dots$$

$$+ n^{-\nu/2} \frac{1}{(\nu+1)!} \sum_{i_1=1}^d \cdots \sum_{i_{\nu+1}=1}^d a_{i_1 \dots i_{\nu+1}} Z^{(i_1)} \dots Z^{(i_{\nu+1})},$$

$$a_{i_1 \dots i_j} = (\partial^j / \partial x^{(i_1)} \dots \partial x^{(i_j)}) A(\mathbf{x})\big|_{\mathbf{x}=\boldsymbol{\mu}},$$

$$\mathbf{Z} = n^{1/2} (\bar{\mathbf{X}} - \boldsymbol{\mu}),$$

and for each $\epsilon, \lambda > 0$,

$$P\Big(|R_n| > n^{-(\nu+1)/2+\epsilon}\Big) = O(n^{-\lambda}).$$

Recall from Theorems 2.1 and 2.2 in Section 2.4 that for $1 \leq j \leq \nu$, π_j is defined uniquely by the property that for each $1 \leq k \leq j + 2$, the kth moment of $S_{n,\nu+1}$ equals

$$\int_{-\infty}^{\infty} x^k \, d\Big\{\Phi(x) + \sum_{j=1}^{\nu} n^{-j/2} \pi_j(x) \, \phi(x)\Big\} + o(n^{-\nu/2}).$$

(Equivalently, the definitions can be given in terms of cumulants; note the expressions for cumulants in terms of moments at (2.6).)   Therefore, it suffices to prove that for $k \geq 1$,

$$E\big(S_{n,\nu+1}^k\big) = \int_{-\infty}^{\infty} x^k \, d\Big\{\Phi(x) + \sum_{j=1}^{\nu} n^{-j/2} r_j(x) \, \phi(x)\Big\} + o(n^{-\nu/2}). \quad (5.32)$$

Observe from (5.31) that for $0 < \epsilon \leq \tfrac{1}{2}$,

$$\sup_{-\infty < x < \infty} \Big| P\big(S_n \leq x \pm n^{-(\nu/2)-\epsilon}\big) - P\big(S_n \leq x\big) \Big| = O(n^{-(\nu/2)-\epsilon}).$$

Therefore,

$$\sup_{-\infty < x < \infty} \left| P(S_n \le x) - P(S_{n,\nu+1} \le x) \right|$$

$$\le \sup_{-\infty < x < \infty} \max_{+,-} \left| P\left(S_n \le x \pm n^{-(\nu/2)-\epsilon}\right) - P(S_n \le x) \right|$$
$$+ P\left(|R_n| > n^{-(\nu/2)-\epsilon}\right)$$
$$= O(n^{-(\nu/2)-\epsilon}).$$

This result and (5.31) imply that for all $0 < \epsilon \le \frac{1}{2}$,

$$\sup_{-\infty < x < \infty} \left| P(S_{n,\nu+1} \le x) - \Phi(x) - \sum_{j=1}^{\nu} n^{-j/2} r_j(x) \phi(x) \right|$$

$$= O(n^{-(\nu/2)-\epsilon}). \tag{5.33}$$

Given $c > 0$, take $\delta = \epsilon/2c$, and note from (5.33) that

$$\sup_{|x| \le n^{\delta}} \left(1 + |x|^c\right) \left| P(S_{n,\nu+1} \le x) - \Phi(x) - \sum_{j=1}^{\nu} n^{-j/2} r_j(x) \phi(x) \right|$$

$$= O(n^{-(\nu/2)-(\epsilon/2)}). \tag{5.34}$$

By Markov's inequality we have for each $\delta, \lambda > 0$,

$$P\left(|S_{n,\nu+1}| > n^{\delta}\right) = O(n^{-\lambda}).$$

(Here we have again used the fact that $\mathbf{X}$ is essentially bounded.) Therefore, the supremum in (5.34) may be taken over all x: for all $c > 0$,

$$\sup_{-\infty < x < \infty} \left(1 + |x|^c\right) \left| P(S_{n,\nu+1} \le x) - \Phi(x) - \sum_{j=1}^{\nu} n^{-j/2} r_j(x) \phi(x) \right|$$

$$= o(n^{-\nu/2}).$$

Hence, integrating by parts,

$$\left| E(S_{n,\nu+1}^k) - \int_{-\infty}^{\infty} x^k \, d\left\{ \Phi(x) + \sum_{j=1}^{\nu} n^{-j/2} r_j(x) \phi(x) \right\} \right|$$

$$\le k \int_{-\infty}^{\infty} |x|^{k-1} \left| P(S_{n,\nu+1} \le x) - \Phi(x) - \sum_{j=1}^{\nu} n^{-j/2} r_j(x) \phi(x) \right| dx$$

$$= o(n^{-\nu/2}).$$

This establishes (5.32) and completes the proof of the theorem.

(iv) Proof of Lemma 5.2

Observe that $\psi(\mathbf{t}) = \chi^\dagger(n^{-1/2}\,\mathbf{t})^n$, where

$$\chi^\dagger(\mathbf{t}) \;=\; E\{\exp(i\mathbf{t}^T\,\mathbf{X}^\dagger)\mid \mathcal{X}\}\,.$$

Since $\|\mathbf{X}^\dagger\| \le 2n^{1/2}$, then

$$|D^{\mathbf{r}}\chi^\dagger(n^{-1/2}\,\mathbf{t})| \;\le\; (2n^{1/2})^{|\mathbf{r}|}\,n^{-|\mathbf{r}|/2} \;=\; 2^{|\mathbf{r}|}\,.$$

Therefore,

$$|D^{\mathbf{r}}\psi(\mathbf{t})| \;\le\; C_{33}\,n^{|\mathbf{r}|}\,|\chi^\dagger(\mathbf{t})|^{n-|\mathbf{r}|}$$

uniformly in $|\mathbf{r}| \le d+1$. It follows that for $n \ge d+1$,

$$a_\delta \;\le\; C_{34}\,n^{d+1}\int_{\|\mathbf{t}\|>C_{15}n^{1/2}} |\chi^\dagger(n^{-1/2}\,\mathbf{t})|^{n-(d+1)}\,\exp(-\|\delta\mathbf{t}\|^{1/2})\,dt\,.$$

If $\mathcal{X} \in \mathcal{E}$ then

$$|\chi^\dagger(\mathbf{t})| \;=\; \left|E\{\exp(i\mathbf{t}^T\,\mathbf{Y})\mid \mathcal{X}\}\right|$$

$$\le\; \left|E\{\exp(i\mathbf{t}^T\,\widehat{\mathbf{V}}^{-1/2}\,\mathbf{X}^*)\mid \mathcal{X}\}\right|$$

$$+\; P\{\|\widehat{\mathbf{V}}^{-1/2}(\mathbf{X}^* - \overline{\mathbf{X}})\| > n^{1/2}\mid \mathcal{X}\}$$

$$\le\; |\chi^*(\widehat{\mathbf{V}}^{-1/2}\,\mathbf{t})| + C_{35}\,n^{-1}\,,$$

and so

$$a_\delta \;\le\; C_{36}\,n^{d+1}\int_{\|\mathbf{t}\|>C_{37}n^{1/2}}\left\{|\chi^*(n^{-1/2}\,\mathbf{t})| + C_{35}\,n^{-1}\right\}^{n-(d+1)}$$

$$\times\; \exp(-C_{38}\,\|\delta\mathbf{t}\|^{1/2})\,dt\,. \tag{5.35}$$

Since the distribution of $\mathbf{X}$ satisfies Cramér's condition, we may choose $\eta_1 \in [\tfrac{1}{2}, 1)$ and $\eta_2 > 0$ such that for all $\epsilon > 0$,

$$\sup_{\|\mathbf{t}\|>\epsilon} |\chi(\mathbf{t})| \;\le\; \max(\eta_1,\; 1 - \eta_2\,\epsilon^2)\,.$$

Given $\epsilon > 0$, let $\eta_3(\epsilon)$ be such that

$$\sup_{\|\mathbf{t}\|>\epsilon} |\chi(\mathbf{t})| \;\le\; \eta_3 \;<\; 1\,,$$

and choose $\eta_4 = \eta_4(\epsilon) > 0$ so small that $\eta_3(1 + \eta_4) < 1$. If $|\chi^*(\mathbf{t})| > \eta_3$ then

$$n^{-1} \;\le\; n^{-1}\eta_3^{-1}\,|\chi^*(\mathbf{t})|\,,$$

in which case we have for $n > 2(d+1)$,

$$\left\{|\chi^*(\mathbf{t})| + C_{35}\,n^{-1}\right\}^{n-(d+1)} \;\le\; \exp(2C_{35}/\eta_3)\,|\chi^*(\mathbf{t})|^{n-(d+1)}\,.$$

If $|\chi^*(\mathbf{t})| \leq \eta_3$ then

$$\left\{|\chi^*(\mathbf{t})| + C_{35}\, n^{-1}\right\}^{n-(d+1)} \;\leq\; \exp(2C_{35}/\eta_3)\, \eta_3^{n-(d+1)}\,.$$

If $|\chi^*(\mathbf{t}) - \chi(\mathbf{t})| \leq \eta_4\, |\chi(\mathbf{t})|$ and $\|\mathbf{t}\| > \epsilon$ then

$$|\chi^*(\mathbf{t})| \;\leq\; \eta_3(1 + \eta_4)\,.$$

If $|\chi^*(\mathbf{t}) - \chi(\mathbf{t})| > \eta_4\, |\chi(\mathbf{t})|$ then for any $m \in [1,\, n-(d+1)]$,

$$|\chi^*(\mathbf{t})|^{n-(d+1)} \;\leq\; \left\{(1 + \eta_4^{-1})\, |\chi^*(\mathbf{t}) - \chi(\mathbf{t})|\right\}^m\,.$$

Combining these estimates we deduce that for any $\epsilon > 0$ we have for all $\|\mathbf{t}\| > \epsilon$,

$$\left\{|\chi^*(\mathbf{t})| + C_{35}\, n^{-1}\right\}^{n-(d+1)} \;\leq\; \exp(2C_{35}/\eta_3)\left[\{\eta_3(1 + \eta_4)\}^{n-(d+1)}\right.$$

$$\left. +\, \left\{(1 + \eta_4^{-1})\, |\chi^*(\mathbf{t}) - \chi(\mathbf{t})|\right\}^m\right]\,. \quad (5.36)$$

Take $\epsilon = C_{37}$,

$$\eta_3 \;=\; \max\left\{\tfrac{1}{2},\, \sup_{\|\mathbf{t}\| > \epsilon} |\chi(\mathbf{t})|\right\},$$

$\eta_4 = \tfrac{1}{4}$ if $\eta_2 = \tfrac{1}{2}$,

$$\eta_4 \;=\; \tfrac{1}{2}\left[\left\{\max\left(\eta_1, 1 - \eta_2\, \epsilon^2\right)\right\}^{-1} - 1\right]$$

if $\eta_2 \neq \tfrac{1}{2}$, and

$$C_{19} \;=\; \max\left\{\tfrac{1}{2}(1 + \eta_1),\, 1 - \tfrac{1}{2}\eta_2\, \epsilon^2\right\} \;\geq\; \eta_3(1 + \eta_4)\,.$$

Substituting (5.36) into (5.35), and defining $C_{20} = C_{36}\, \exp(2C_{35}/\eta_3)$ and $C_{21} = 1 + \eta_4^{-1}$, we obtain (5.24). This completes the proof of Lemma 5.2.

5.3 Edgeworth Expansions of Coverage Probability

Our aim in this section is to provide a rigorous justification of the results developed in Section 3.5, particularly Proposition 3.1. From that flow most of our conclusions about the manner in which various orders of correctness influence the coverage properties of confidence intervals. We shall develop formulae for coverage error up to terms of size n^{-1}, with a remainder of $o(n^{-1})$. Higher-order expansions, useful for describing coverage properties of symmetric intervals or iterated bootstrap intervals (see Sections 3.6 and 3.11), may be established rigorously by similar means. Throughout our development we assume the "smooth function model" described in

Section 2.4 and subsection 5.2.1. In particular, $\theta_0 = g(\boldsymbol{\mu})$, $\hat{\theta} = g(\overline{\mathbf{X}})$, and $\hat{\sigma} = h(\overline{\mathbf{X}})$, where g, h are smooth functions from $\mathbb{R}^d$ to $\mathbb{R}$, $\overline{\mathbf{X}}$ denotes the mean of a d-variate random sample $\mathcal{X} = \{\mathbf{X}_1, \ldots, \mathbf{X}_n\}$, and $\boldsymbol{\mu} = E(\mathbf{X})$ is the population mean.

To begin, we recall the coverage problem introduced in Section 3.5. Let

$$\mathcal{I}_1 = \mathcal{I}_1(\alpha) = \left(-\infty, \ \hat{\theta} + n^{-1/2}\,\hat{\sigma}(z_\alpha + \hat{c}_\alpha) \right) \qquad (5.37)$$

denote a one-sided confidence interval, where $z_\alpha = \Phi^{-1}(\alpha)$ and

$$\hat{c}_\alpha = n^{-1/2}\,\hat{s}_1(z_\alpha) + n^{-1}\,\hat{s}_2(z_\alpha) + \ldots$$

for polynomials $\hat{s}_j$. We wish to develop an Edgeworth expansion of the probability $P(\theta_0 \in \mathcal{I}_1)$. In many cases of practical interest, $\hat{s}_j$ is obtained by substituting sample moments for population moments in a polynomial s_j, which is of degree $j+1$ and is odd or even according to whether j is even or odd. (This follows from properties of Cornish-Fisher expansions, described in Sections 2.5 and 3.3.) We claim that under moment conditions on $\mathbf{X}$, there exist random variables M_1, M_2 that are linear combinations of sample moments of the form

$$M = n^{-1} \sum_{i=1}^{n} \prod_{j=1}^{d} \left(X_i^{(j)} \right)^{l_j}, \qquad (5.38)$$

and fixed constants b_1, b_2, such that for $\epsilon, \delta > 0$ sufficiently small,

$$P\left[\sup_{n^{-\epsilon} \le \alpha \le 1 - n^{-\epsilon}} \left| \hat{c}_\alpha - \left\{ n^{-1/2}\,s_1(z_\alpha) + n^{-1}\,s_2(z_\alpha) \right\} \right. \right.$$
$$\left. \left. - n^{-1/2}\left\{ b_1(M_1 - EM_1) + b_2\,z_\alpha^2(M_2 - EM_2) \right\} \right| > n^{-1-\delta} \right]$$
$$= o(n^{-1}). \qquad (5.39)$$

(In (5.38), $l_1, \ldots, l_d$ denote nonnegative integers.)

To appreciate why, observe that it is typically the case that for some $\epsilon, \delta > 0$,

$$P\left\{ \sup_{n^{-\epsilon} \le \alpha \le 1 - n^{-\epsilon}} \left| \hat{c}_\alpha - n^{-1/2}\,\hat{s}_1(z_\alpha) - n^{-1}\,\hat{s}_2(z_\alpha) \right| > n^{-1-2\delta} \right\}$$
$$= o(n^{-1}). \qquad (5.40)$$

For example, if $\mathcal{I}_1$ is a percentile or percentile-t confidence interval then $\hat{c}_\alpha = -\hat{u}_{1-\alpha} - z_\alpha$ or $\hat{c}_\alpha = -\hat{v}_{1-\alpha} - z_\alpha$, respectively, and then (5.40) follows immediately from Theorem 5.2. (See subsection 3.5.2 for discussion of forms that $\hat{c}_\alpha$ can take.) If M is given by (5.38) then for any specified

$\eta > 0$, it may be proved by Markov's inequality that under sufficiently stringent moment conditions on $\mathbf{X}$,

$$P\big(|M - EM| > n^{-(1/2)+\eta}\big) \;=\; o(n^{-1})\,.$$

The same result holds if M is replaced by the linear combination M_i. Furthermore, since s_1 is an even polynomial of degree 2,

$$s_1(x) \;=\; a_1\mu_1 + a_2\mu_2 x^2 \qquad \text{and} \qquad \hat{s}_1(x) \;=\; a_1\hat{\mu}_1 + a_2\hat{\mu}_2 x^2\,,$$

where the a_i's are constants, $\hat{\mu}_i$ is a sample estimate of a population moment μ_i, and $a_i(\hat{\mu}_i - \mu_i) = b_i(M_i - EM_i) + O_p(n^{-1})$, for approximate choices of the constant b_i and the linear combination M_i of moments. Arguing thus, and noting that $|z_\alpha| \le C_1(\log n)^{1/2}$ uniformly in $n^{-\epsilon} \le \alpha \le 1 - n^{-\epsilon}$, we may prove that if $E(\|\mathbf{X}\|^l) < \infty$ for sufficiently large l then there exist M_1, M_2, constants b_1, b_2, and $\epsilon, \delta > 0$, such that

$$P\Bigg\{ \sup_{n^{-\epsilon} \le \alpha \le 1-n^{-\epsilon}} \Big| \hat{s}_1(z_\alpha) - s_1(z_\alpha) - b_1(M_1 - EM_1) \\ - b_2\, z_\alpha^2 (M_2 - EM_2) \Big| > n^{-(1/2)-2\delta} \Bigg\} = o(n^{-1})\,, \quad (5.41)$$

$$P\Bigg\{ \sup_{n^{-\epsilon} \le \alpha \le 1-n^{-\epsilon}} \big| \hat{s}_2(z_\alpha) - s_2(z_\alpha) \big| > n^{-2\delta} \Bigg\} \;=\; o(n^{-1})\,. \quad (5.42)$$

The desired result (5.39) follows on combining (5.40)–(5.42).

We are now in a position to state and prove rigorously a version of Proposition 3.1. Let $\mathcal{I}_1$ denote the confidence interval defined in (5.37). Put $g_j(\mathbf{x}) = (\partial/\partial x^{(j)})\, g(\mathbf{x})$, and observe that if M_1, M_2 are linear combinations of sample moments of the form (5.38) then for constants c_1, c_2 not depending on n,

$$E\Bigg[\bigg\{ \sum_{j=1}^d (\bar{\mathbf{X}} - \boldsymbol{\mu})^{(j)}\, g_j(\boldsymbol{\mu}) \bigg\} (M_i - EM_i) \Bigg] \;=\; n^{-1} c_i\,, \qquad i = 1,2\,.$$

Define

$$a_\alpha \;=\; \sigma^{-1}\big(b_1\, c_1 + b_2\, c_2\, z_\alpha^2 \big)\,, \tag{5.43}$$

where b_1, b_2 are as in (5.39). If any of the components comprising the linear combination M_1 or M_2 is not an element of $\bar{\mathbf{X}}$, it may be adjoined by lengthening each data vector so that the new vectors include the appropriate products of earlier terms. We shall assume that this lengthening has already been carried out, in which case each M_i is a linear combination of elements of $\bar{\mathbf{X}}$. Then we have the following version of Proposition 3.1.

THEOREM 5.3. *Assume that g and h each have four bounded derivatives in a neighbourhood of μ, that $E(\|\mathbf{X}\|^l) < \infty$ for sufficiently large l, that Cramér's condition (5.5) holds, that $\hat{c}_\alpha$ admits the approximation (5.39) for polynomials s_1 and s_2, and that the distribution of $n^{1/2}(\hat{\theta} - \theta_0)/\hat{\sigma}$ admits the Edgeworth expansion*

$$P\{n^{1/2}(\hat{\theta} - \theta_0)/\hat{\sigma} \leq x\}$$
$$= \Phi(x) + n^{-1/2}\, q_1(x)\, \phi(x) + n^{-1}\, q_2(x)\, \phi(x) + o(n^{-1}) \quad (5.44)$$

for polynomials q_1 and q_2. Then for the same ϵ appearing in (5.39), and with a_α defined by (5.43),

$$\sup_{n^{-\epsilon} \leq \alpha \leq 1 - n^{-\epsilon}} \left| P\{\theta_0 \in \mathcal{I}_1(\alpha)\} - \alpha \right.$$
$$\left. - n^{-1/2}\, r_1(z_\alpha)\, \phi(z_\alpha) - n^{-1}\, r_2(z_\alpha)\, \phi(z_\alpha) \right| = o(n^{-1}), \quad (5.45)$$

where

$$r_1 = s_1 - q_1$$

and

$$r_2(z_\alpha) = q_2(z_\alpha) + s_2(z_\alpha) - \tfrac{1}{2}\, z_\alpha\, s_1(z_\alpha)^2$$
$$+ s_1(z_\alpha)\{z_\alpha\, q_1(z_\alpha) - q_1'(z_\alpha)\} - a_\alpha\, z_\alpha.$$

Existence of the expansion (5.44) is not really an assumption, since it is guaranteed to hold uniformly in x under the other conditions of Theorem 5.3; see Theorem 2.2. Our formula for a_α is equivalent to that given in (3.35) in Chapter 3, the only difference being that in (5.43) we have spelled out the way in which it depends on z_α and the constants b_k, c_k. Thus, (5.45) implies the expansion (3.40) of Proposition 3.1, up to a remainder of $o(n^{-1})$.

We do not specify the value of l in the theorem since our proof makes no attempt to be economical in terms of moment conditions.

It is not possible, in general, to extend the range of α in (5.45) from $n^{-\epsilon} \leq \alpha \leq 1 - n^{-\epsilon}$ to $0 < \alpha < 1$. This follows from properties noted earlier in connection with confidence intervals constructed by explicit skewness correction; see Section 3.8. In particular, there exist examples where all the conditions in Theorem 5.3 are satisfied but, owing to the fact that $z_\alpha + \hat{c}_\alpha$ is not monotone increasing in α,

$$\limsup_{\alpha \uparrow 1} P\{\theta_0 \in \mathcal{I}_1(\alpha)\}$$

is bounded below 1 as $n \to \infty$. The interval $\mathcal{I}$ discussed in the second-last paragraph of Section 3.8, and also the accelerated bias-corrected interval $(-\infty, \hat{y}_{\mathrm{ABC}, \alpha})$ introduced in subsection 3.10.3, can have this problem.

However, if $z_\alpha + \hat{c}_\alpha$ is monotone in α, and if (5.45) holds for some $\epsilon > 1$, then we may replace the range $n^{-\epsilon} \leq \alpha \leq 1 - n^{-\epsilon}$ by $0 < \alpha < 1$. To appreciate why, note that if $\beta = n^{-\epsilon} = o(n^{-1})$ then

$$\sup_{0 < \alpha \leq \beta} \left| P\{\theta_0 \in \mathcal{I}_1(\alpha)\} - \alpha - n^{-1/2} r_1(z_\alpha)\,\phi(z_\alpha) - n^{-1} r_2(z_\alpha)\,\phi(z_\alpha) \right|$$

$$= \sup_{0 < \alpha \leq \beta} P\{\theta_0 \in \mathcal{I}_1(\alpha)\} + o(n^{-1})$$

$$\leq P\{\theta_0 \in \mathcal{I}_1(\beta)\} + o(n^{-1}) = o(n^{-1}),$$

the last identity following from (5.45); and similarly,

$$\sup_{1-\beta \leq \alpha < 1} \left| P\{\theta_0 \in \mathcal{I}_1(\alpha)\} - \alpha \right.$$
$$\left. - n^{-1/2} r_1(z_\alpha)\,\phi(z_\alpha) - n^{-1} r_2(z_\alpha)\,\phi(z_\alpha) \right| = o(n^{-1}).$$

In specific cases, such as that of the percentile method (where $\hat{c}_\alpha = -\hat{u}_{1-\alpha} - z_\alpha$), or the "other percentile method" ($\hat{c}_\alpha = \hat{u}_\alpha - z_\alpha$), or percentile-$t$ ($\hat{c}_\alpha = -v_{1-\alpha} - z_\alpha$), or bias correction, it is not difficult to establish (5.39) for $\epsilon > 1$, assuming sufficiently many moments of the distribution of $\mathbf{X}$. In those instances, (5.45) holds uniformly in $0 < \alpha < 1$.

We conclude this section with a proof of Theorem 5.3, for which the following lemma is particularly helpful. Write $\mathbf{V}$ for the variance matrix of $\mathbf{X}$, let $\{\kappa_r\}$ denote the multivariate cumulant sequence of $\mathbf{X}$, and let $P_j(-\Phi_{0,\mathbf{V}} : \{\kappa_r\})$ be the signed measure with density $P_j(-\phi_{0,\mathbf{V}} : \{\kappa_r\})$; see part (i) of subsection 5.2.3 for more details. Write $(\partial\mathcal{R})^\epsilon$ for the set of points distant no more than ϵ from the boundary $\partial\mathcal{R}$ of $\mathcal{R}$.

LEMMA 5.4. *If $E(\|\mathbf{X}\|^{\nu+2}) < \infty$ for an integer $\nu \geq 0$, and if Cramér's condition (5.5) holds, then*

$$P\{n^{1/2}(\overline{\mathbf{X}} - \boldsymbol{\mu}) \in \mathcal{R}\} = \sum_{j=0}^{\nu} n^{-j/2} P_j\big(-\Phi_{0,\mathbf{V}} : \{\kappa_r\}\big)(\mathcal{R}) + o(n^{-\nu/2})$$

uniformly in any class R of sets $\mathcal{R} \subseteq \mathbb{R}^d$ satisfying

$$\sup_{\mathcal{R} \in R} \int_{(\partial\mathcal{R})^\eta} \exp\big(-\tfrac{1}{2}\|\mathbf{x}\|^2\big)\,d\mathbf{x} = O(\eta) \tag{5.46}$$

as $\eta \downarrow 0$.

This result is almost the simplest form of Edgeworth expansion for a sum of independent random vectors; see Bhattacharya and Rao (1976, p. 215).

Since M_i is a linear combination of elements of $\overline{\mathbf{X}}$ then we may write $M_i = \mathbf{v}_i^{\mathrm{T}} \overline{\mathbf{X}}$ for fixed vectors $\mathbf{v}_i$, $i = 1, 2$. Let δ be as in (5.39), and define $p(\alpha) = P\{\theta_0 \in \mathcal{I}_1(\alpha)\}$,

$$U = n^{1/2}(\hat{\theta} - \theta_0)\hat{\sigma}^{-1} + n^{-1/2}\Big\{b_1(M_1 - EM_1) + b_2 z_\alpha^2(M_2 - EM_2)\Big\},$$

$$A(\mathbf{x}) = \{g(\mathbf{x}) - g(\boldsymbol{\mu})\}\, h(\mathbf{x})^{-1} + n^{-1}\Big\{b_1\mathbf{v}_1^{\mathrm{T}}(\mathbf{x} - \boldsymbol{\mu}) + b_2 z_\alpha^2 \mathbf{v}_2^{\mathrm{T}}(\mathbf{x} - \boldsymbol{\mu})\Big\},$$

$$y_\pm = -\Big\{z_\alpha + n^{-1/2}\, s_1(z_\alpha) + n^{-1}\, s_2(z_\alpha)\Big\} \pm n^{-1-\delta},$$

$$\mathcal{R}_\pm = \mathcal{R}_\pm(\alpha, n) = \Big\{\mathbf{x} \in \mathbb{R}^d : n^{1/2}\, A(\boldsymbol{\mu} + n^{-1/2}\,\mathbf{x}) > y_\pm\Big\}$$

and

$$p_\pm = P(U > y_\pm) = P\{n^{1/2}(\overline{\mathbf{X}} - \boldsymbol{\mu}) \in \mathcal{R}_\pm\}. \tag{5.47}$$

In view of (5.39),

$$p(\alpha) \le p_-(\alpha) + o(n^{-1}), \qquad p(\alpha) \ge p_+(\alpha) + o(n^{-1}) \tag{5.48}$$

uniformly in $n^{-\epsilon} \le \alpha \le 1 - n^{-\epsilon}$. It is readily shown by Taylor expanding the function A about $\boldsymbol{\mu}$ that for sufficiently large n_0, the class of sets

$$R = \Big\{\mathcal{R}_\pm(\alpha, n) : n \ge n_0, \ \ n^{-\epsilon} \le \alpha \le 1 - n^{-\epsilon}\Big\}$$

has property (5.46). Hence by Lemma 5.4 and formula (5.47),

$$p_\pm(\alpha) = \sum_{j=0}^{\nu} n^{-j/2}\, P_j(-\Phi_{0,\mathbf{V}} : \{\kappa_{\mathbf{r}}\})(\mathcal{R}_\pm) + o(n^{-1}) \tag{5.49}$$

uniformly in $n^{-\epsilon} \le \alpha \le 1 - n^{-\epsilon}$.

Bearing in mind the definition of P_j given in part (i) of subsection 5.2.3, we may Taylor expand the series in P_j in (5.49) to obtain

$$p_\pm(\alpha) = \alpha + n^{-1/2}\, r_1(z_\alpha)\, \phi(z_\alpha) + n^{-1}\, r_2(z_\alpha)\, \phi(z_\alpha) + o(n^{-1}) \tag{5.50}$$

uniformly in $n^{-\epsilon} \le \alpha \le 1 - n^{-\epsilon}$, where r_1 and r_2 are polynomials. This brings us to a version of (5.31) in the proof of Theorem 5.1, after which point it remains only to identify the polynomials r_1 and s_2. That part of the present proof may be conducted along the same lines as part (iii) of the proof of Theorem 5.1, showing that r_1 and r_2 are determined by the cumulant argument used in subsection 3.5.2 to prove Proposition 3.1. Theorem 5.3 now follows from (5.48) and (5.50).

5.4 Linear Regression

We discussed the case of simple linear regression in subsections 4.3.3 and 4.3.4 of Chapter 4, and showed that in several key respects it is different from the more traditional circumstances encountered in Chapter 3. In particular, a percentile-t confidence interval for the slope parameter has coverage error $O(n^{-2})$, rather than $O(n^{-1})$. However, the method of proof of results such as this is very similar to that in the traditional case (see Sections 5.2 and 5.3), the only important difference being that we now work with weighted sums of independent and identically distributed random variables rather than with unweighted sums. In the present section we outline the necessary changes to proofs, using the case of slope estimation as an example.

We begin by recalling the basic ingredients of simple linear regression. The simple linear regression model is

$$Y_i = c + x_i d + \epsilon_i, \qquad 1 \le i \le n,$$

where c, d, x_i, Y_i, ϵ_i are scalars, c and d are unknown constants representing intercept and slope respectively, the ϵ_i's are independent and identically distributed random variables with zero mean and variance σ^2, and the x_i's are given design points. The usual least-squares estimates of d and σ^2 are

$$\hat{d} = \sigma_x^{-2} n^{-1} \sum_{i=1}^{n} (x_i - \bar{x})(Y_i - \bar{Y}), \qquad \hat{\sigma}^2 = n^{-1} \sum_{i=1}^{n} \hat{\epsilon}_i^2,$$

where

$$\sigma_x^2 = n^{-1} \sum_{i=1}^{n} (x_i - \bar{x})^2, \qquad \hat{\epsilon}_i = Y_i - \bar{Y} - (x_i - \bar{x}) \hat{d}.$$

Since $\hat{d}$ has mean d and variance $n^{-1} \sigma_x^{-2} \sigma^2$ then inference about d would typically be based on the statistic $T = n^{1/2}(\hat{d} - d) \sigma_x / \hat{\sigma}$, whose distribution may be estimated by the bootstrap, as follows.

Let $\mathcal{X}^* = \{\epsilon_1^*, \ldots, \epsilon_n^*\}$ denote a resample drawn randomly, with replacement, from the collection $\mathcal{X} = \{\hat{\epsilon}_1, \ldots, \hat{\epsilon}_n\}$, and put

$$Y_i^* = \hat{c} + x_i \hat{d} + \epsilon_i^* = \bar{Y} + (x_i - \bar{x}) \hat{d} + \epsilon_i^*,$$

where $\hat{c} = \bar{Y} - \bar{x} \hat{d}$. Define $\bar{Y}^* = n^{-1} \sum Y_i^*$,

$$\hat{d}^* = \sigma_x^{-2} n^{-1} \sum_{i=1}^{n} (x_i - \bar{x})(Y_i^* - \bar{Y}^*),$$

$$\hat{\epsilon}_i^* = Y_i^* - \bar{Y}^* - (x_i - \bar{x}) \hat{d}^*, \qquad \hat{\sigma}^{*2} = n^{-1} \sum_{i=1}^{n} \hat{\epsilon}_i^{*2}.$$

The conditional distribution of $T^* = n^{1/2}(\hat{d}^* - \hat{d})\,\sigma_x/\hat{\sigma}^*$, given $\mathcal{X}$, approximates the unconditional distribution of T.

To see how we might develop Edgeworth expansions of the distributions of T and T^*, first define $v_i = (x_i - \bar{x})/\sigma_x$,

$$U_i^{(1)} = \epsilon_i/\sigma\,, \quad U_i^{(2)} = v_i\,\epsilon_i/\sigma\,, \quad U_i^{(3)} = \epsilon_i^2\,\sigma^{-2} - 1\,,$$

$$U_i^{(1)^*} = \epsilon_i^*/\hat{\sigma}\,, \quad U_i^{(2)^*} = v_i\,\epsilon_i^*/\hat{\sigma}\,, \quad U_i^{(3)^*} = \epsilon_i^{*2}\,\hat{\sigma}^{-2} - 1\,,$$

$$\mathbf{U}_i = \left(U_i^{(1)}, U_i^{(2)}, U_i^{(3)}\right)^{\mathrm{T}}\,, \quad \mathbf{U}_i^* = \left(U_i^{(1)^*}, U_i^{(2)^*}, U_i^{(3)^*}\right)^{\mathrm{T}}\,,$$

$$\bar{\mathbf{U}} = n^{-1}\sum_{i=1}^{n}\mathbf{U}_i = \left(\bar{U}^{(1)}, \bar{U}^{(2)}, \bar{U}^{(3)}\right)^{\mathrm{T}}\,,$$

$$\bar{\mathbf{U}}^* = n^{-1}\sum_{i=1}^{n}\mathbf{U}_i^* = \left(\bar{U}^{(1)^*}, \bar{U}^{(2)^*}, \bar{U}^{(3)^*}\right)^{\mathrm{T}}\,,$$

$$A(u^{(1)}, u^{(2)}, u^{(3)}) = u^{(2)}/\left(1 + u^{(3)} - u^{(1)^2} - u^{(2)^2}\right)^{1/2}\,.$$

Then

$$T = n^{1/2}\,A(\bar{\mathbf{U}})\,, \qquad T^* = n^{1/2}\,A(\bar{\mathbf{U}}^*)\,.$$

This makes it clear that T and T^* may be expressed as smooth functions of means of independent random variables. Therefore, we may use minor modifications of the methods employed to prove Theorems 2.2 and 5.1 to establish Edgeworth expansions. Indeed, if q_j and $\hat{q}_j$, $j \geq 1$, denote the polynomials introduced in subsection 4.3.3 of Chapter 4, then the following result may be proved.

THEOREM 5.4. *Assume that* $E(|\epsilon|^{2\nu+4}) < \infty$ *for an integer* $\nu \geq 0$, *that Cramér's condition*

$$\limsup_{|t|\to\infty} |E\{\exp(it\,\epsilon)\}| < 1 \tag{5.51}$$

holds, and that $\{x_i\}$ *represents a sequence of independent realizations of a random variable* X *with* $E(|X|^{\nu+2}) < \infty$ *and* $\mathrm{var}(X) > 0$. *Then for a class of sequences* $\{x_i\}$ *arising with probability one,*

$$\sup_{-\infty < w < \infty} \left| P(T \leq w) - \Phi(w) - \sum_{j=1}^{\nu} n^{-j/2}\,q_j(w)\,\phi(w) \right|$$

$$= O(n^{-\nu/2}) \tag{5.52}$$

as $n \to \infty$. *If* $\lambda > 0$ *is given, and if the moment conditions on* ϵ *and* X *are strengthened to* $E(|\epsilon|^{2l}) < \infty$ *and* $E(|X|^l) < \infty$ *for a sufficiently large*

$l = l(\lambda)$, then for a constant $C > 0$,

$$P\left\{ \sup_{-\infty < w < \infty} \left| P(T^* \leq w \mid \mathcal{X}) - \Phi(w) \right. \right.$$
$$\left. \left. - \sum_{j=1}^{\nu} n^{-j/2} \hat{q}_j(w) \phi(w) \right| > C n^{-(\nu+1)/2} \right\} = O(n^{-\lambda}) \quad (5.53)$$

as $n \to \infty$.

Taking $\lambda > 1$ in (5.53) we may deduce from the Borel-Cantelli lemma that

$$\sup_{-\infty < w < \infty} \left| P(T^* \leq w \mid \mathcal{X}) - \Phi(w) - \sum_{j=1}^{\nu} n^{-j/2} \hat{q}_j(w) \phi(w) \right|$$
$$= O(n^{-(\nu+1)/2}) \quad (5.54)$$

with probability one. Result (5.54) is an explicit form of (4.29), under specified regularity conditions, and likewise (5.52) is an explicit form of (4.27). Expansions of coverage probability, such as that in (4.41), may be derived by doing no more than paralleling the arguments in Section 5.3.

We conclude by outlining the proof of Theorem 5.4. Let $\mathbf{V}_i$ denote the variance matrix and $\{\kappa_{\mathbf{r},i}\}$ the cumulant sequence of $\mathbf{U}_i$, and put

$$\overline{\mathbf{V}} = n^{-1} \sum_{i=1}^{n} \mathbf{V}_i, \qquad \bar{\kappa}_r = n^{-1} \sum_{i=1}^{n} \kappa_{\mathbf{r},i}.$$

Define the signed measure $P_j(-\Phi_{\mathbf{0},\overline{\mathbf{V}}} : \{\bar{\kappa}_{\mathbf{r}}\})$ as in part (i) of subsection 5.2.3. The following analogue of Lemma 5.4 may be proved by an argument similar to that which produced Lemma 5.4.

LEMMA 5.5. *Assume that* $E(|\epsilon|^{2\nu+4}) < \infty$ *for an integer* $\nu \geq 0$, *that Cramér's condition (5.51) holds, and that* $\{x_i\}$ *represents a sequence of independent realizations of a random variable* X *with* $E(|X|^{\nu+2}) < \infty$ *and* $\mathrm{var}(X) > 0$. *Then for a class of sequences* $\{x_i\}$ *arising with probability one,*

$$P(n^{1/2} \overline{\mathbf{U}} \in \mathcal{R}) = \sum_{j=0}^{\nu} n^{-j/2} P_j(-\Phi_{\mathbf{0},\bar{\mathbf{v}}} : \{\bar{\kappa}_{\mathbf{r}}\})(\mathcal{R}) + o(n^{-\nu/2}) \quad (5.55)$$

uniformly in any class R *of sets* $\mathcal{R} \subseteq \mathbb{R}^3$ *satisfying*

$$\sup_{\mathcal{R} \in R} \int_{(\partial\mathcal{R})^\eta} \exp\left(-\tfrac{1}{2} \|\mathbf{x}\|^2 \right) d\mathbf{x} = O(\eta)$$

as $\eta \downarrow 0$.

To prove (5.52), take

$$\mathcal{R} = \mathcal{R}(n, x) = \{\mathbf{y} \in \mathbb{R}^d : n^{1/2} A(n^{-1/2} \mathbf{y}) \leq x\}$$

in (5.55), and Taylor expand

$$P_j(-\Phi_{0,\bar{\mathbf{V}}} : \{\bar{\kappa}_r\})(\mathcal{R}(n, x))$$

about

$$P_j(-\Phi_{0,\bar{\mathbf{V}}} : \{\bar{\kappa}_r\})\,(\mathcal{R}(\infty, x))\,,$$

using techniques from part (iii) of subsection 5.2.3 to identify terms in the resulting expansion. The proof of (5.53) is very similar to that of Theorem 5.1 in Section 5.2, with only minor modifications to take account of the fact that the random series involved are now *weighted* sums of independent and identically distributed random variables.

5.5 Nonparametric Density Estimation

5.5.1 Introduction

Let $\mathcal{X} = \{X_1, \ldots, X_n\}$ denote a random sample from a population with density f. In Section 4.4 we developed Edgeworth expansions for the distribution of kernel-type nonparametric density estimators of f, in both the Studentized and non-Studentized cases. All the statistics involved may be represented very simply in terms of smooth functions of sums of independent and identically distributed random variables. For example, the Studentized estimator has the form

$$T \;=\; T(x) \;=\; \{\hat{f}(x) - E\hat{f}(x)\}/\{(nh)^{-1}\,\hat{f}_2(x) - n^{-1}\,\hat{f}(x)^2\}^{1/2}\,, \quad (5.56)$$

where $\hat{f}(x) = \hat{f}_1(x)$,

$$\hat{f}_j(x) \;=\; (nh)^{-1} \sum_{i=1}^{n} K\{(x - X_i)/h\}^j\,, \qquad (5.57)$$

and h denotes the bandwidth. Therefore, in principle, the techniques described earlier in this section go through with little change, and Edgeworth expansions may be developed for both T and its bootstrap approximant.

However, there is one technical difficulty, that of Cramér's condition. It is not immediately obvious that the random variable

$$Y \;=\; K\{(x - X)/h\}$$

will satisfy an appropriate version of Cramér's condition. Indeed, if K is the so-called uniform kernel,

$$K(y) \;=\; \begin{cases} \frac{1}{2} & \text{if } |y| \leq 1\,, \\ 0 & \text{otherwise}\,, \end{cases}$$

then Cramér's condition fails. This feature, and the fact that the orders of magnitude of terms in Edgeworth expansions for nonparametric density estimators are nonstandard, indicate that we should spend some time recounting rigorous theory for the expansions described in Section 4.4. The present section is devoted to that task. Subsection 5.5.2 states a variety of expansions from Section 4.4, under explicit regularity conditions, and subsections 5.5.3 and 5.5.4 provide proofs.

5.5.2 Main Results

We need the following notation from Section 4.4. Define $\hat{f}_j$ and T as in (5.56) and (5.57), and put

$$\mu_j = \mu_j(x) = E\{\hat{f}_j(x)\},$$

$$\mu_{ij} = \mu_{ij}(x) = h^{-1} E\Big([K\{(x-X)/h\} - EK\{(x-X)/h\}]^i$$
$$\times [K\{(x-X)/h\}^2 - EK\{(x-X)/h\}^2]^j\Big),$$

$$S = S(x) = \{\hat{f}(x) - \mu_1(x)\}/\{(nh)^{-1}\mu_2(x) - n^{-1}\mu_1(x)^2\}^{1/2},$$

$$p_1(y) = -\tfrac{1}{6}\mu_{20}^{-3/2}\mu_{30}(y^2-1),$$

$$p_2(y) = -\tfrac{1}{24}\mu_{20}^{-2}\mu_{40}\,y(y^2-3) - \tfrac{1}{72}\mu_{20}^{-3}\mu_{30}^2(y^4 - 10y^2 + 15),$$

$$q_{10}(y) = \tfrac{1}{2}\mu_{20}^{-3/2}\mu_{11} - \tfrac{1}{6}\mu_{20}^{-3/2}(\mu_{30} - 3\mu_{11})(y^2-1),$$

$$q_{20}(y) = -\mu_{20}^{-3}\mu_{30}^2\,y - \big(\tfrac{2}{3}\mu_{20}^{-3}\mu_{30}^2 - \tfrac{1}{12}\mu_{20}^{-2}\mu_{40}\big)y(y^2-3)$$
$$\qquad - \tfrac{1}{18}\mu_{20}^{-3}\mu_{30}^2\,y(y^4 - 10y^2 + 15),$$

$$q_{11}(y) = -\mu_1\mu_{20}^{-1/2}\,y^2.$$

Let $\mathcal{X}^* = \{X_1^*, \ldots, X_n^*\}$ denote a resample drawn randomly, with replacement, from $\mathcal{X}$, and define

$$\hat{f}_j^*(x) = (nh)^{-1}\sum_{i=1}^n K\{(x-X_i^*)/h\}^j, \qquad \hat{f}^* = \hat{f}_1^*,$$

$$\hat{\mu}_j = \hat{\mu}_j(x) = \hat{f}_j(x),$$

$$\hat{\mu}_{ij} = \hat{\mu}_{ij}(x) = h^{-1} E\Big([K\{(x-X^*)/h\} - E\{K((x-X^*)/h) \mid \mathcal{X}\}]^i$$
$$\times [K\{(x-X^*)/h\} - E\{K((x-X^*)/h) \mid \mathcal{X}\}]^j \mid \mathcal{X}\Big),$$

$$S^* = S^*(x) = \{\hat{f}^*(x) - \hat{f}(x)\}/\{(nh)^{-1}\hat{f}_2(x) - n^{-1}\hat{f}_1(x)^2\}^{1/2},$$

$$T^* = T^*(x) = \{\hat{f}^*(x) - \hat{f}(x)\}/\{(nh)^{-1}\hat{f}_2^*(x) - n^{-1}\hat{f}_1^*(x)^2\}^{1/2}.$$

Let $\hat{p}_j$ and $\hat{q}_{ij}$ have the same definitions as p_j and q_{ij}, except that μ_k and μ_{kl} are replaced by $\hat{\mu}_k$ and $\hat{\mu}_{kl}$ throughout.

The following regularity conditions are needed. The first guarantees a version of Cramér's condition, the second demands a minimal amount of smoothness of f, and the third asks that h satisfy the usual conditions for strong consistency of f:

> K has compact support $[a, b]$; for some decomposition $a = y_0 < y_1 < \cdots < y_m = b$, K' exists and is bounded on each interval (y_{j-1}, y_j) and is either strictly positive or strictly negative there; $\int K = 1$; $\qquad$ (5.58)

> $f(y) = (d/dy) P(X \le y)$ exists in a neighbourhood of x, is continuous at x, and satisfies $f(x) > 0$; $\qquad$ (5.59)

> $h = h(n) \to 0 \quad$ and $\quad nh/\log n \to \infty$. $\qquad$ (5.60)

Condition (5.58) excludes the uniform kernel, which fails to satisfy our bivariate version of Cramér's condition (see Lemma 5.6 in subsection 5.5.3), and that fact renders our proof inapplicable. However, an Edgeworth expansion for $S(x)$ in the case of the uniform kernel may be derived by methods which are routine for lattice-valued random variables (e.g., Gnedenko and Kolmogorov 1954, p. 212ff), and extended to $T(x)$ by noting that for the uniform kernel,

$$T(x) \;=\; (nh)^{1/2}\{\hat{f}(x) - \mu_1(x)\}/[\hat{f}(x)\{1 - h\hat{f}(x)\}]^{1/2}.$$

We are now in a position to state our main theorems about Edgeworth expansions, which make rigorous the work in subsections 4.4.3 and 4.4.4. The first expansion applies to the distributions of S and T, the second to the conditional distributions of S^* and T^*. We give here only the limited expansions used in Section 4.4. Longer expansions are readily developed, using the general Theorem 5.8 proved in subsection 5.5.3.

THEOREM 5.5. *Assume conditions* (5.58)–(5.60). *Then*

$$\sup_{-\infty < y < \infty} \left| P(S \le y) - \Phi(y) - \sum_{j=1}^{2} (nh)^{-j/2} p_j(y)\, \phi(y) \right|$$

$$= O\{(nh)^{-3/2}\}, \qquad (5.61)$$

$$\sup_{-\infty < y < \infty} \left| P(T \le y) - \Phi(y) - \sum_{j=1}^{2} (nh)^{-j/2} q_j(y)\, \phi(y) - (h/n)^{1/2} q_3(y)\, \phi(y) \right|$$

$$= O\{(nh)^{-3/2} + n^{-1}\}. \qquad (5.62)$$

THEOREM 5.6. *Assume conditions* (5.58)–(5.60). *Then*

$$\sup_{-\infty < y < \infty} \left| P(S^* \leq y \mid \mathcal{X}) - \Phi(y) - \sum_{j=1}^{2} (nh)^{-j/2} \hat{p}_j(y) \phi(y) \right|$$

$$= O\{(nh)^{-3/2}\}, \quad (5.63)$$

$$\sup_{-\infty < y < \infty} \left| P(T^* \leq y \mid \mathcal{X}) - \Phi(y) \right.$$

$$\left. - \sum_{j=1}^{2} (nh)^{-j/2} \hat{q}_j(y) \phi(y) - (h/n)^{1/2} \hat{q}_3(y) \phi(y) \right|$$

$$= O\{(nh)^{-3/2} + n^{-1}\}, \quad (5.64)$$

with probability one.

Formulae (5.61)–(5.64) are versions of the Edgeworth expansions (4.83), (4.86), (4.91), and (4.92), respectively, stated here under explicit regularity conditions.

Finally, we state rigorous forms of the coverage error expansions in subsection 4.4.5. Define $\hat{u}_\alpha$, $\hat{v}_\alpha$ by

$$P(S^* \leq \hat{u}_\alpha \mid \mathcal{X}) = P(T^* \leq \hat{v}_\alpha \mid \mathcal{X}) = \alpha.$$

Put $\hat{\sigma}(x)^2 = \hat{f}_2(x) - h\hat{f}_1(x)^2$, and recall that one-sided percentile and percentile-*t* confidence intervals for $E\hat{f}(x)$ are given by

$$\widehat{I}_1 = \left(-\infty, \ \hat{f}(x) - (nh)^{-1/2} \hat{\sigma}(x) \hat{u}_{1-\alpha}\right),$$

$$\widehat{J}_1 = \left(-\infty, \ \hat{f}(x) - (nh)^{-1/2} \hat{\sigma}(x) \hat{v}_{1-\alpha}\right),$$

respectively. Our next theorem provides Edgeworth expansions of the respective coverage probabilities.

Let

$$\gamma_1(x) = \mu_{20}(x)^{3/2} \mu_{30}(x),$$

$$\gamma_2(x) = \mu_{20}(x)^{-2} \mu_{40}(x) - \tfrac{3}{2} \mu_{20}(x)^{-3} \mu_{30}(x)^2.$$

THEOREM 5.7. *Assume conditions* (5.58)–(5.60). *Then*

$$P\{E\hat{f}(x) \in \widehat{I}_1\} = \alpha - (nh)^{-1/2} \tfrac{1}{2} \gamma_1(x) z_\alpha^2 \phi(z_\alpha)$$

$$+ O\{(nh)^{-1} + (h/n)^{1/2}\}, \quad (5.65)$$

$$P\{E\hat{f}(x) \in \widehat{J}_1\} = \alpha - (nh)^{-1} \tfrac{1}{6} \gamma_2(x) z_\alpha (2z_\alpha^2 + 1) \phi(z_\alpha)$$

$$+ O\{(nh)^{-3/2} + n^{-1}\}, \quad (5.66)$$

uniformly in $0 < \alpha < 1$, *as* $n \to \infty$.

Results (5.65) and (5.66) are identical to (4.94) and (4.95), respectively.

5.5.3 Proof of Theorems 5.5 and 5.6

We begin by introducing essential notation. Let

$$Y_r = K\{(x - x)/h\}^r - E[K\{(x - x)/h\}^r]$$

and write $\mathbf{V} = (v_{rs})$ for the 2×2 matrix with $v_{11} = \mu_{20}$, $v_{12} = \mu_{11}$, $v_{22} = \mu_{02}$. Put $\mathbf{t} = (t_1, t_2)^T$ and define the polynomial $P_{jk}(t_1, t_2)$ by the formal expansion

$$\exp\left[u^{-2} \sum_{k=0}^{\infty} (-v)^k (k+1)^{-1} \left\{ \sum_{j=2}^{\infty} i^j (j!)^{-1} u^j h^{-1} E(t_1 Y_1 + t_2 Y_2)^j \right\}^{k+1} \right]$$

$$= \exp(-\tfrac{1}{2} \mathbf{t}^T \mathbf{V} \mathbf{t}) \left\{ 1 + \sum_{j=1}^{\infty} \sum_{k=0}^{j/2} u^j v^k P_{jk}(t_1, t_2) \right\}.$$

Let Q_{jk} be the signed measure whose Fourier-Stieltjes transform is $\exp(-\tfrac{1}{2} \mathbf{t}^T \mathbf{V} \mathbf{t}) P_{jk}(\mathbf{t})$,

$$\int_{\mathbb{R}^2} \exp(i\mathbf{t}^T \mathbf{y}) \, dQ_{jk}(\mathbf{y}) = \exp(-\tfrac{1}{2} \mathbf{t}^T \mathbf{V} \mathbf{t}) P_{jk}(\mathbf{t}).$$

The density of Q_{jk} may be written as $\rho_{jk}(\mathbf{y}) \, \phi_{0,\mathbf{V}}(\mathbf{y})$, where ρ_{jk} is a polynomial whose coefficients are rational polynomials in the moments μ_{rs}, and $\phi_{0,\mathbf{V}}$ is the bivariate normal $N(\mathbf{0}, \mathbf{V})$ density. Let $\Phi_{0,\mathbf{V}}$ denote the $N(\mathbf{0}, \mathbf{V})$ probability measure.

Let $\widehat{\mathbf{V}}$, $\widehat{Q}_{jk}$, $\hat{\rho}_{jk}$ be the versions of $\mathbf{V}$, Q_{jk}, ρ_{jk} respectively, obtained on replacing μ_{rs} by $\hat{\mu}_{rs}$ (for each r, s) wherever it appears in the respective formulae. Define

$$S_{nr} = (nh)^{-1/2} \sum_{j=1}^{n} \left[K\{(x - X_j)/h\}^r - EK\{(x - X_j)/h\}^r \right],$$

$$S_{nr}^* = (nh)^{-1/2} \sum_{j=1}^{n} \left[K\{(x - X_j^*)/h\}^r - h\hat{f}_r(x) \right],$$

$$\mathbf{S}_n = (S_{n1}, S_{n2})^T, \qquad \mathbf{S}_n^* = (S_{n1}^*, S_{n2}^*)^T.$$

Write $\partial \mathcal{R}$ for the boundary of a set $\mathcal{R} \subseteq \mathbb{R}^2$ and $(\partial \mathcal{R})^\epsilon$ for the set of all points distant no more than ϵ from $\partial \mathcal{R}$. We first prove a precursor to Theorems 5.5 and 5.6.

THEOREM 5.8. *Assume the conditions of Theorem 5.5. Let R denote a class of Borel sets $\mathcal{R} \subseteq \mathbb{R}^2$ that satisfy*

$$\sup_{\mathcal{R} \in R} \int_{(\partial \mathcal{R})^\epsilon} \exp\left(-\tfrac{1}{2} \|\mathbf{x}\|^2 \right) d\mathbf{x} = O(\epsilon)$$

as $\epsilon \downarrow 0$. Then for each integer $\nu \geq 1$,

$$\sup_{\mathcal{R} \in R} \left| P(\mathbf{S}_n \in \mathcal{R}) - \Phi_{0,\mathbf{v}}(\mathcal{R}) - \sum_{j=1}^{\nu} \sum_{k=0}^{j/2} (nh)^{-j/2} h^k Q_{jk}(\mathcal{R}) \right|$$
$$= O\{(nh)^{-(\nu+1)/2}\}, \tag{5.67}$$

$$\sup_{\mathcal{R} \in R} \left| P(\mathbf{S}_n^* \in \mathcal{R} \mid \mathcal{X}) - \Phi_{0,\hat{\mathbf{v}}}(\mathcal{R}) - \sum_{j=1}^{\nu} \sum_{k=0}^{j/2} (nh)^{-j/2} h^k \hat{Q}_{jk}(\mathcal{R}) \right|$$
$$= O\{(nh)^{-(\nu+1)/2}\},$$

the latter result holding with probability one.

Our proof of Theorem 5.8 is prefaced by three lemmas.

LEMMA 5.6. *Under conditions (5.58) and (5.59) we have for each $\epsilon > 0$,*

$$\sup_{|t_1|+|t_2|>\epsilon} \left| \int_{-\infty}^{\infty} \exp\{it_1 K(u) + it_2 K^2(u)\} d_u F(x - hu) \right| \leq 1 - C(x, \epsilon) h$$

for all sufficiently small h, where $C(x, \epsilon) > 0$.

Proof.

Let a, b be as in condition (5.58), and observe that

$$I(t_1, t_2)$$
$$= -\int_{-\infty}^{\infty} \exp\{it_1 K(u) + it_2 K^2(u)\} d_u F(x - hu)$$
$$= 1 - h \int_a^b f(x - hu) \, du + hf(x) \int_a^b \exp\{it_1 K(u) + it_2 K^2(u)\} \, du$$
$$\quad + h \int_a^b \exp\{it_1 K(u) + it_2 K^2(u)\} \{f(x - hu) - f(x)\} \, du.$$

We claim that the conditions imposed on K imply that for each

$$-\infty < c < d < \infty, \qquad \text{and each} \qquad \epsilon > 0$$

there exists $\epsilon' > 0$ such that

$$\sup_{|t_1|+|t_2|>\epsilon} \left| (d-c)^{-1} \int_c^d \exp\{it_1 K(u) + it_2 K^2(u)\} \, du \right| < 1 - 3\epsilon'. \tag{5.68}$$

Accepting this for the time being, take $c = a$ and $d = b$, and choose h so small that

$$\int_a^b \left| f(x - hu) - f(x) \right| du \; < \; (b - a) f(x) \, \epsilon' \,.$$

Then

$$
\begin{aligned}
|I(t_1, t_2)| \; \leq \; & 1 - h(b - a) \, f(x)(1 - \epsilon') \\
& + h(b - a) \, f(x)(1 - 3\epsilon') + h(b - a) \, f(x)\epsilon' \\
= \; & 1 - h(b - a) f(x)\epsilon' \,,
\end{aligned}
$$

which gives Lemma 5.6.

Finally we check (5.68). Observe that

$$
\int_a^b \exp\left\{ it_1 \, K(u) + it_2 \, K^2(u) \right\} du
$$
$$
= \sum_{j=1}^m \int_{K(y_{j-1})}^{K(y_j)} \exp\left(it_1 \, u + it_2 \, u^2 \right) dK^{-1}(u) \,,
$$

where K^{-1} is interpreted in the obvious manner. Define

$$L(u) \; = \; \frac{d}{du} K^{-1}(u) \; = \; 1/K'\{K^{-1}(u)\} \,,$$

and note that

$$\left| \int_{K(y_{j-1})}^{K(y_j)} |L(u)| \, du \right| \; = \; y_j - y_{j-1} < \infty \,.$$

Therefore L is integrable on the interval $I_j = (a_j, b_j)$ where endpoints are $K(y_j)$ and $K(y_{j-1})$. Result (4.68) will follow if we prove that for each $\epsilon > 0$ and each j,

$$\sup_{|t_1| + |t_2| > \epsilon} \left| \int_{a_j}^{b_j} \exp\left(it_1 \, u + it_2 \, u^2 \right) \right| \; < \; \left| \int_{a_j}^{b_j} L(u) \, du \right| \,.$$

Since L may be approximated arbitrarily closely by a step function, it suffices to prove that for each $c < d$,

$$\sup_{|t_1| + |t_2| > \epsilon} \left| (d - c)^{-1} \int_c^d \exp\left(it_1 \, u + it_2 \, u^2 \right) du \right| \; < \; 1 \,.$$

Noting that $|e^{itx}| \leq 1$ we see that this result will follow if we prove that

$$\limsup_{|t_1| + |t_2| \to \infty} \left| (d - c)^{-1} \int_c^d \exp\left(it_1 \, u + it_2 \, u^2 \right) du \right| \; = \; 0 \,,$$

which may be accomplished using the standard argument for establishing the Riemann-Lebesgue lemma (see, e.g., Whittaker and Watson 1927, pp. 172-174).

LEMMA 5.7. *Let* $\mathbf{Z}$ *be a bivariate* $N(\mathbf{0}, \mathbf{I})$ *random variable independent of* $X_1, X_2, \ldots$, *and let* p *denote the probability density of* $\mathbf{S}_n + n^{-c}\mathbf{Z}$. *Assume the conditions of Theorem 5.5. Then for each pair of positive integers* (a, b) *there exists* $c_0(a, b) > 0$ *such that for all* $c \geq c_0(a, b)$,

$$\sup_{\mathbf{y} \in \mathbb{R}^2} (1 + \|\mathbf{y}\|)^a \left| p(\mathbf{y}) - \phi_{\mathbf{0},\mathbf{v}}(\mathbf{y}) - \sum_{j=1}^{b} \sum_{k=0}^{j/2} (nh)^{-j/2} h^k \rho_{jk}(\mathbf{y}) \phi_{\mathbf{0},\mathbf{v}}(\mathbf{y}) \right|$$

$$= O\{(nh)^{-(b+1)/2}\}.$$

Proof.

The inequality (5.17) may be complemented by the following result. For each integer $a \geq 0$ there exists a constant $C_1(a) > 0$ such that for any function $g : \mathbb{R}^d \to \mathbb{R}$ with Fourier transform $\check{g}$,

$$\sup_{\mathbf{y} \in \mathbb{R}^d} (1 + \|\mathbf{y}\|)^a |g(\mathbf{y})| \leq C_1(a) \sup{}' \int_{\mathbb{R}^d} |D^{\mathbf{r}} \check{g}(\mathbf{t})| \, d\mathbf{t},$$

where $\sup'$ denotes the supremum over integer vectors $\mathbf{r} \geq \mathbf{0}$ such that no more than one element of $\mathbf{r}$ is nonzero and any nonzero element lies in the interval $[1, a]$. Applying this inequality in the case $d = 2$ we see that it suffices to prove that with

$$g(\mathbf{y}) = p(\mathbf{y}) - \phi_{\mathbf{0},\mathbf{v}}(\mathbf{y}) - \sum_{j=1}^{b} \sum_{k=0}^{j/2} (nh)^{-j/2} h^k \rho_{jk}(\mathbf{y}) \phi_{\mathbf{0},\mathbf{v}}(\mathbf{y}), \quad (5.69)$$

we have

$$\sup{}' \int_{\mathbb{R}^2} |D^{\mathbf{r}} \check{g}(\mathbf{t})| \, d\mathbf{t} = O\{(nh)^{-(b+1)/2}\}, \quad (5.70)$$

provided c is sufficiently large.

For notational simplicity we shall treat only the case $\mathbf{r} = (r^{(1)}, r^{(2)}) = (0, 0)$, proving that

$$\int_{\mathbb{R}^2} |\check{g}(\mathbf{t})| \, d\mathbf{t} = O\{(nh)^{-(b+1)/2}\} \quad (5.71)$$

for c sufficiently large. There is no important technical difference between proving this result and (5.70) for *all* nonnegative integer pairs $(r^{(1)}, r^{(2)})$, since the function K is bounded. However, requisite notation for the case $(r^{(1)}, r^{(2)}) \neq (0, 0)$ is rather cumbersome.

The Fourier transform of p equals

$$\psi\{\mathbf{t}/(nh)^{1/2}\}^n \exp(-\tfrac{1}{2} n^{-2c} \|\mathbf{t}\|^2), \quad (5.72)$$

where

$$\psi(\mathbf{t}) = E\{\exp(it_1 Y_1 + it_2 Y_2)\}. \quad (5.73)$$

Since
$$E\big(|Y_1|^m\big) + E\big(|Y_2|^m\big) \leq C_2(m)h$$

for each $m \geq 1$ and $0 < h \leq 1$, by expanding the exponential function in (5.73) as a power series we may deduce that

$$\left| \psi(\mathbf{t}) - \sum_{j=0}^m i^j (j!)^{-1} E(t_1 Y_1 + t_2 Y_2)^j \right| \leq C_3(m)\, h\big(|t_1|^{m+1} + |t_2|^{m+1}\big).$$

In consequence, noting that $E(Y_1) = E(Y_2) = 0$,

$$n \log \psi\{\mathbf{t}/(nh)^{1/2}\}$$

$$= nh \sum_{k=0}^\infty (-h)^k\, (k+1)^{-1} \left\{ \sum_{j=2}^{b+2} i^j (j!)^{-1}\, (nh)^{-j/2}\, h^{-1}\, E(t_1\, Y_1 + t_2\, Y_2)^j \right\}^{k+1}$$

$$+\, R_{1b}(\mathbf{t}),$$

where, if $\|\mathbf{t}\| \leq \epsilon_1 (nh)^{1/2}$ and $\epsilon_1 = \epsilon_1(b)$ is sufficiently small,

$$|R_{1b}(\mathbf{t})| \leq C_4(b)\, (nh)^{-(b+1)/2}\, \big(|t_1|^{b+3} + |t_2|^{b+3}\big).$$

Hence if $\|\mathbf{t}\| \leq \epsilon_2\, (nh)^{1/2}$ and $\epsilon_2 = \epsilon_2(b)$ is sufficiently small,

$$\psi\{\mathbf{t}/(nh)^{1/2}\}^n = \exp\big[n \log \psi\{\mathbf{t}/(nh)^{1/2}\}\big]$$

$$= \beta_b(\mathbf{t}) + R_{2,b}(\mathbf{t}), \tag{5.74}$$

where

$$\beta_b(\mathbf{t}) = \exp(-\tfrac{1}{2}\mathbf{t}^T\mathbf{V}\mathbf{t})\left\{ 1 + \sum_{j=1}^b \sum_{k=0}^{j/2} (nh)^{-j/2}\, h^k\, P_{jk}(t_1, t_2) \right\}$$

is the Fourier transform of

$$\phi_{\mathbf{0},\mathbf{v}}(\mathbf{y}) + \sum_{j=1}^b \sum_{k=0}^{j/2} (nh)^{-j/2}\, h^k\, \rho_{jk}(\mathbf{y})\, \phi_{\mathbf{0},\mathbf{v}}(\mathbf{y})$$

and

$$|R_{2,b}(\mathbf{t})| \leq C_5(b)\, (nh)^{-(b+1)/2}\, \exp\big\{ - C_6(b)\, (t_1^2 + t_2^2) \big\}.$$

Substituting the results from (5.74) down into (5.72) we see that if $c \geq (b+1)/4$ and $\|\mathbf{t}\| \leq \epsilon_2 (nh)^{1/2}$ then the Fourier transform $\check{g}$ of the function g defined at (5.69) satisfies

$$|\check{g}(\mathbf{t})| \leq C_7(b)\, (nh)^{-(b+1)/2}\, \exp\big\{ - C_6(b)\, (t_1^2 + t_2^2) \big\}.$$

Therefore,

$$\int_{\|\mathbf{t}\|\le\epsilon_2(nh)^{1/2}} |\breve{g}(\mathbf{t})|\, dt \;=\; O\{(nh)^{-(b+1)/2}\}. \tag{5.75}$$

Furthermore,

$$\int_{\|\mathbf{t}\|>n^{2c}} |\breve{g}(\mathbf{t})|\, dt \;\le\; \int_{\|\mathbf{t}\|>n^{2c}} \exp(-\tfrac{1}{2}\, n^{-2c}\, \|\mathbf{t}\|^2)\, dt$$

$$= \; n^{2c} \int_{\|\mathbf{t}\|>n^c} \exp\left(-\tfrac{1}{2}\, \|\mathbf{t}\|^2\right)\, dt \;=\; O(n^{-C}) \tag{5.76}$$

for all $C > 0$. Formula (5.71) follows from (5.75), (5.76), and the result

$$\int_{\epsilon_2(nh)^{1/2}\le\|\mathbf{t}\|\le n^{2c}} |\breve{g}(\mathbf{t})|\, dt \;=\; O\{(nh)^{-(b+1)/2}\}, \tag{5.77}$$

which we prove next.

The left-hand side of (5.77) is dominated by

$$\left\{ \sup_{\|\mathbf{t}\|>\epsilon_2(nh)^{1/2}} |\breve{g}(\mathbf{t})| \right\} \int_{\|\mathbf{t}\|\le n^{2c}} dt$$

$$\le \; \left\{ \sup_{\|\mathbf{t}\|>\epsilon_2} |\psi(\mathbf{t})| \right\}^n \int_{\|\mathbf{t}\|\le n^{2c}} dt + O(n^{-C}), \tag{5.78}$$

for all $C > 0$. By Lemma 5.6,

$$|\psi(\mathbf{t})| \;=\; \left| \int_{-\infty}^{\infty} \exp\{it_1\, K(u) + it_2\, K^2(u)\}\, d_u\, F(x - hu) \right|$$

$$\le \; 1 - C_8(\epsilon_2)h$$

for all $\|\mathbf{t}\| > \epsilon_2$ and all sufficiently small h. Therefore, the quantity at (5.78) is dominated by

$$\{1 - C_8\,(\epsilon_2)h\}^n\, C_9\, n^{4c} + O(n^{-C})$$

$$\le \; C_9\, n^{4c} \exp\{-C_8\,(\epsilon_2)\, nh\} + O(n^{-C}) \;=\; O(n^{-C})$$

for all $C > 0$, since $nh/\log n \to \infty$. This proves (5.77) and completes the proof of Lemma 5.7.

LEMMA 5.8. *Let $\mathbf{Z}$ be a bivariate $N(\mathbf{0}, \mathbf{I})$ random variable independent of $X_1, X_2, \dots$ and of the resampling, and let $\hat{p}$ denote the probability density of $\mathbf{S}_n^* + n^{-c}\mathbf{Z}$ conditional on $\mathcal{X}$. Assume the conditions of Theorem 5.5.*

Then for each pair of positive integers (a, b) there exists $c_0(a, b) > 0$ such that for all $c \geq c_0(a, b)$,

$$\sup_{\mathbf{y} \in \mathbb{R}^2} (1 + \|\mathbf{y}\|)^a \left| \hat{p}(\mathbf{y}) - \phi_{0,\mathbf{v}}(\mathbf{y}) - \sum_{j=1}^{b} \sum_{k=0}^{j/2} (nh)^{-j/2} h^k \hat{\rho}_{jk}(\mathbf{y}) \phi_{0,\hat{\mathbf{v}}}(\mathbf{y}) \right|$$
$$= O\{(nh)^{-(b+1)/2}\}$$

with probability one.

The proof of Lemma 5.8 is very similar to that of Lemma 5.7, and so will not be given here. The differences between the proofs have already been illustrated in the proof of Theorem 5.1.

Proof of Theorem 5.8.

We establish only (5.67), since the second part of the theorem may be proved by similar arguments. Taking $a = 2$ in Lemma 5.7, and computing probabilities by integrating relevant densities over sets, we deduce that for all $c \geq c_0(2, b)$,

$$\sup_{\mathcal{R} \in \mathcal{B}} \left| P(\mathbf{S}_n + n^{-c} \mathbf{Z} \in \mathcal{R}) - \Phi_{0,\mathbf{v}}(\mathcal{R}) - \sum_{j=1}^{b} \sum_{k=0}^{j/2} (nh)^{-j/2} h^k Q_{jk}(\mathcal{R}) \right|$$
$$= O\{(nh)^{-(b+1)/2}\} \tag{5.79}$$

where $\mathcal{B}$ is the set of all Borel subsets of $\mathbb{R}^2$. Put $\delta = \delta(n) = n^{-c/2}$. Now,

$$\left| P(\mathbf{S}_n \in \mathcal{R}) - P(\mathbf{S}_n + n^{-c} \mathbf{Z} \in \mathcal{R}) \right|$$
$$\leq P\{\mathbf{S}_n + n^{-c} \mathbf{Z} \in (\partial \mathcal{R})^\delta\} + P(\|\mathbf{Z}\| > n^{c/2}).$$

We may deduce from (5.79) that if $c \geq b + 1$,

$$\sup_{\mathcal{R} \in \mathcal{R}} P\{\mathbf{S}_n + n^{-c} \mathbf{Z} \in (\partial \mathcal{R})^\delta\} = O\{(nh)^{-(b+1)/2} + n^{-c/2}\},$$

and, of course, $P(\|\mathbf{Z}\| > n^{c/2}) = O(n^{-C})$ for all $C > 0$. Therefore,

$$\sup_{\mathcal{R} \in \mathcal{R}} \left| P(\mathbf{S}_n \in \mathcal{R}) - P(\mathbf{S}_n + n^{-c} \mathbf{Z} \in \mathcal{R}) \right| = O\{(nh)^{-(b+1)/2}\}.$$

The desired result now follows from a second application of (5.79).

Proof of Theorems 5.5 and 5.6.

Since $S(x) = S_{n1}$ and $S^*(x) = S_{n1}^*$, the first part of each of Theorems 5.5 and 5.6 is contained in Theorem 5.8. To obtain the second part of Theorem 5.5, note that $T(x) = g_1(S_{n1}, S_{n2})$ where $g_1(u, v) = u/g_2(u, v)^{1/2}$

and

$$g_2(u, v) = \mu_2(x) + (nh)^{-1/2} v - h\{\mu_1(x) + (nh)^{-1/2} u\}^2 .$$

Hence, with

$$\mathcal{R}(y) = \{(u, v) : g_2(u, v) > 0, \quad g_1(u, v) \le y\},$$

we have $P\{T(x) \le y\} = P\{S_n \in \mathcal{R}(y)\}$. The second part of Theorem 5.5 may now be proved from the first part of Theorem 5.8, by using the argument following (5.29) in the first part of Theorem 5.1. Likewise, the second part of Theorem 5.6 follows from the second part of Theorem 5.8. Since the Edgeworth expansions are nonstandard, we conclude with a brief account of how terms in the second expansion in Theorem 5.5 may be identified.

Define $s = (nh)^{1/2}$,

$$\hat{f}_r(x) = (nh)^{-1} \sum_{j=1}^{n} K^r\{(x - X_j)/h\}, \qquad \mu_r(x) = E\{\hat{f}_r(x)\},$$

$$\sigma^2(x) = \mu_2(x) - h\mu_1^2(x) = \mu_{20}, \qquad \rho(x) = \mu_1(x)/\sigma(x),$$

$$\hat{\sigma}^2(x) = \hat{f}_2(x) - h\hat{f}_1^2(x), \qquad \Delta_j(x) = s\{\hat{f}_j(x) - \mu_j(x)\}/\sigma(x)^j .$$

Dropping the argument x, we have

$$\hat{\sigma}^2 = \sigma^2\{1 + s^{-1}\Delta_2 - 2hs^{-1}\rho\Delta_1 + O_p(n^{-1})\},$$

whence

$$\hat{\sigma}^{-1} = \sigma^{-1}\{1 - \tfrac{1}{2}s^{-1}\Delta_2 + \tfrac{3}{8} s^{-2}\Delta_2^2 + hs^{-1}\rho\Delta_1 + O_p(s^{-3} + n^{-1})\}.$$

Therefore,

$$T = \Delta_1\sigma\hat{\sigma}^{-1}$$

$$= \Delta_1\{1 - \tfrac{1}{2}s^{-1}\Delta_2 + \tfrac{3}{8} s^{-2}\Delta_2^2 + hs^{-1}\rho\Delta_1\} + O_p(s^{-3} + n^{-1}).$$

Raising T to a power, taking expectations, and ignoring terms of order $s^{-3} + n^{-1}$ or smaller, we deduce that

$$E(T) = -\tfrac{1}{2}s^{-1} E(\Delta_1\Delta_2) + hs^{-1}\rho E(\Delta_1^2),$$

$$E(T^2) = E(\Delta_1^2) - s^{-1} E(\Delta_1^2\Delta_2) + s^{-2} E(\Delta_1^2\Delta_2^2),$$

$$E(T^3) = E(\Delta_1^3) - \tfrac{3}{2}s^{-1} E(\Delta_1^3\Delta_2) + 3hs^{-1}\rho E(\Delta_1^4),$$

$$E(T^4) = E(\Delta_1^4) - 2s^{-1} E(\Delta_1^4\Delta_2) + 3s^{-2} E(\Delta_1^4\Delta_2^2).$$

Now,

$$E(\Delta_1^2) = 1, \qquad\qquad E(\Delta_1 \Delta_2) = \sigma^{-3}\mu_{11},$$

$$E(\Delta_1^3) = s^{-1}\sigma^{-3}\mu_{30}, \qquad E(\Delta_1^2\Delta_2) = s^{-1}\sigma^{-4}\mu_{21},$$

$$E(\Delta_1^4) = \sigma^{-4}(3\mu_{20}^2 + s^{-2}\mu_{40}) + O(n^{-1}),$$

$$E(\Delta_1^3\Delta_2) = \sigma^{-5}3\mu_{20}\mu_{11} + O(s^{-2}),$$

$$E(\Delta_1^2\Delta_2^2) = \sigma^{-6}(2\mu_{11}^2 + \mu_{20}\mu_{02}) + O(s^{-2}),$$

$$E(\Delta_1^4\Delta_2) = s^{-1}\sigma^{-6}(4\mu_{11}\mu_{30} + 6\mu_{20}\mu_{21}) + O(s^{-3}),$$

$$E(\Delta_1^4\Delta_2^2) = \sigma^{-8}(3\mu_{20}^2\mu_{02} + 12\mu_{11}^2\mu_{20}) + O(s^{-2}).$$

Using these formulae, noting that $\sigma^2 = \mu_{20}$, ignoring terms of order $s^{-3} + n^{-1}$ or smaller, and employing identities expressing cumulants in terms of moments (see (2.6)), we see that the first four cumulants of T are

$$\kappa_1(T) = -\tfrac{1}{2}s^{-1}\mu_{20}^{-3/2}\mu_{11} + hs^{-1}\rho, \tag{5.80}$$

$$\kappa_2(T) = 1 + s^{-2}\left(\tfrac{7}{4}\mu_{20}^{-3}\mu_{11}^2 + \mu_{20}^{-2}\mu_{02} - \mu_{20}^{-2}\mu_{21}\right),$$

$$\kappa_3(T) = s^{-1}\mu_{20}^{-3/2}(\mu_{30} - 3\mu_{11}) + 6hs^{-1}\rho, \tag{5.81}$$

$$\kappa_4(T) = s^{-2}\Big(\mu_{20}^{-2}\mu_{40} - 6\mu_{20}^{-3}\mu_{11}\mu_{30} - 6\mu_{20}^{-2}\mu_{21}$$
$$+ 18\mu_{20}^{-3}\mu_{11}^2 + 3\mu_{20}^{-2}\mu_{02}\Big).$$

All higher-order cumulants are of order s^{-3} or smaller, by an analogue of Theorem 2.1.

If the nonnegative integers i, j, k, l satisfy $i + 2j = k + 2l$, then

$$\mu_{ij} - \mu_{kl} = O(h).$$

Therefore, ignoring terms of order $s^{-3} + n^{-1}$ or smaller,

$$\kappa_2(T) = 1 + s^{-2}\tfrac{7}{4}\mu_{20}^{-3}\mu_{30}^2, \tag{5.82}$$

$$\kappa_4(T) = s^{-2}\left(12\mu_{20}^{-3}\mu_{30}^2 - 2\mu_{20}^{-2}\mu_{40}\right). \tag{5.83}$$

Combining (5.80)–(5.83) we see that the characteristic function of T satisfies

$$E(e^{itT})$$

$$= e^{-t^2/2} \exp\left[s^{-1}\left\{ -\tfrac{1}{2}\mu_{20}^{-3/2}\mu_{11}it + h\rho it \right.\right.$$
$$\left. + \tfrac{1}{6}\mu_{20}^{-3/2}(\mu_{30} - 3\mu_{11})(it)^3 + h\rho(it)^3 \right\}$$
$$+ s^{-2}\left\{ \tfrac{7}{8}\mu_{20}^{-3}\mu_{30}^2(it)^2 \right.$$
$$\left.\left. + \left(\tfrac{1}{2}\mu_{20}^{-3}\mu_{30}^2 - \tfrac{1}{12}\mu_{20}^{-2}\mu_{40}\right)(it)^4 \right\} \right] + O(n^{-1})$$

$$= e^{-t^2/2}\left[1 + s^{-1}\left\{ -\tfrac{1}{2}\mu_{20}^{-3/2}\mu_{11}\,it + \tfrac{1}{6}\mu_{20}^{-3/2}(\mu_{30} - 3\mu_{11})(it)^3 \right\} \right.$$
$$+ hs^{-1}\rho\{it + (it)^3\}$$
$$+ s^{-2}\left\{ \mu_{20}^{-3}\mu_{30}^2\,(it)^2 + \left(\tfrac{2}{3}\mu_{20}^{-3}\mu_{30}^2 - \tfrac{1}{12}\mu_{20}^{-2}\mu_{40}\right)(it)^4 \right.$$
$$\left.\left. + \tfrac{1}{18}\mu_{20}^{-3}\mu_{30}^2\,(it)^6 \right\} \right] + O(s^{-3} + n^{-1}).$$

The desired Edgeworth expansion for $P\{T(x) \le y\}$ follows, in a formal sense, on inverting this characteristic function.

5.5.4 Proof of Theorem 5.7

The argument needed to prove Theorem 5.7 closely parallels that used to establish Theorem 5.3 in Section 5.3, and so will not be given in detail. However, to indicate why the terms in the expansions have the form claimed for them we shall outline the derivation of the second part of Theorem 5.7. Put $s = (nh)^{1/2}$,

$$\Delta_j(x) = s\{\hat{f}_j(x) - \mu_j(x)\}/\sigma(x)^j,$$
$$\Delta = \Delta_3(x) - \tfrac{3}{2}\mu_3(x)\sigma(x)^{-3}\Delta_2(x).$$

Routine Taylor expansion gives

$$\hat{\mu}_{20}^{-3/2}\hat{\mu}_{11} - \mu_{20}^{-3/2}\mu_{11} = s^{-1}\Delta + O_p(s^{-2} + hs^{-1}),$$
$$\hat{\mu}_{20}^{-3/2}\hat{\mu}_{30} - \mu_{20}^{-3/2}\mu_{30} = s^{-1}\Delta + O_p(s^{-2} + hs^{-1}).$$

Hence,

$$\hat{q}_1(y) - q_1(y) = s^{-1}\tfrac{1}{6}(2y^2 + 1)\Delta + O_p(s^{-2} + hs^{-1}).$$

Therefore, writing v_α for the solution of $P(T \le v_\alpha) = \alpha$, we have

$$\hat{v}_\alpha - v_\alpha = -s^{-1}\{\hat{q}_1(z_\alpha) - q_1(z_\alpha)\} + O_p(s^{-3} + n^{-1})$$
$$= -s^{-2}\tfrac{1}{6}(2y^2 + 1)\Delta + O_p(s^{-3} + n^{-1}),$$

whence

$$P(T \le \hat{v}_\alpha) = P\Big\{T + s^{-2}\tfrac{1}{6}(2z_\alpha^2 + 1)\Delta \le v_\alpha + O_p(s^{-3} + n^{-1})\Big\}$$
$$= P(T' \le v_\alpha) + O(s^{-3} + n^{-1}), \tag{5.84}$$

where $T' = T + s^{-2}\tfrac{1}{6}(2z_\alpha^2 + 1)\Delta$. Here we have used the delta method. All but the second of the first four cumulants of T and T' are identical up to (but not including) terms of order $s^{-3} + n^{-1}$; and

$$\mathrm{var}(T') = \mathrm{var}(T) + s^{-2}\tfrac{1}{3}(2z_\alpha^2 + 1)\gamma_2(x) + O(s^{-4} + n^{-1}).$$

Therefore, comparing Edgeworth expansions for T and T',

$$P(T' \le t) = P(T \le t) - s^{-2}\tfrac{1}{6}(2z_\alpha^2 + 1)\gamma_2(x)\,t\,\phi(t) + O(s^{-3} + n^{-1}).$$

Taking $t = v_\alpha = z_\alpha + O(s^{-1})$ and using (5.84), we deduce that

$$P(T \le \hat{v}_\alpha) = \alpha - s^{-2}\tfrac{1}{6}z_\alpha(2z_\alpha^2 + 1)\gamma_2(x)\,\phi(z_\alpha) + O(s^{-3} + n^{-1}).$$

Hence, changing α to $1 - \alpha$ and noting that $z_{1-\alpha} = -z_\alpha$, we obtain

$$P\{E\hat{f}(x) \in \hat{J}_1\} = P(T > \hat{v}_{1-\alpha})$$
$$= \alpha - s^{-2}\tfrac{1}{6}z_\alpha(2z_\alpha^2 + 1)\gamma_2(x)\,\phi(z_\alpha) + O(s^{-3} + n^{-1}),$$

as required.

5.6 Bibliographical Notes

Rigorous theory for bootstrap Edgeworth expansions dates from work of Singh (1981), Bickel and Freedman (1980), Beran (1982), Babu and Singh (1983, 1984, 1985), and Hall (1986a, 1988a). Development of expansions of coverage probability started with Hall (1986a), and our work in Sections 5.2 and 5.3 is based on that contribution. The case of simple linear regression, discussed briefly in Section 5.4, is treated in greater detail by Hall (1988d). The theory for nonparametric density estimation in Section 5.5 is taken from Hall (1991b).

Appendix I

Number and Sizes of Atoms of Nonparametric Bootstrap Distribution

Let $\mathcal{X}^*$ denote a same-size resample drawn, with replacement, from a given sample $\mathcal{X}$, and let $\hat{\theta}^*$ be the value of a statistic computed from $\mathcal{X}^*$. In most cases of practical interest, each distinct $\mathcal{X}^*$ (without regard for order) gives rise to a distinct $\hat{\theta}^*$. For example, if $\hat{\theta}^*$ is a mean or a variance, and if $\mathcal{X}$ is drawn at random from a continuous distribution, then the statement made in the previous sentence is true for almost all samples $\mathcal{X}$, i.e., with probability one. In this circumstance the number of different possible values of $\hat{\theta}^*$ equals (with probability one) the number of different possible resamples $\mathcal{X}^*$.

If the sample $\mathcal{X}$ is of size n, and if all elements of $\mathcal{X}$ are distinct, then the number of different possible resamples $\mathcal{X}^*$ equals the number, $N(n)$, of distinct ways of placing n indistinguishable objects into n numbered boxes, the boxes being allowed to contain any number of objects. To appreciate why, write $\mathcal{X} = \{X_1, \dots, X_n\}$ and let m_i denote the number of times X_i is repeated in $\mathcal{X}^*$. The number of different possible resamples equals the number of different ways of choosing the ordered n-vector $(m_1, \dots, m_n)$ such that each $m_i \geq 0$ and $m_1 + \dots + m_n = n$. Think of m_i as the number of objects in box i.

Calculation of $N(n)$ is an old but elementary combinatorial problem. In fact, $N(n) = \binom{2n-1}{n}$. One proof of this formula runs as follows. Place $2n - 1$ dots in a line. Mark a cross on exactly $n - 1$ of the dots, and put into box i a number of objects equal to the number of dots remaining between the $(i - 1)$th and ith crosses, $1 \leq i \leq n$. (Cross 0 represents the far left of the arrangement, and cross n denotes the far right.) Each arrangement of n indistinguishable objects into n numbered boxes may be represented uniquely in this manner. Therefore, the total number $N(n)$ of arrangements equals the number of ways of placing the $n - 1$ crosses among the $2n - 1$ dots; this is $\binom{2n-1}{n-1}$, or equivalently, $\binom{2n-1}{n}$. See Roberts (1984, p. 42) for a similar treatment of a more general problem.

For an alternative proof, note that the number $N(n)$ of ordered sequences $(m_1, \ldots, m_n)$ with each $m_i \geq 0$ and $m_1 + \ldots + m_n = n$, equals the coefficient of x^n in $(1 + x + x^2 + \ldots)^n = (1 - x)^{-n}$. That coefficient is $(-1)^n \binom{-n}{n}$, which is identical to $\binom{2n-1}{n}$.

Not all of the $\binom{2n-1}{n}$ atoms of the bootstrap distribution of $\hat{\theta}^*$ have equal mass. To compute probabilities, let $\mathcal{X}^*(m_1, \ldots, m_n)$ denote the resample drawn from $\mathcal{X} = \{X_1, \ldots, X_n\}$ in which X_i is repeated precisely m_i times, for $1 \leq i \leq n$. The chance of drawing X_i out of $\mathcal{X}$ on any given draw equals n^{-1}. Therefore, the chance that n draws (with replacement after each draw) result in $\mathcal{X}^*(m_1, \ldots, m_n)$ equals the multinomial probability

$$\binom{n}{m_1, \ldots, m_n} (n^{-1})^{m_1} \ldots (n^{-1})^{m_n} = \frac{n!}{n^n m_1! \ldots m_n!}.$$

If $\hat{\theta}^*(m_1, \ldots, m_n)$ denotes the value of the statistic $\hat{\theta}^*$ when the resample is $\mathcal{X}^*(m_1, \ldots, m_n)$, then in the circumstances discussed in the first paragraph of this appendix,

$$P\{\hat{\theta}^* = \hat{\theta}^*(m_1, \ldots, m_n) \mid \mathcal{X}\} = \frac{n!}{n^n m_1! \ldots m_n!},$$

$$\text{all } m_i \geq 0 \quad \text{with } m_1 + \ldots + m_n = n,$$

(A.1)

for almost all samples $\mathcal{X}$. These probabilities represent the atoms of the bootstrap distribution of $\hat{\theta}^*$.

The right-hand side of formula (A.1) is at its largest when each $m_i = 1$, in which case $\mathcal{X}^*$ is identical to $\mathcal{X}$. Hence, the most likely resample to be drawn is the original sample, this event occurring with probability $p_n = n!/n^n$. The modal atom of the bootstrap distribution of $\hat{\theta}^*$ is therefore $\hat{\theta}$, the original statistic computed from $\mathcal{X}$. While $\mathcal{X}$ is more likely to arise than any other resample, its probability is very small for even small n, being only 3.6×10^{-4} when $n = 10$. The probability decreases exponentially quickly with increasing n. That property reflects two features of the bootstrap distribution of $\hat{\theta}^*$: the total number of atoms increases rapidly with increasing sample size, and the bootstrap distribution of $\hat{\theta}^*$ is "smudged" very widely over these different atoms.

Table A.1 lists values of $N(n)$ and p_n for different values of n. As n increases, $N(n) \sim (n\pi)^{-1/2} 2^{2n-1}$ and $p_n \sim (2n\pi)^{1/2} e^{-n}$, demonstrating that $N(n)$ increases exponentially quickly while p_n decreases exponentially quickly.

If B bootstrap simulations are conducted then the chance that no resamples are repeated is not less than

$$(1 - p_n)(1 - 2p_n) \ldots \{1 - (B-1)p_n\} \geq 1 - \tfrac{1}{2} B(B-1) p_n, \quad \text{(A.2)}$$

TABLE A.1. Number of atoms, $\binom{2n-1}{n}$, of bootstrap distribution, together with probability, $n!/n^n$, of most likely or modal atom. In each case the modal value is $\hat{\theta}$, the original statistic computed from the sample.

n	number of atoms	probability of most likely atom
2	3	0.5
3	10	0.2222
4	35	0.0940
5	126	0.0384
6	462	1.5×10^{-2}
7	1,716	6.1×10^{-3}
8	6,435	2.4×10^{-3}
9	24,310	9.4×10^{-4}
10	92,378	3.6×10^{-4}
12	1,352,078	5.4×10^{-5}
15	7.8×10^7	3.0×10^{-6}
20	6.9×10^{10}	2.3×10^{-8}

provided $(B-1)\,p_n < 1$. As n and B increase,

$$\tfrac{1}{2} B(B-1)\,p_n \; \sim \; B^2 (n\pi/2)^{1/2}\, e^{-n}\,.$$

Therefore, if B increases with n more slowly than $n^{-1/4}\, e^{n/2}$ — in particular, if $B = O(n^c)$ for any fixed $c > 0$ — then the probability that one or more repeats occur in the B values of $\hat{\theta}^*$, converges to zero as $n \to \infty$. The number of resampling operations has to increase with sample size at an exponentially fast rate before repetitions become a problem.

To be more specific, if $n = 20$ and $B = 2{,}000$ then the chance of no repetitions is not less than

$$1 - \tfrac{1}{2} B(B-1)\,p_n \; = \; 0.954\,.$$

This is a convenient case to remember: with a sample size of 20, and with as many as 2,000 resampling operations, the chance of repeating the same resample is less than 1 in 20. Therefore, for many practical purposes the bootstrap distribution of a statistic $\hat{\theta}^*$ may be regarded as continuous.

For small values of n, say $n \leq 8$, it is often feasible to calculate a bootstrap estimate exactly, by computing all the atoms of the bootstrap algorithm. Fisher and Hall (1991) discuss algorithms for exact bootstrap calculations. However, for larger samples, say $n \geq 10$, Monte Carlo simulation is usually the only practical method.

Appendix II

Monte Carlo Simulation

II.1 Introduction

In many problems of practical interest, the nonparametric bootstrap is employed to estimate an expected value. For example, if $\hat{\theta}$ is an estimate of an unknown quantity θ then we might wish to estimate bias, $E(\hat{\theta} - \theta)$, or the distribution function of $\hat{\theta}$, $E\{I(\hat{\theta} \leq x)\}$. Generally, suppose we wish to estimate $E(U)$, where U is a random variable that (in the notation of Chapter 1) will often be a functional of both the population distribution function F_0 and the empirical distribution function F_1 of the sample $\mathcal{X}$,

$$U = f(F_0, F_1).$$

Let F_2 denote the empirical distribution function of a resample $\mathcal{X}^*$ drawn randomly, with replacement, from $\mathcal{X}$, and put $U^* = f(F_1, F_2)$. As noted in Chapter 1,

$$\hat{u} = E\{f(F_1, F_2) \mid F_1\}$$

$$= E(U^* \mid \mathcal{X})$$

is "the bootstrap estimate" of $u = E(U)$.

In the bias example considered above, we would have $U = \hat{\theta} - \theta$ and $U^* = \hat{\theta}^* - \hat{\theta}$, where $\hat{\theta}^*$ is the version of $\hat{\theta}$ computed for $\mathcal{X}^*$ rather than $\mathcal{X}$. In the case of a distribution function, $U = I(\hat{\theta} \leq x)$ and $U^* = I(\hat{\theta}^* \leq x)$.

Our aim in this Appendix is to describe some of the available methods for approximating $\hat{u}$ by Monte Carlo simulation and to provide a little theory for each. The methods that we treat are uniform resampling, linear approximation, the centring method, balanced resampling, antithetic resampling, and importance resampling, and are discussed in Sections II.2–II.7, respectively. This account is not exhaustive; for example, we do not treat Richardson extrapolation (Bickel and Yahav 1988), computation by saddlepoint methods (Davison and Hinkley 1988; Reid 1988), or balanced importance resampling (Hall 1990e). Section II.8 will briefly describe the problem of quantile estimation, which does not quite fit into the format of approximating $\hat{u} = E(U^* \mid \mathcal{X})$.

II.2 Uniform Resampling

Since $\hat{u}$ is defined in terms of uniform resampling — that is, random resampling with replacement, in which each sample value is drawn with the same probability n^{-1} — then uniform resampling is the most obvious approach to simulation. Conditional on $\mathcal{X}$, draw B independent resamples $\mathcal{X}_1^*, \ldots, \mathcal{X}_B^*$ by resampling uniformly, and let $\hat{\theta}_b^*$ denote the version of $\hat{\theta}$ computed for $\mathcal{X}_b^*$ rather than $\mathcal{X}$. Then

$$\hat{u}_B^* = B^{-1} \sum_{b=1}^{B} U_b^*$$

is a Monte Carlo approximation to $\hat{u}$. With probability one, conditional on $\mathcal{X}$, $\hat{u}_B^*$ converges to $\hat{u}_\infty^* = \hat{u}$ as $B \to \infty$.

We refer to $\hat{u}_B^*$ as an approximation rather than an estimate. The bootstrap estimate $\hat{u}$ is based on $B = \infty$, and only its approximate form is concerned with finite B's. Thus, we draw a distinction between the statistical problem of estimating u and the numerical problem of approximating $\hat{u}$. While this view might seem semantic, it does help to distinguish between different approaches to inference. For example, the bootstrap approach to hypothesis testing described in Section 4.6 is concerned only with the case $B = \infty$, since it uses the "full" (i.e., $B = \infty$) bootstrap estimates of quantiles and critical points. Employing a finite value of B would produce an approximation to bootstrap hypothesis testing. On the other hand, the method of Monte Carlo hypothesis testing (Barnard 1963; Hope 1968; and Marriott 1979) is concerned intrinsically with the case of finite B. For example, it is common for $B = 99$ to be recommended for testing at the 5% level. Thus, a Monte Carlo hypothesis test is an approximate bootstrap hypothesis test, but a bootstrap test is not a Monte Carlo test. See also Hall and Titterington (1989).

The expected value of $\hat{u}_B^*$, conditional on $\mathcal{X}$, equals $\hat{u}$:

$$E(\hat{u}_B^* \mid \mathcal{X}) = B^{-1} \sum_{b=1}^{B} E(U_b^* \mid \mathcal{X}) = B^{-1} \sum_{b=1}^{B} \hat{u} = \hat{u}.$$

Therefore, $\hat{u}_B^*$ is an unbiased approximation to $\hat{u}$, and the performance of $\hat{u}_B^*$ may be reasonably described in terms of conditional variance,

$$\mathrm{var}(\hat{u}_B^* \mid \mathcal{X}) = B^{-1} \mathrm{var}(U^* \mid \mathcal{X}) = B^{-1} \left\{ E(U^{*2} \mid \mathcal{X}) - \hat{u}^2 \right\}.$$

In many problems of practical importance, $\mathrm{var}(U^* \mid \mathcal{X})$ is asymptotic to either a constant or a constant multiple of n^{-1}, as $n \to \infty$. For example, if we are estimating a distribution function then U^* is an indicator function, and so $U^{*2} = U^*$, whence

$$\text{var}(U^* \mid \mathcal{X}) = \hat{u}(1 - \hat{u}) \longrightarrow u_0(1 - u_0),$$

where u_0 (typically a value of a Normal distribution function) equals the limit as $n \to \infty$ of $\hat{u}$. If we are estimating bias and $\hat{\theta} = g(\overline{\mathbf{X}})$ is a smooth function of a d-variate sample mean, then $U^* = \hat{\theta}^* - \hat{\theta} = g(\overline{\mathbf{X}}^*) - g(\overline{\mathbf{X}})$, where $\overline{\mathbf{X}}^*$ is the resample mean. By Taylor expansion,

$$U^* = g(\overline{\mathbf{X}}^*) - g(\overline{\mathbf{X}}) \approx \sum_{j=1}^d (\overline{\mathbf{X}}^* - \overline{\mathbf{X}})^{(j)} g_j(\overline{\mathbf{X}}), \qquad (\text{A.3})$$

where $x^{(j)} = (\mathbf{x})^{(j)}$ denotes the jth element of a d-vector $\mathbf{x}$ and

$$g_j(\mathbf{x}) = (\partial/\partial x^{(j)}) \, g(\mathbf{x}).$$

Thus,

$$\text{var}(U^* \mid \mathcal{X}) = \text{var}\left\{ \sum_{j=1}^d (\overline{\mathbf{X}}^* - \overline{\mathbf{X}})^{(j)} g_j(\overline{\mathbf{X}}) \mid \mathcal{X} \right\} + O(n^{-2})$$

$$= n^{-1} \hat{\sigma}^2 + O(n^{-2}),$$

where

$$\hat{\sigma}^2 = n^{-1} \sum_{i=1}^n \left\{ \sum_{j=1}^d (\mathbf{X}_i - \overline{\mathbf{X}})^{(j)} g_j(\overline{\mathbf{X}}) \right\}^2$$

$$\longrightarrow \sigma^2 = E\left\{ \sum_{j=1}^d (\mathbf{X} - \boldsymbol{\mu})^{(j)} g_j(\boldsymbol{\mu}) \right\}^2$$

(as $n \to \infty$) and $\boldsymbol{\mu} = E(\mathbf{X})$ is the population mean.

These two examples describe situations that are typical of a great many that arise in statistical problems. When the target $\hat{u}$ is a distribution function or a quantile, the conditional variance of the uniform bootstrap approximant is roughly equal to CB^{-1} for large n and large B; and when the target is the expected value of a smooth function of a mean, the variance is approximately $CB^{-1}n^{-1}$. In both cases, C is a constant not depending on B or n. Efficient approaches to Monte Carlo approximation can reduce the value of C in the case of estimating a distribution function, or increase the power of n^{-1} (say, from n^{-1} to n^{-2}) in the case of estimating the expected value of a smooth function of a mean. Most importantly, they usually do not increase the power of B^{-1}. Therefore, generally speaking, even the more efficient of Monte Carlo methods have mean squared error that decreases like B^{-1} as B increases, for a given sample.

II.3 Linear Approximation

We motivate linear approximation by considering the bias estimation example of the previous section. Suppose our aim is to approximate

$$\hat{u} = E(U^* \mid \mathcal{X}),$$

where

$$U^* = \hat{\theta}^* - \hat{\theta} = g(\bar{\mathbf{X}}^*) - g(\bar{\mathbf{X}})$$

and g is a smooth function of d variables. Let us extend the Taylor expansion (A.3) to another term,

$$U^* = \sum_{j=1}^{d} (\bar{\mathbf{X}}^* - \bar{\mathbf{X}})^{(j)} g_j(\bar{\mathbf{X}})$$

$$+ \tfrac{1}{2} \sum_{j=1}^{d} \sum_{k=1}^{d} (\bar{\mathbf{X}}^* - \bar{\mathbf{X}})^{(j)} (\bar{\mathbf{X}}^* - \bar{\mathbf{X}})^{(k)} g_{jk}(\bar{\mathbf{X}}) + \ldots, \qquad \text{(A.4)}$$

where $g_{j_1 \ldots j_r}(\mathbf{x}) = (\partial^r / \partial x^{(j_1)} \ldots \partial x^{(j_r)}) g(\mathbf{x})$. As we noted in Section II.2, the conditional variance of U^* is determined asymptotically by the variance of the first term on the right-hand side of (A.4), which is the linear component in the Taylor expansion. Now, our aim is to approximate $E(U^* \mid \mathcal{X})$, and the linear component does not contribute anything to that expectation:

$$E\left\{ \sum_{j=1}^{d} (\bar{\mathbf{X}}^* - \bar{\mathbf{X}})^{(j)} g_j(\bar{\mathbf{X}}) \mid \mathcal{X} \right\}$$

$$= \sum_{j=1}^{d} E\{(\bar{\mathbf{X}}^* - \bar{\mathbf{X}})^{(j)} \mid \mathcal{X}\} g_j(\bar{\mathbf{X}}) = 0. \qquad \text{(A.5)}$$

Therefore, it makes sense to remove the linear component; that is, base the uniform resampling approximation on

$$V^* = U^* - \sum_{j=1}^{d} (\bar{\mathbf{X}}^* - \bar{\mathbf{X}})^{(j)} g_j(\bar{\mathbf{X}}),$$

instead of on U^*. Conditional on $\mathcal{X}$, draw B independent resamples $\mathcal{X}_1^*, \ldots, \mathcal{X}_B^*$ by resampling uniformly (exactly as in Section II.2) and put

$$V_b^* = U_b^* - \sum_{j=1}^{d} (\bar{\mathbf{X}}_b^* - \bar{\mathbf{X}})^{(j)} g_j(\bar{\mathbf{X}})$$

$$= g(\bar{\mathbf{X}}_b^*) - g(\bar{\mathbf{X}}) - \sum_{j=1}^{d} (\bar{\mathbf{X}}_b^* - \bar{\mathbf{X}})^{(j)} g_j(\bar{\mathbf{X}}),$$

$1 \leq b \leq B$, where $\overline{\mathbf{X}}_b^*$ denotes the mean of the resample $\mathcal{X}_b^*$. Define

$$\hat{v}_B^* = B^{-1} \sum_{b=1}^{B} V_b^* .$$

Then $\hat{v}_B^*$ is the linear approximation method approximation of $\hat{u}$. With probability one, conditional on $\mathcal{X}$, $\hat{v}_B^* \rightarrow \hat{u}$ as $B \rightarrow \infty$.

In view of (A.5), $E(V^* \mid \mathcal{X}) = E(U^* \mid \mathcal{X}) = \hat{u}$, and so $\hat{v}_B^*$ is an unbiased approximant of $\hat{u}$. The conditional variance of $\hat{v}_B^*$ is dominated by that of the first remaining term in the Taylor expansion, just as in the case of $\hat{u}_B^*$,

$$\text{var}\big(\hat{v}_B^* \mid \mathcal{X}\big) = B^{-1} \text{var}(V^* \mid \mathcal{X}),$$

and

$\text{var}(V^* \mid \mathcal{X})$

$$= \text{var}\left\{ \tfrac{1}{2} \sum_{j=1}^{d} \sum_{k=1}^{d} (\overline{\mathbf{X}}^* - \overline{\mathbf{X}})^{(j)} (\overline{\mathbf{X}}^* - \overline{\mathbf{X}})^{(k)} g_{jk}(\overline{\mathbf{X}}) \mid \mathcal{X} \right\} + O(n^{-3})$$

$$= n^{-2} \hat{\beta} + O(n^{-3}), \tag{A.6}$$

with probability one as $n \rightarrow \infty$, where[1]

$$\hat{\beta} = \tfrac{1}{2} \sum_{j_1=1}^{d} \sum_{k_1=1}^{d} \sum_{j_2=1}^{d} \sum_{k_2=1}^{d} g_{j_1 k_1}(\overline{\mathbf{X}}) \, g_{j_2 k_2}(\overline{\mathbf{X}}) \, \hat{\sigma}^{j_1 j_2} \, \hat{\sigma}^{k_1 k_2} , \tag{A.7}$$

$$\hat{\sigma}^{jk} = n^{-1} \sum_{i=1}^{n} (\mathbf{X}_i - \overline{\mathbf{X}})^{(j)} (\mathbf{X}_i - \overline{\mathbf{X}})^{(k)} . \tag{A.8}$$

Therefore, the order of magnitude of the variance has been reduced from $B^{-1} n^{-1}$ (in the case of $\hat{u}_B^*$) to $B^{-1} n^{-2}$ (in the case of $\hat{v}_B^*$). The numerical value of the reduction, for a given problem and sample, will depend on values of the first and second derivatives of g, as well as on the higher-order terms that our asymptotic argument has ignored.

More generally, we may approximate U^* by an arbitrary number of terms in the Taylor expansion (A.4), and thereby compute a general "polynomial approximation" to $\hat{u}$. For example, if $m \geq 1$ is an integer then we may define

$$W_b^* = U_b^* - \sum_{r=1}^{m} \frac{1}{r!} \sum_{j_1=1}^{d} \cdots \sum_{j_r=1}^{d} (\overline{\mathbf{X}}_b^* - \overline{\mathbf{X}})^{(j_1)} \ldots (\overline{\mathbf{X}}_b^* - \overline{\mathbf{X}})^{(j_r)} g_{j_1 \ldots j_r}(\overline{\mathbf{X}})$$

---

[1]In Hall (1989a), the factor $\tfrac{1}{2}$ was inadvertently omitted from the right-hand side of (A.7).

(a generalization of V_b^*),

$$\tilde{w} = \sum_{r=1}^{m} \frac{1}{r!} \sum_{j_1=1}^{d} \cdots \sum_{j_r=1}^{d} E\{(\overline{\mathbf{X}}^* - \overline{\mathbf{X}})^{(j_1)} \ldots (\overline{\mathbf{X}}^* - \overline{\mathbf{X}})^{(j_r)} \mid \mathcal{X}\} g_{j_1 \ldots j_r}(\overline{\mathbf{X}}),$$

and

$$\hat{w}_B^* = B^{-1} \sum_{b=1}^{B} W_b^* + \tilde{w}.$$

Then $\hat{w}_B^*$ is an unbiased approximation of $\hat{u}$. Of course, the approximation is only practicable if we can compute $\tilde{w}$, which is a linear form in the first m central sample moments, with coefficients equal to the derivatives of g. In the special case $m = 1$, $\hat{w}_B^*$ reduces to our linear approximation $\hat{v}_B^*$, and there $\tilde{w} = 0$. The conditional variance of $\hat{w}_B^*$ is of order $B^{-1} n^{-(m+1)}$ as $n \to \infty$. Of course, $\hat{w}_B^* \to \hat{u}$ as $B \to \infty$, for fixed $\mathcal{X}$. If the function g is analytic then for fixed $\mathcal{X}$ and B, $\hat{w}_B^* \to \hat{u}$ as $m \to \infty$. Each of these limits is attained with probability one, conditional on $\mathcal{X}$.

Monte Carlo methods based on functional approximation have been discussed by Oldford (1985) and Davison, Hinkley, and Schechtman (1986).

II.4 Centring Method

To motivate the centring method approximation, recall that the linear approximation method produces the approximant

$$\hat{v}_B^* = B^{-1} \sum_{b=1}^{B} V_b^*$$

$$= B^{-1} \sum_{b=1}^{B} \left\{ g(\overline{\mathbf{X}}_b^*) - g(\overline{\mathbf{X}}) - \sum_{j=1}^{d} (\overline{\mathbf{X}}_b^* - \overline{\mathbf{X}})^{(j)} g_j(\overline{\mathbf{X}}) \right\}$$

$$= \hat{u}_B^* - \sum_{j=1}^{d} (\overline{\overline{\mathbf{X}}}^* - \overline{\mathbf{X}})^{(j)} g_j(\overline{\mathbf{X}}),$$

where $\hat{u}_B^*$ is the uniform resampling approximation of $\hat{u}$ and

$$\overline{\overline{\mathbf{X}}}^* = B^{-1} \sum_{b=1}^{B} \overline{\mathbf{X}}_b^*$$

is the grand mean of all the resamples. Now,

$$\sum_{j=1}^{d} (\overline{\overline{\mathbf{X}}}^* - \overline{\mathbf{X}})^{(j)} g_j(\overline{\mathbf{X}}) \approx g(\overline{\overline{\mathbf{X}}}^*) - g(\overline{\mathbf{X}}),$$

by Taylor expansion, and so

$$\hat{v}_B \;\approx\; \hat{u}_B^* - \{g(\overline{\mathbf{X}}_{\cdot}^*) - g(\overline{\mathbf{X}})\} \;=\; \hat{x}_B^*\,,$$

where

$$\hat{x}_B^* \;=\; B^{-1} \sum_{b=1}^{B} g(\overline{\mathbf{X}}_b^*) - g(\overline{\mathbf{X}}_{\cdot}^*)\,.$$

We call $\hat{x}_B^*$ the centring method approximation to $\hat{u}$. It differs from $\hat{u}_B^*$ in that the mean of the $g(\overline{\mathbf{X}}_b^*)$'s is now centred at $g(\overline{\mathbf{X}}_{\cdot}^*)$ rather than $g(\overline{\mathbf{X}})$; recall that

$$\hat{u}_B^* \;=\; B^{-1} \sum_{b=1}^{B} g(\overline{\mathbf{X}}_b^*) - g(\overline{\mathbf{X}})\,.$$

The centring method approximation was suggested by Efron (1988).

As expected, $\hat{x}_B^* \to \hat{u}$ as $B \to \infty$, conditional on $\mathcal{X}$. The approximant $\hat{x}_B^*$ is not unbiased for $\hat{u}$, although the bias is generally very low, of smaller order than the error about the mean. Indeed, it may be proved that

$$E(\hat{x}_B^* \mid \mathcal{X}) - \hat{u} \;=\; -(Bn)^{-1}\,\hat{\alpha} + O\{(Bn)^{-2}\}\,, \qquad (A.9)$$

$$\mathrm{var}(\hat{x}_B^* \mid \mathcal{X}) \;=\; (Bn^2)^{-1}\,\hat{\beta} + O\{(Bn)^{-2} + (Bn^3)^{-1}\}\,, \qquad (A.10)$$

with probability one as $n \to \infty$, where

$$\hat{\alpha} \;=\; \tfrac{1}{2} \sum_{j=1}^{d} \sum_{k=1}^{d} g_{jk}(\overline{\mathbf{X}})\,\hat{\sigma}^{jk}\,, \qquad (A.11)$$

and $\hat{\beta}$, $\hat{\sigma}^{jk}$ are given by (A.7) and (A.8), respectively. See Hall (1989a). Note particularly that by A.6) and (A.10), the conditional asymptotic variances of $\hat{v}_B^*$ and $\hat{x}_B^*$ are identical. Since $\hat{v}_B^*$ is an unbiased approximant and the bias of $\hat{x}_B^*$ is negligible relative to the error about the mean (order $B^{-1}\,n^{-1}$ relative to $B^{-1/2}\,n^{-1}$), then the approximations $\hat{v}_B^*$ and $\hat{x}_B^*$ have asymptotically equivalent mean squared error.

II.5 Balanced Resampling

If we could ensure that the grand mean of the bootstrap resamples was identical to the sample mean, i.e.,

$$\overline{\mathbf{X}}_{\cdot}^* \;=\; \overline{\mathbf{X}}\,, \qquad (A.12)$$

then the uniform approximation $\hat{u}_B^*$, the linear approximation $\hat{v}_B^*$ and the centring approximation $\hat{x}_B^*$ would all be identical. The only practical way of guaranteeing (A.12) is to resample in such a way that each data point $\mathbf{X}_i$ occurs the same number of times in the union of the resamples $\mathcal{X}_b^*$.

To achieve this end, write down each of the sample values $\mathbf{X}_1, \ldots, \mathbf{X}_n$ B times, in a string of length Bn; then randomly permute the elements of this string; and finally, divide the permuted string into B chunks of length n, putting all the sample values lying between positions $(b-1)n + 1$ and bn of the permuted string into the bth resample $\mathcal{X}_b^\dagger$, for $1 \le b \le B$. This is balanced resampling, and amounts to random resampling subject to the constraint that $\mathbf{X}_i$ appears just B times in $\cup_b \mathcal{X}_b^\dagger$. Balanced resampling was introduced by Davison, Hinkley, and Schechtman (1986), and high-order balance has been discussed by Graham, Hinkley, John, and Shi (1990). See also Ogbonmwan and Wynn (1986, 1988). An algorithm for performing balanced resampling has been described by Gleason (1988). The method of Latin hypercube sampling (McKay, Beckman, and Conover 1979; Stein 1987), used for Monte Carlo simulation in a nonbootstrap setting, is closely related to balanced resampling.

The balanced resampling approximation of $\hat{u} = E\{g(\overline{\mathbf{X}}^*) \mid \mathcal{X}\}$ is

$$\hat{u}_B^\dagger = B^{-1} \sum_{b=1}^{B} g(\overline{\mathbf{X}}_b^\dagger),$$

where $\overline{\mathbf{X}}_b^\dagger$ denotes the mean of $\mathcal{X}_b^\dagger$. Once again, $\hat{u}_B^\dagger \to \hat{u}$ as $B \to \infty$, with probability one conditional on $\mathcal{X}$. The balanced resampling approximation shares the asymptotic bias and variance formulae of the centring approximation $\hat{x}_B^*$, introduced in Section II.4,

$$E(\hat{u}_B^\dagger \mid \mathcal{X}) - \hat{u} = -(Bn)^{-1}\hat{\alpha} + O(B^{-1}n^{-2}), \qquad (A.13)$$

$$\mathrm{var}(\hat{u}_B^\dagger \mid \mathcal{X}) = (Bn^2)^{-1}\hat{\beta} + O\{(Bn)^{-2} + (Bn^3)^{-1}\}, \qquad (A.14)$$

with probability one as $n \to \infty$, where $\hat{\alpha}$, $\hat{\beta}$ are given by (A.11) and (A.7), respectively; compare (A.9) and (A.10). In particular, in the context of bias estimation for a smooth function of a mean, balanced resampling reduces the orders of magnitude of variance and mean squared error by a factor of n^{-1}. Formulae (A.13) and (A.14) are not entirely trivial to prove, and the asymptotic equivalence of bias and variance formulae for the centring method and balanced resampling is not quite obvious; see Hall (1989a).

In Sections II.3 and II.4, and so far in the present section, we have treated only the case of approximating a smooth function of a mean. The methods of linear approximation and centring do not admit a wide range of other applications. For example, linear approximation relies on Taylor expansion, and that demands a certain level of smoothness of the statistic U. However, balanced resampling is not constrained in this way, and in principle applies to a much wider range of problems, including distribution function and quantile estimation. In those cases the extent of improvement of variance and mean squared error is generally by a constant factor, not by a factor n^{-1}.

Suppose that U is an indicator function of the form $U = I(S \leq x)$ or $U = I(T \leq x)$, where $S = n^{1/2}(\hat{\theta} - \theta)/\sigma$ and $T = n^{1/2}(\hat{\theta} - \theta)/\hat{\sigma}$ are statistics that are asymptotically Normal $N(0, 1)$. The bootstrap versions are

$$U^* = I(S^* \leq x) \qquad \text{and} \qquad U^* = I(T^* \leq x),$$

respectively, where

$$S^* = n^{1/2}(\hat{\theta}^* - \hat{\theta})/\hat{\sigma} \qquad \text{and} \qquad T^* = n^{1/2}(\hat{\theta}^* - \hat{\theta})/\hat{\sigma}^*.$$

(The pros and cons of pivoting are not relevant to the present discussion.) To construct a balanced resampling approximation to $\hat{u} = E(U^* \mid \mathcal{X})$, first draw B balanced resamples $\mathcal{X}_b^\dagger$, $1 \leq b \leq B$, as described two paragraphs earlier. Let $\hat{\theta}_b^\dagger$, $\hat{\sigma}_b^\dagger$ denote the versions of $\hat{\theta}$, $\hat{\sigma}$ respectively computed from $\mathcal{X}_b^\dagger$ instead of $\mathcal{X}$, and put

$$S_b^\dagger = n^{1/2}(\hat{\theta}_b^\dagger - \hat{\theta})/\hat{\sigma}, \qquad T_b^\dagger = n^{1/2}(\hat{\theta}_b^\dagger - \hat{\theta})/\hat{\sigma}_b^\dagger,$$

and

$$U_b^\dagger = I(S_b^\dagger \leq x)$$

or $I(T_b^\dagger \leq x)$, depending on whether $U = I(S \leq x)$ or $I(T \leq x)$. (We define $T_b^\dagger = c$, for an arbitrary but fixed constant c, if $\sigma_b^\dagger = 0$.) The balanced resampling approximation to $\hat{u}$ is

$$\hat{u}_B^\dagger = B^{-1} \sum_{b=1}^{B} U_b^\dagger.$$

Recall from Section II.2 that the uniform resampling approximation $\hat{u}_B^*$ is unbiased for $\hat{u}$, in the sense that $E(\hat{u}_B^* \mid \mathcal{X}) = \hat{u}$, and has variance

$$\text{var}(\hat{u}_B^* \mid \mathcal{X}) = B^{-1} \hat{u}(1 - \hat{u}) \sim B^{-1} \Phi(x)\{1 - \Phi(x)\}$$

as $n \to \infty$. The balanced resampling approximant is slightly biased, although the bias is low relative to the error about the mean,

$$E(\hat{u}_B^\dagger \mid \mathcal{X}) - \hat{u} = O(B^{-1})$$

with probability one. The asymptotic variance of $\hat{u}_B^\dagger$ is less than that of $\hat{u}_B^*$ by a constant factor $\rho(x)^{-1} < 1$, since

$$\text{var}(\hat{u}_B^\dagger \mid \mathcal{X}) \sim B^{-1} \left[\Phi(x)\{1 - \Phi(x)\} - \phi(x)^2 \right];$$

see Hall (1990a). Thus, the asymptotic efficiency of balanced resampling relative to uniform resampling, in this context, is

$$\rho(x) \;=\; \Phi(x)\{1 - \Phi(x)\}\big[\,\Phi(x)\{1 - \Phi(x)\} - \phi(x)^2\,\big]^{-1},$$

which reaches a maximum at $x = 0$ and decreases to 1 as $|x| \uparrow \infty$. See Hall (1990a).

The same asymptotic efficiencies apply to the case of quantile estimation, which we discuss in more detail in Section II.8. Figure A.1 graphs the efficiencies.

II.6 Antithetic Resampling

The method of antithetic resampling dates back at least to Hammersley and Morton (1956) and Hammersley and Mauldon (1956). See Snijders (1984) for a recent account in connection with Monte Carlo estimation of probabilities. Antithetic resampling may be described as follows. Suppose we have two estimates $\hat\theta_1$ and $\hat\theta_2$ of the same parameter θ, with identical means and variances but negative covariance. Assume that the costs of computing $\hat\theta_1$ and $\hat\theta_2$ are identical. Define $\hat\theta_3 = \frac{1}{2}(\hat\theta_1 + \hat\theta_2)$. Then $\hat\theta_3$ has the same mean as either $\hat\theta_1$ or $\hat\theta_2$, but less than half the variance, since

$$\mathrm{var}(\hat\theta_3) \;=\; \tfrac{1}{4}\{\mathrm{var}\,\hat\theta_1 + \mathrm{var}\,\hat\theta_2 + 2\mathrm{cov}(\hat\theta_1,\hat\theta_2)\}$$

$$\;=\; \tfrac{1}{2}\{\mathrm{var}(\hat\theta_1) + \mathrm{cov}(\hat\theta_1,\,\hat\theta_2)\} \;<\; \tfrac{1}{2}\mathrm{var}(\hat\theta_1).$$

Since the cost of computing $\hat\theta_3$ is scarcely more than twice the cost of computing either $\hat\theta_1$ or $\hat\theta_2$, but the variance is more than halved, there is an advantage from the viewpoint of cost-effectiveness in using $\hat\theta_3$, rather than either $\hat\theta_1$ or $\hat\theta_2$, to estimate θ. Obviously, the advantage increases with increasing negativity of the covariance, all other things being equal.

To appreciate how this idea may be applied to the case of resampling, let U^* denote the version of a statistic U computed from a (uniform) resample $\mathcal{X}^* = \{X_1^*,\ldots,X_n^*\}$ rather than the original sample $\mathcal{X} = \{X_1,\ldots,X_n\}$. Let π be an arbitrary but fixed permutation of the integers $1,\ldots,n$, and let $j_1,\ldots,j_n$ be the random integers such that $X_i^* = X_{j_i}$ for $1 \le i \le n$. Define $X_i^{**} = X_{\pi(j_i)}$, $1 \le i \le n$, and put $\mathcal{X}^{**} = \{X_1^{**},\ldots,X_n^{**}\}$. That is, $\mathcal{X}^{**}$ is the (uniform) resample obtained by replacing each appearance of X_k in $\mathcal{X}^*$ by $X_{\pi(k)}$. If U^{**} denotes the version of U computed from $\mathcal{X}^{**}$ instead of $\mathcal{X}$, then U^* and U^{**} have the same distributions, conditional on $\mathcal{X}$. In particular, they have the same conditional mean and variance. If we choose the permutation π in such a way that the conditional covariance of U^* and U^{**} is negative, we may apply the antithetic argument to the pair (U^*, U^{**}). That is, the approximant

$$U^@ = \tfrac{1}{2}(U^* + U^{**})$$

will have the same conditional mean as U^* but less than half the conditional variance of U^*.

If the X_i's are scalars then in many cases of practical interest, the "asymptotically optimal" permutation π (which asymptotically minimizes the covariance of U^* and U^{**}), is that which takes the largest X_i into the smallest X_i, the second largest X_i into the second smallest X_i, and so on. That is, if we index the X_i's such that $X_1 \leq \ldots \leq X_n$, then $\pi(i) = n-i+1$ for $1 \leq i \leq n$. For example, this is true when $U = g(\bar{X}) - g(\mu)$, where g is a smooth function, and also when

$$U = I\left[n^{1/2}\{g(\bar{X}) - g(\mu)\} \leq x \right]. \tag{A.15}$$

The asymptotically optimal permutation π in the case of a d-dimensional sample, where $U = g(\bar{\mathbf{X}}) - g(\boldsymbol{\mu})$ or

$$U = I\left[n^{1/2}\{g(\bar{\mathbf{X}}) - g(\boldsymbol{\mu})\} \leq x \right], \tag{A.16}$$

is that one that reverses the order of the quantities

$$Y_i = \sum_{j=1}^{d} (\mathbf{X}_i - \bar{\mathbf{X}})^{(j)}\, g^{(j)}(\bar{\mathbf{X}}), \qquad 1 \leq i \leq n.$$

That is, if we index the $\mathbf{X}_i$'s such that $Y_1 \leq \ldots \leq Y_n$, then $\pi(i) = n-i+1$. We shall call π the antithetic permutation. These results remain true if we Studentize the arguments of the indicator functions at (A.15) and (A.16); the pros and cons of pivoting do not have a role to play here. The reader is referred to Hall (1989b) for details.

The method of antithetic resampling may be used in the following way to approximate the bootstrap estimate $\hat{u} = E(U^* \mid \mathcal{X})$. Draw B independent, uniform resamples $\mathcal{X}_b^*$, $1 \leq b \leq B$; by applying the antithetic permutation, convert $\mathcal{X}_b^*$ into the corresponding antithetic resample $\mathcal{X}_b^{**}$; compute the versions of U_b^* and U_b^{**} for $\mathcal{X}_b^*$ and $\mathcal{X}_b^{**}$, respectively; and define

$$\hat{u}_B^@ = \tfrac{1}{2} B^{-1} \sum_{b=1}^{B} (U_b^* + U_b^{**}).$$

Note that $\hat{u}_B^@$ is an unbiased approximant of $\hat{u}$, in the sense that

$$E\left(\hat{u}_B^@ \mid \mathcal{X} \right) = \hat{u}.$$

The conditional variance of $\hat{u}_B^@$ is given by

$$\text{var}\left(\hat{u}_B^@ \mid \mathcal{X} \right) = (2B)^{-1}\left\{ \text{var}(U^* \mid \mathcal{X}) + \text{cov}(U^*, U^{**} \mid \mathcal{X}) \right\},$$

where (U^*, U^{**}) denotes a generic pair (U_b^*, U_b^{**}). In general, as $n \to \infty$,

$$\text{var}\big(\hat{u}_B^{@} \mid \mathcal{X}\big) = q\,\text{var}(\hat{u}_B^* \mid \mathcal{X}) + o\{\text{var}(\hat{u}_B^* \mid \mathcal{X})\},$$

where $0 \leq q < 1$.

The exact value of q depends on the situation, it being greater for less symmetric distributions. For example, in the case of distribution function estimation where $\hat{u} = P(S^* \leq x \mid \mathcal{X})$ or $P(T^* \leq x \mid \mathcal{X})$, the value of

$$q = q(x) = \lim_{n \to \infty} \lim_{B \to \infty} \big\{\text{var}(\hat{u}_B^{@} \mid \mathcal{X})/\text{var}(\hat{u}_B^* \mid \mathcal{X})\big\}$$

is an increasing function of both $|x|$ and the skewness of the sampling distribution. The minimum value $q = 0$ can only arise in this context when both $x = 0$ and the sampling distribution has zero skewness.

Thus, generally speaking, antithetic resampling reduces variance by a constant factor. In this respect, balanced resampling is superior in the case of approximating the conditional mean of a smooth function, since it reduces variance by a factor that converges to zero as sample size increases. Again, the reader is referred to Hall (1989b) for details.

II.7 Importance Resampling

II.7.1 Introduction

The method of importance resampling is a standard technique for improving the efficiency of Monte Carlo approximations. See Hammersley and Handscomb (1964, p. 60ff). It was first suggested in the context of bootstrap resampling by Johns (1988) and Davison (1988). The account in this section differs from Johns' and Davison's.

We show that the improvements offered by importance resampling are asymptotically negligible in the contexts of bootstrap bias estimation and variance estimation, but that they can be substantial when distribution functions or quantiles are the subject of interest. Hinkley and Shi (1989), Hall (1991a), and Do and Hall (1991) have also discussed importance resampling.

Our aim in this section is to give a general account of importance resampling, so that its application to a wide variety of problems such as bias estimation, variance estimation, and quantile estimation may be considered. On grounds of efficiency relative to uniform resampling, there is little reason to consider importance resampling in the cases of bias and variance estimation, but that negative result is not clear without a reasonably general analysis of the method. Subsection II.7.2 will introduce and develop the principle of importance resampling, and subsections II.7.3 and II.7.4 will apply the principle to bias estimation, variance estimation, and distribution function estimation. The empirical approach to importance resampling described in subsection II.7.4 below is drawn from Do and Hall (1991).

II.7.2 Concept of Importance Resampling

Let $\mathcal{X} = \{X_1, \ldots, X_n\}$ denote the sample from which a resample will be drawn. (This notation is only for the sake of convenience, and in no way precludes a multivariate sample.) Under importance resampling, each X_i is assigned a probability p_i of being selected on any given draw, where $\sum p_i = 1$. Sampling is conducted with replacement, so that the chance of drawing a resample of size n in which X_i appears just m_i times $(1 \leq i \leq n)$ is given by a multinomial formula,

$$\frac{n!}{m_1! \ldots m_n!} \prod_{i=1}^{n} p_i^{m_i}.$$

Of course, $\sum m_i = n$. Taking $p_i = n^{-1}$ for each i, we obtain the uniform resampling method of Section II.2.

The name "importance" derives from the fact that resampling is designed to take place in a manner that ascribes more importance to some sample values than others. The aim is to select the p_i's so that the value assumed by a bootstrap statistic is relatively likely to be close to the quantity whose value we wish to approximate.

There are two parts to the method of importance resampling: first, a technique for passing from a sequence of importance resamples to an approximation of a quantity that would normally be defined in terms of a uniform resample; and second, a method for computing the appropriate values of p_i for the importance resampling algorithm. One would usually endeavour to choose the p_i's so as to minimize the error, or variability, of the approximation.

We know from Appendix I that there are $N = \binom{2n-1}{n}$ different possible resamples. Let these be $\mathcal{X}_1, \ldots, \mathcal{X}_N$, indexed in any order, and let m_{ji} denote the number of times X_i appears in $\mathcal{X}_j$. The probability of obtaining $\mathcal{X}_j$ after n resampling operations, under uniform resampling or importance resampling, equals

$$\pi_j = \frac{n!}{m_{j1}! \ldots m_{jn}!} n^{-n}.$$

or

$$\pi'_j = \frac{n!}{m_{j1}! \ldots m_{jn}!} \prod_{i=1}^{n} p_i^{m_{ji}} = \pi_j \prod_{i=1}^{n} (np_i)^{m_{ji}}, \qquad (A.17)$$

respectively. Let U be the statistic of interest, a function of the original sample. We wish to construct a Monte Carlo approximation to the bootstrap estimate $\hat{u}$ of the mean of U, $u = E(U)$.

Let $\mathcal{X}^*$ denote a resample drawn by uniform resampling, and write U^* for the value of U computed for $\mathcal{X}^*$. Of course, $\mathcal{X}^*$ will be one of the $\mathcal{X}_j$'s. Write u_j for the value of U^* when $\mathcal{X}^* = \mathcal{X}_j$. In this notation,

$$\hat{u} = E(U^* \mid \mathcal{X}) = \sum_{j=1}^{N} u_j \pi_j = \sum_{j=1}^{N} u_j \pi'_j \prod_{i=1}^{n} (np_i)^{-m_{ji}}, \qquad (A.18)$$

the last identity following from (A.17).

Let $\mathcal{X}^\dagger$ denote a resample drawn by importance resampling, write $U^\dagger$ for the value of U computed from $\mathcal{X}^\dagger$, and let $M_i^\dagger$ be the number of times X_i appears in $\mathcal{X}^\dagger$. Then by (A.18),

$$\hat{u} = E\left\{ U^\dagger \prod_{i=1}^{n} (np_i)^{-M_i^\dagger} \mid \mathcal{X} \right\}.$$

Therefore, it is possible to approximate $\hat{u}$ by importance resampling. In particular, if $\mathcal{X}_b^\dagger$, $1 \leq b \leq B$, denote independent resamples drawn by importance resampling, if $U_b^\dagger$ equals the value of U computed for $\mathcal{X}_b^\dagger$, and if $M_{bi}^\dagger$ denotes the number of times X_i appears in $\mathcal{X}_b^\dagger$, then the importance resampling approximant of $\hat{u}$ is given by

$$\hat{u}_B^\dagger = B^{-1} \sum_{b=1}^{B} U_b^\dagger \prod_{i=1}^{n} (np_i)^{-M_{bi}^\dagger}.$$

This approximation is unbiased, in the sense that $E\left(\hat{u}_B^\dagger \mid \mathcal{X}\right) = \hat{u}$. Note too that conditional on $\mathcal{X}$, $\hat{u}_B^\dagger \to \hat{u}$ with probability one as $B \to \infty$.

If we take each $p_i = n^{-1}$ then $\hat{u}_B^\dagger$ is just the usual uniform resampling approximant $\hat{u}_B^*$. We wish to choose $p_1, \ldots, p_n$ to optimize the performance of $\hat{u}_B^\dagger$. Since $\hat{u}_B^\dagger$ is unbiased, the performance of $\hat{u}_B^\dagger$ may be described in terms of variance,

$$\mathrm{var}\left(\hat{u}_B^\dagger \mid \mathcal{X}\right) = B^{-1} \mathrm{var}\left\{ U_b^\dagger \prod_{i=1}^{n} (np_i)^{-M_{bi}^\dagger} \mid \mathcal{X} \right\}$$

$$= B^{-1}(\hat{v} - \hat{u}^2), \qquad (A.19)$$

where

$$\hat{v} \;=\; \hat{v}(p_1,\dots,p_n) \;=\; E\!\left[\left\{U_B^\dagger \prod_{i=1}^{n}(np_i)^{-M_{bi}^\dagger}\right\}^2 \;\middle|\; \mathcal{X}\right]$$

$$= \sum_{j=1}^{N} \pi_j' \, u_j^2 \prod_{i=1}^{n}(np_i)^{-2m_{ji}}$$

$$= \sum_{j=1}^{N} \pi_j \, u_j^2 \prod_{i=1}^{n}(np_i)^{-m_{ji}}$$

$$= E\!\left\{U^{*2} \prod_{i=1}^{n}(np_i)^{-M_i^*} \;\middle|\; \mathcal{X}\right\}. \tag{A.20}$$

On the last line, M_i^* denotes the number of times X_i appears in the uniform resample $\mathcal{X}^*$. Ideally we would like to choose $p_1,\dots,p_n$ so as to minimize $\hat{v}(p_1,\dots,p_n)$, subject to $\sum p_i = 1$. Subsections II.7.3 and II.7.4 below treat special cases of this problem.

II.7.3 Importance Resampling for Approximating Bias, Variance, Skewness, etc.

Let $\hat{\theta}$ denote an estimate of an unknown parameter θ_0, calculated from the sample $\mathcal{X}$, and write $\hat{\theta}^*$ for the version of $\hat{\theta}$ computed from a uniform resample $\mathcal{X}^*$. Take $U = (\hat{\theta} - \theta_0)^k$ and $U^* = (\hat{\theta}^* - \hat{\theta})^k$, where k is a positive integer. In the cases $k = 1, 2,$ and 3, estimation of $u = u_k = E(U)$ is essentially equivalent to estimation of bias, variance, and skewness, respectively, of which the bootstrap estimate is $\hat{u} = \hat{u}_k = E(U^* \mid \mathcal{X})$. There are some minor differences of detail; for example, u_2 strictly equals mean squared error rather than variance, the latter being $u_2 - u_1^2$. However, the importance resampling approximation to $\hat{u}_2 - \hat{u}_1^2$ has the same asymptotic properties as the approximation to $\hat{u}_2$, and so it is appropriate to focus attention on approximating $\hat{u}_2$. We shall prove that there are no asymptotic advantages in using importance resampling rather than uniform resampling to approximate $\hat{u}_k$.

Again we assume the "smooth function model" introduced in Section 2.4. Thus, the sample values $\mathbf{X}_i$ are d-variate with population mean $\boldsymbol{\mu} = E(\mathbf{X})$, and θ_0, $\hat{\theta}$ may be represented by $\theta_0 = g(\boldsymbol{\mu})$, $\hat{\theta} = g(\overline{\mathbf{X}})$, where g is a smooth real-valued function of d variables. Define $\delta_i = -\log(np_i)$,

$$\hat{\sigma}^2 \;=\; n^{-1} \sum_{i=1}^{n} \left\{ \sum_{j=1}^{d} (\mathbf{X}_i - \overline{\mathbf{X}})^{(j)} \, g_j(\overline{\mathbf{X}}) \right\}^2 ,$$

$$\epsilon_i \;=\; n^{-1/2} \, \hat{\sigma}^{-1} \sum_{j=1}^{d} (\mathbf{X}_i - \overline{\mathbf{X}})^{(j)} \, g_j(\overline{\mathbf{X}}) . \tag{A.21}$$

Write M_i^* for the number of times $\mathbf{X}_i$ appears in $\mathcal{X}^*$. Then

$$U^* = \left\{g(\overline{\mathbf{X}}^*) - g(\overline{\mathbf{X}})\right\}^k \simeq \left\{\sum_{j=1}^{d} (\overline{\mathbf{X}}^* - \overline{\mathbf{X}})^{(j)} g_j(\overline{\mathbf{X}})\right\}^k$$

$$= \left(n^{-1/2}\, \hat{\sigma} \sum_{i=1}^{n} M_i^*\, \epsilon_i\right)^k,$$

whence by (A.20),

$$\hat{v} = E\left\{U^{*2} \prod_{i=1}^{n} (np_i)^{-M_i^*} \mid \mathcal{X}\right\}$$

$$\sim (n^{-1/2}\, \hat{\sigma})^{2k}\, E\left\{\left(\sum_{i=1}^{n} M_i^*\, \epsilon_i\right)^{2k} \prod_{i=1}^{n} (np_i)^{-M_i^*} \mid \mathcal{X}\right\}$$

$$= (n^{-1/2}\, \hat{\sigma})^{2k}\, E\left(V_1^{2k}\, e^{V_2} \mid \mathcal{X}\right), \tag{A.22}$$

where $V_1 = \sum M_i^*\, \epsilon_i$, $V_2 = \sum M_i^*\, \delta_i$.

Conditional on $\mathcal{X}$, the vector $(M_1^*, \dots, M_n^*)$ has a multinomial distribution with all probabilities equal to n^{-1}, and so

$$E(M_1^* \mid \mathcal{X}) = 1, \qquad \mathrm{var}(M_1^* \mid \mathcal{X}) = E(M_1^*\, M_2^* \mid \mathcal{X}) = 1 - n^{-1}.$$

It is now readily proved by a central limit theorem for weighted multinomials (for example, in Holst 1972) that conditional on $\mathcal{X}$, (V_1, V_2) is asymptotically Normally distributed with mean $(0, \sum \delta_i)$, variances $(1, \sum \delta_i^2 - n^{-1}(\sum \delta_i)^2)$, and covariance $\sum \epsilon_i \delta_i$. Here we have used the fact that $\sum \epsilon_i = 0$ and $\sum \epsilon_i^2 = 1$. The constraint $\sum p_i = 1$ implies that $\sum \delta_i = \frac{1}{2} \sum \delta_i^2 + o(1)$ and $n^{-1}(\sum \delta_i)^2 \to 0$. Therefore, defining $s^2 = \sum \delta_i^2$ and $\rho = s^{-1} \sum \epsilon_i \delta_i$, we have

$$\hat{v} \sim (n^{-1/2}\, \hat{\sigma})^{2k}\, E\left(N_1^{2k}\, e^{N_2} \mid \mathcal{X}\right)$$

$$= (n^{-1/2}\, \hat{\sigma})^{2k}\, E\{(N_1 + s\rho^2)^{2k} \mid \mathcal{X}\}\, e^{s^2}, \tag{A.23}$$

where conditional on $\mathcal{X}$, (N_1, N_2) is bivariate Normal with means $(0, \frac{1}{2} s^2)$, variances $(1, s^2)$, and correlation ρ. The right-hand side of (A.23) is minimized by taking $s = 0$, i.e., $\delta_i = 0$ for each i, which is equivalent to demanding that each $p_i = n^{-1}$.

Therefore, the minimum asymptotic variance of the importance resampling approximant occurs when each $p_i = n^{-1}$, and in view of (A.19) and (A.23) is given by

$$B^{-1}(\hat{v} - \hat{u}^2) \sim B^{-1}(n^{-1/2}\, \hat{\sigma})^{2k} \left\{E(N^{2k}) - (EN^k)^2\right\},$$

where N has the Standard Normal distribution.

II.7.4 Importance Resampling for a Distribution Function

In the case of estimating a distribution function there can be a significant advantage in choosing nonidentical p_i's, the amount of improvement depending on the argument of the distribution function. To appreciate the extent of improvement we shall consider the case of estimating the distribution function of a Studentized statistic,

$$T = n^{1/2}(\hat{\theta} - \theta_0)/\hat{\sigma},$$

assuming the smooth function model. Other cases, such as that where $S = n^{1/2}(\hat{\theta} - \theta_0)/\sigma$ is the subject of interest, are similar; the issue of Studentizing does not play a role here.

Use notation introduced in subsection II.7.3 above, and take

$$U^* = I(T^* \le x)$$

where

$$T^* = n^{1/2}(\hat{\theta}^* - \hat{\theta})/\hat{\sigma}^*,$$

$$\hat{\sigma}^{*2} = n^{-1}\sum_{i=1}^{n}\left\{\sum_{j=1}^{d}(\mathbf{X}_i^* - \overline{\mathbf{X}}^*)^{(j)}\, g_j(\overline{\mathbf{X}}^*)\right\}^2.$$

(We agree to define $T^* = c$, for an arbitrary but fixed constant c, in the event that $\hat{\sigma}^* = 0$.) Now, $\hat{u} = P(T^* \le x \mid \mathcal{X})$, and for this particular x we wish to choose $p_1, \ldots, p_n$ to minimize

$$\hat{v} = \hat{v}(p_1, \ldots, p_n) = E\left\{I(T^* \le x)\prod_{i=1}^{n}(np_i)^{-M_i^*}\mid \mathcal{X}\right\}.$$

There are two practical solutions to this problem, one of them asymptotic and the other empirical. We first treat the asymptotic solution.

Define ϵ_i, δ_i, M_i^*, V_1, V_2, N_1, N_2, s, and ρ as in the previous subsection, and note that $T^* \simeq \sum M_i^* \epsilon_i$. Then, by arguments similar to those in (A.22) and (A.23),

$$\hat{v} \sim E\left\{I\left(\sum_{i=1}^{n} M_i^* \epsilon_i \le x\right)\prod_{i=1}^{n}(np_i)^{-M_i^*}\mid \mathcal{X}\right\}$$

$$= E\left\{I(V_1 \le x)\,e^{V_2}\mid \mathcal{X}\right\}$$

$$\sim E\left\{I(N_1 \le x)\,e^{N_2}\mid \mathcal{X}\right\}$$

$$= \Phi(x - s\rho)\,e^{s^2}.$$

The values of s and ρ that minimize $\Phi(x - s\rho)\, e^{s^2}$ are $(s, \rho) = \pm(A, 1)$, where $A = A(x) > 0$ is chosen to minimize $\Phi(x - A)\, e^{A^2}$. Taking $\delta_i = A\epsilon_i + C$, where C is chosen to ensure that $\Sigma\, p_i = n^{-1}\Sigma\, e^{-\delta_i} = 1$, we see that $s \to A$ and $\rho \to 1$.

Therefore, the minimum asymptotic variance of the importance resampling approximant occurs when

$$p_i = \frac{e^{-A\epsilon_i}}{\sum_{j=1}^{n} e^{-A\epsilon_j}}, \qquad 1 \le i \le n, \tag{A.24}$$

where ϵ_i is given by (A.21). This minimum variance is

$$B^{-1}(\hat{v} - \hat{u}^2) \sim B^{-1}\big\{ \Phi(x - A)\, e^{A^2} - \Phi(x)^2 \big\} \tag{A.25}$$

since $\hat{u} \to \Phi(x)$ as $n \to \infty$.

Writing $\hat{v}$ for $\hat{v}(p_1, \dots, p_n)$, we see that the extent of improvement offered by this approach is given by

$$r(x) = \frac{\text{variance of } \hat{u}_B^* \;\; (\text{under uniform resampling})}{\text{variance of } \hat{u}_B^\dagger \;\; (\text{under importance resampling})}$$

$$= \frac{B^{-1}(\hat{u} - \hat{u}^2)}{B^{-1}(\hat{v} - \hat{u}^2)} = \frac{\hat{u} - \hat{u}^2}{\hat{v} - \hat{u}^2}$$

$$\sim \frac{\Phi(x)\{1 - \Phi(x)\}}{\Phi(x - A)\, e^{A^2} - \Phi(x)^2}\,.$$

(To obtain the second identity, note that the conditional variance of $\hat{u}_B^*$ equals $B^{-1}(\hat{u} - \hat{u}^2)$ since $\hat{v} = \hat{u}$ when each $p_i = n^{-1}$. To obtain the last asymptotic relation, use (A.25) and observe that $\hat{u} - \hat{u}^2 \to \Phi(x)\{1 - \Phi(x)\}$.) Values of $A = A(x)$ that maximize $r(x)$, and of $r(x)$ for this choice of A, are given in Table A.2.

Note that r is a strictly decreasing function, with $r(-\infty) = \infty$, $r(0) = 1.7$, and $r(+\infty) = 1$. Therefore, importance resampling can be considerably more efficacious for negative x than positive x. If we wish to approximate $\widehat{G}(x) = P(T^* \le x \mid \mathcal{X})$ for a value of $x > 0$, it is advisable to work throughout with $-T^*$ rather than T^* and use importance resampling to calculate

$$P(-T^* \le -x \mid \mathcal{X}) = 1 - \widehat{G}(x)\,.$$

Next we describe an empirical approach to selecting the optimal p_i's for importance resampling. Divide the program of B simulations into two parts, involving B_1 and B_2 simulations respectively, where $B_1 + B_2 = B$.

TABLE A.2. Optimal values of $A = A(x)$ and values of asymptotic
efficiency $r(x)$ for selected x (taken to be quantiles of the Standard
Normal distribution).

$\Phi(x)$	$A(x)$	$r(x)$
0.005	2.6561	69.0
0.01	2.5704	38.0
0.025	2.1787	17.6
0.05	1.8940	10.0
0.1	1.5751	5.8
0.5	0.6120	1.7
0.9	0.1150	1.1
0.95	0.0602	1.1
0.975	0.0320	1.0
0.99	0.0139	1.0
0.995	0.0074	1.0

Conduct the first B_1 simulations by uniform resampling, and thereby approximate $\hat{v}$ by

$$\hat{v}_{B_1} = \hat{v}_{B_1}(p_1, \ldots, p_n) = B_1^{-1} \sum_{b=1}^{B_1} I(T_b^* \leq x) \prod_{i=1}^{n} (np_i)^{-M_{bi}^*},$$

where T_b^* denotes the value of T^* computed for the uniform resample $\mathcal{X}_b^*$, $1 \leq b \leq B_1$. Choose $p_1, \ldots, p_n$ to minimize $\hat{v}_{B_1}$, subject to $\sum p_i = 1$. Conduct the remaining B_2 simulations by importance resampling, using the p_i's that minimize $\hat{v}_{B_1}$. Approximate $\hat{u} = \widehat{G}(x) = P(T^* \leq x \mid \mathcal{X})$ from the B_1 uniform resamples and the B_2 importance resamples, obtaining $\tilde{G}_1(x)$ and $\tilde{G}_2(x)$ where

$$\tilde{G}_1(x) = \hat{u}_{B_1} = B_1^{-1} \sum_{b=1}^{B_1} I(T_b^* \leq x),$$

$$\tilde{G}_2(x) = B_2^{-1} \sum_{b=1}^{B_2} I(T_b^{**} \leq x) \prod_{i=1}^{n} (np_i)^{-M_{bi}^{**}},$$

and T_b^{**} is the value of T computed from the bth importance resample $\mathcal{X}_b^{**}$. Finally, combine these approximations to obtain

$$\tilde{G}(x) = \beta \tilde{G}_1(x) + (1 - \beta) \tilde{G}_2(x),$$

where

$$\beta = \max \left[\left\{ \min_p \hat{v}_{B_1}(p) - \hat{u}_{B_1}^2 \right\} \left\{ \hat{u}_{B_1} + \min_p \hat{v}_{B_1}(p) - 2\hat{u}_{B_1}^2 \right\}^{-1}, \, 0 \right].$$

(The latter is an estimate of $\mathrm{var}(\tilde{G}_2 \mid \mathcal{X}) / \{\mathrm{var}(\tilde{G}_2 \mid \mathcal{X}) + \mathrm{var}(\tilde{G}_1 \mid \mathcal{X})\}$.)

II.8 Quantile Estimation

Here we consider the problem of estimating the αth quantile, ξ_α, of the distribution of a random variable R such as $S = n^{1/2}(\hat{\theta} - \theta_0)/\sigma$ or $T = n^{1/2}(\hat{\theta} - \theta_0)/\hat{\sigma}$. We define ξ_α to be the solution of the equation

$$P(R \le \xi_\alpha) = \alpha.$$

Now, the bootstrap estimate of ξ_α is the solution $\hat{\xi}_\alpha$ of

$$P(R^* \le \hat{\xi}_\alpha \mid \mathcal{X}) = \alpha, \tag{A.26}$$

where

$$R^* = S^* = n^{1/2}(\hat{\theta}^* - \hat{\theta})/\hat{\sigma} \qquad \text{(if } R = S)$$

or

$$R^* = T^* = n^{1/2}(\hat{\theta}^* - \hat{\theta})/\hat{\sigma}^* \qquad \text{(if } R = T).$$

The equation (A.26) usually cannot be solved exactly, although, as pointed out in Appendix I, the error will be an exponentially small function of n. For the sake of definiteness we shall define

$$\hat{\xi}_\alpha = \hat{H}^{-1}(\alpha) = \inf\{x : \hat{H}(x) \ge \alpha\}, \tag{A.27}$$

where $\hat{H}(x) = P(R^* \le x \mid \mathcal{X})$.

Approximation of the function $\hat{H}(x)$ falls neatly into the format described in earlier sections, where we discussed Monte Carlo methods for approximating a conditional expectation of the form $\hat{u} = E(U^* \mid \mathcal{X})$. Take $U^* = I(R^* \le x)$ and apply any one of the methods of uniform resampling, balanced resampling, antithetic resampling, or importance resampling to obtain an approximation $\hat{H}_B(x)$ to $\hat{H}(x)$. Then, following the prescription suggested by (A.27), define

$$\hat{\xi}_{\alpha,B} = \hat{H}_B^{-1}(\alpha) = \inf\{x : \hat{H}_b(x) \ge \alpha\}. \tag{A.28}$$

For each of these methods, the asymptotic performance of $\hat{\xi}_{\alpha,B}$ as an approximation to $\hat{\xi}_\alpha$ may be represented in terms of the performance of $\hat{H}_B(\hat{\xi}_\alpha)$ as an approximation to $\hat{H}(\hat{\xi}_\alpha) = \alpha$, by using the method of Bahadur representation,

$$\hat{\xi}_{\alpha,B} - \hat{\xi}_\alpha = \{\alpha - \hat{H}_B(\hat{\xi}_\alpha)\} \phi(z_\alpha)^{-1} \{1 + o_p(1)\},$$

where of course ϕ is the Standard Normal density and z_α the α-level Standard Normal quantile. See Hall (1990a, 1991a). (Note from subsection II.7.4 that if $\alpha > 0.5$ and we are using importance resampling, it is efficacious to approximate instead the $(1-\alpha)$th quantile of $-T^*$, and then change sign.)

This makes it clear that the asymptotic gain in performance obtained by using an efficient resampling algorithm to approximate $\widehat{H}^{-1}(\alpha)$ is equivalent to the asymptotic improvement when approximating $\widehat{H}(x)$ at the point $x = \hat{\xi}_\alpha$. In the case of importance resampling, where optimal choice of the resampling probabilities p_i depends very much on the value of $\hat{\xi}_\alpha$, one implication of this result is that we may select the p_i's as though we were approximating $\widehat{H}(z_\alpha)$ (bearing in mind that $\hat{\xi}_\alpha \to z_\alpha$ as $n \to \infty$). Alternatively, we might construct an approximation $\tilde{\xi}_\alpha$ of $\hat{\xi}_\alpha$ via a pilot program of uniform resampling, and then select the p_i's as though we were approximating $\widehat{H}(\tilde{\xi}_\alpha)$.

A graph of the efficiency, relative to uniform resampling, of a distribution or quantile approximation by Monte Carlo means, is typically a bell-shaped or similar curve that asymptotes to unity in the tails. This means that the efficiency is greater towards the centre of the distribution than in the tails, where it is most often needed for constructing confidence intervals or hypothesis tests. An exception to this rule is importance resampling, where the efficiency curve is cup-shaped, lowest at the centre of the distribution and diverging to $+\infty$ in either tail; see Figure A.1. That observation demonstrates the considerable potential of importance resampling for quantile estimation.

Formula (A.28) generally produces an approximation $\hat{\xi}_{\alpha,B}$ that is identical to one of the resampled values of R^*. In the case of uniform resampling, careful choice of B can enhance the coverage accuracy of a confidence interval obtained by using this quantile approximation. Let $R_1^*, \dots, R_B^*$ denote the B values of R^* obtained by uniform resampling, and suppose we choose B and ν such that

$$\nu/(B+1) = \alpha.$$

In the case $\alpha = 0.95$ this would involve taking $B = 99, 199, \dots$ instead of $B = 100, 200, \dots$, respectively. Since the B values of R^* divide the real line into $B+1$, not B, parts, then it makes sense to select $\hat{\xi}_{\alpha,B}$ as the νth largest of $R_1^*, \dots, R_B^*$.

This definition of $\hat{\xi}_{\alpha,B}$ may be justified more fully as follows. We wish to estimate ξ_α, which has the property

$$P(R \le \xi_\alpha) = \alpha = \nu/(B+1).$$

If we use the approximant $\hat{\xi}_{\alpha,B}$ in place of ξ_α in a formula for a one-sided

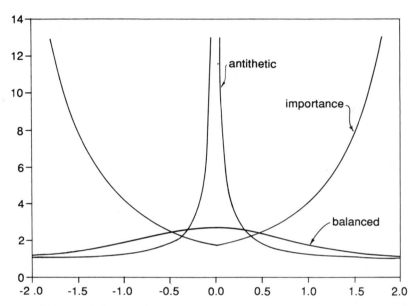

FIGURE A.1. Asymptotic efficiencies of balanced resampling, antithetic resampling, and importance resampling for approximating the bootstrap distribution function

$$\widehat{H}(x) = P(R^* \leq x \mid \mathcal{X}),$$

where $R^* = S^*$ or T^*.   The horizontal axis represents x.   The same efficiencies apply to approximation of the bootstrap quantile $\hat{\xi}_\alpha = \widehat{H}^{-1}(\alpha)$, when $\alpha = \Phi(x)$. Each curve represents a plot of the function

$$e(x) = \lim_{n \to \infty} \lim_{B \to \infty} [\mathrm{var}\{\widehat{H}_B^*(x) \mid \mathcal{X}\}/\mathrm{var}\{\widehat{H}_B(x) \mid \mathcal{X}\}],$$

where $\widehat{H}_B^*(x)$ denotes the approximation after B simulations of uniform resampling and $\widehat{H}_B(x)$ is the approximation after B simulations of balanced resampling, antithetic resampling, or importance resampling respectively. In the case of balanced resampling, $e(x) = \rho(x)$, where the latter is defined in Section II.5. For antithetic resampling, $e(x) = q(x)^{-1}$, where $q(x)$ is defined in Section II.6. (We have plotted the most favourable case for antithetic resampling, in which the sampling distribution is symmetric.) For importance resampling, $e(x) = r(x)$, where the latter is defined in subsection II.7.4. The methods of balanced and importance resampling may be combined to produce a new method that improves on both the other two; see Hall (1990e).

confidence interval of the form $(\hat{\theta} - \hat{t}, \infty)$, then the coverage probability of that interval will be $P(R \leq \hat{\xi}_{\alpha,B})$ rather than α. To appreciate the size of the error $P(R \leq \hat{\xi}_{\alpha,B}) - \alpha$, define $\hat{p} = P(R^* \leq R \mid \mathcal{X})$. Conditional on $\mathcal{X}$, let the random variable N have the binomial $Bi(B, \hat{p})$ distribution. (We put a circumflex on $\hat{p}$ simply to indicate that $\hat{p}$ is a random variable.) Let $R^*_{(1)} \leq R^*_{(2)} \leq \dots \leq R^*_{(B)}$ denote the values $R^*_1, \dots, R^*_B$ arranged in order of increasing size. Then $\hat{\xi}_{\alpha,B} = R^*_{(\nu)}$ is our uniform resampling approximation to $\hat{\xi}_\alpha$, and

$$P(R \leq \hat{\xi}_{\alpha,B} \mid \mathcal{X}) = P(\text{at most } \nu - 1 \text{ out of } T^*_1, \dots, T^*_B \text{ are } < R \mid \mathcal{X})$$

$$= P(N \leq \nu - 1 \mid \mathcal{X})$$

$$= \sum_{j=0}^{\nu-1} \binom{B}{j} \hat{p}^j (1 - \hat{p})^{B-}.$$

Therefore,

$$P(R \leq \hat{\xi}_{\alpha,B}) = E\{P(R \leq \hat{\xi}_{\alpha,B} \mid \mathcal{X})\}$$

$$= \sum_{j=0}^{\nu-1} \binom{B}{j} \int_0^1 p^j (1 - p)^{B-j} \, dP(\hat{p} \leq p). \qquad (A.29)$$

To elucidate formula (A.29), recall that R denotes either $n^{1/2}(\hat{\theta} - \theta_0)/\sigma$ or $n^{1/2}(\hat{\theta} - \theta_0)/\hat{\sigma}$, and that $\hat{H}(x) = P(R^* \leq x \mid \mathcal{X})$ denotes the bootstrap distribution function of R^*. If $R = n^{1/2}(\hat{\theta} - \theta_0)/\hat{\sigma}$ then

$$P(\hat{p} \leq p) = P\{\hat{F}(R) \leq p\} = P\{R \leq \hat{F}^{-1}(p)\} = P(R \leq \hat{\xi}_p)$$

$$= P\left\{\theta_0 \in (\hat{\theta} - n^{-1/2}\hat{\sigma}\hat{\xi}_p, \infty)\right\}.$$

That is, $P(\hat{p} \leq p)$ equals the coverage probability of a nominal p-level percentile-t confidence interval for θ_0, and so

$$P(\hat{p} \leq p) = p + n^{-1} A_n(p),$$

where A_n is a function of bounded variation. See subsection 3.5.4. In the case $R = n^{1/2}(\hat{\theta} - \theta_0)/\sigma$ we have

$$P(\hat{p} \leq p) = P\left\{\theta_0 \in (\hat{\theta} - n^{-1/2}\sigma\hat{\xi}_p, \infty)\right\} = p + n^{-1} A_n(p),$$

for another function A_n of bounded variation. Therefore, by (A.29),

$$P(R \leq \hat{\xi}_{\alpha,B}) = \sum_{j=0}^{\nu-1} \binom{B}{j} \int p^j (1-p)^{B-j} \, d\{p + A_n(p)\}$$

$$= \nu(B+1)^{-1} + n^{-1}\rho_n(B,\nu), \qquad (A.30)$$

where

$$\rho_n(B,\nu) = \int K(p) \, dA_n(p),$$

$$K(p) = \sum_{j=0}^{\nu-1} \binom{B}{j} p^j (1-p)^{B-j}.$$

(Note that $\int K(p) \, dp = \nu(B+1)^{-1}$, and that $\rho_n(B,\nu)$ is bounded uniformly in B and ν.) Formula (A.30) makes it clear that $\nu(B+1)^{-1}$ should reasonably be regarded as the nominal value of the probability $P(R \leq \hat{\xi}_{\alpha,B})$, and so provides a theoretical underpinning for the definition of $\hat{\xi}_{\alpha,B}$ given two paragraphs earlier.

We claim that

$$\sup_{B,\nu} |\rho_n(B,\nu)| \leq \sup_p |A_n(p)|, \qquad (A.31)$$

whence it follows via (A.30) that the worst error between true coverage and nominal coverage, i.e.,

$$\sup_{B,\nu} \left| P(R \leq \hat{\xi}_{\alpha,B}) - \nu(B+1)^{-1} \right|,$$

is no greater than

$$\sup_{\alpha} \left| P(R \leq \hat{\xi}_{\alpha}) - \alpha \right| = O(n^{-1}).$$

To verify (A.31), let $\pi = \pi(u)$ be the solution of the equation $K(p) = u$, for $0 < u < 1$. Since $K(p)$ decreases from 1 to 0 as p increases from 0 to 1,

$$\rho_n(B,\nu) = \int_0^1 K(p) \, dA_n(p) = \int_0^1 dA_n(p) \int_0^{K(p)} du$$

$$= \int_0^1 du \int_0^{\pi(u)} dA_n(p) = \int_0^1 A_n\{\pi(u)\} \, du,$$

whence (A.31) is immediate. The reader is referred to Hall (1986b) for further details.

It is worth reiterating the main conclusion of this argument: if ν and B are chosen so that $\nu(B+1)^{-1} = \alpha$, then the coverage error of the bootstrap confidence interval $(\hat{\theta} - n^{-1/2}\hat{\sigma}\hat{\xi}_{\alpha,B}, \infty)$ will equal $O(n^{-1})$ *uniformly in* B. For example, if for all values of n we do just $B = 19$ simulations of T^*, and always take $\hat{\xi}_{0.95,B} = T^*_{19}$ (the largest of the simulated values), then since $19/(19+1) = 0.95$, the coverage error of the confidence interval $(\hat{\theta} - n^{-1/2}\,\hat{\sigma}\,\hat{\xi}_{0.95,B}, \infty)$ will equal $O(n^{-1})$ even though B stays fixed as $n \to \infty$. Of course, this result does not suggest that small numbers of simulations are appropriate when confidence intervals are being constructed. However, it does indicate that if B and ν are chosen correctly then the error arising from doing too few simulations will not be in terms of coverage accuracy, but rather will arise from erratic or unreliable positioning of the endpoint of the confidence interval, perhaps leading to an overly long interval.

Appendix III

Confidence Pictures

The arguments throughout this monograph are of the frequentist type. We are aware of the Bayesian criticism of the general theory of confidence intervals (e.g., Robinson 1982, p. 124). However, as indicated in the Preface, our monograph has a particular didactic purpose — to present those aspects of the bootstrap and Edgeworth expansion that are related and that shed light on one another. Edgeworth expansions are particularly well suited to exploring the frequentist theory of bootstrap methods. We do not claim to present a comprehensive account of the bootstrap, least of all from a Bayesian viewpoint.

We are also aware that there are criticisms that can be levelled at confidence intervals from a frequentist viewpoint. One of these, which we find particularly persuasive, is that the notion of a confidence interval is somewhat outdated and unsophisticated. Confidence intervals date from an era where the most powerful computational device was a hand-driven mechanical calculator. In that context they are a reasonable way of describing our uncertainty as to the position of a true parameter value, but in the 1990s it would be fair to ask whether alternative, more sophisticated techniques might not convey more information in an equally palatable form. One such device is a "confidence picture", by means of which one may present empirical, graphical evidence about the relative likelihood of the true parameter value lying in different regions.

To define a confidence picture, observe first that an α-level percentile-t confidence interval for the parameter θ is given by $(\hat{\theta} - n^{-1/2} \hat{\sigma} \hat{v}_\alpha, \infty)$, where $\hat{v}_\alpha = \widehat{G}^{-1}(\alpha)$ and

$$\widehat{G}(x) = P\{n(\hat{\theta}^* - \hat{\theta})/\hat{\sigma}^* \leq x \mid \mathcal{X}\}.$$

See Example 1.2 in Section 1.3, and Chapter 3. Of course, $\widehat{G}$ is an estimator of the distribution function G, defined by

$$G(x) = P\{n^{1/2}(\hat{\theta} - \theta)/\hat{\sigma} \leq x\}.$$

By applying a kernel density estimator to simulated values of $n^{1/2}(\hat{\theta}^* - \hat{\theta})/\hat{\sigma}$, which would be needed anyway to construct a Monte Carlo approximation to $\widehat{G}$, we may produce an estimator $\hat{g}$ of the density $g = G'$. A confidence picture is a graph of the pair

$$\left(\hat{\theta} - n^{-1/2} \hat{\sigma} x, \hat{g}(x)\right), \qquad -\infty < x < \infty,$$

and provides an impression of the relative uncertainty with which we might ascribe different values to θ, given the data. It also conveys an easily-interpreted picture of the skewness of our information about where the true value of θ lies. One or more confidence intervals may be superimposed on the confidence picture, to provide additional information; see Figure A.2.

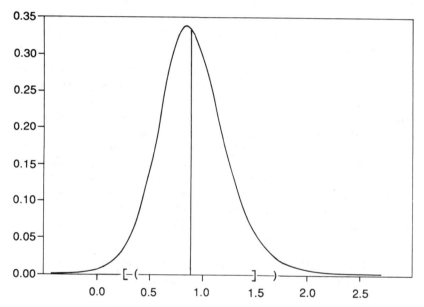

FIGURE A.2. Confidence picture in the case of estimating the mean of an exponential distribution. For the purpose of this example we draw a sample of size $n = 10$ from the exponential distribution with density $f(x) = e^{-x}$, $x > 0$. We took $\hat{\theta} = \bar{X}$, the sample mean, and $\hat{\sigma}^2 = n^{-1} \sum (X_i - \bar{X})^2$, the sample variance. We simulated $n^{1/2}(\hat{\theta}^* - \hat{\theta})/\hat{\sigma}^*$ a total of $B = 1,000$ times, and computed $\hat{g}$ from these data using a kernel estimator with Standard Normal kernel and bandwidth $h = \{4/(3n)\}^{1/5} \hat{\sigma}$. (The latter is asymptotically optimal when sampling from a Normal $N(0, \hat{\sigma}^2)$ distribution.)   The figure depicts the locus of the point $(\hat{\theta} - n^{-1/2}\hat{\sigma} x, \hat{g}(x))$ as x, indicated on the horizontal axis, varies over the range $-\infty < x < \infty$. Symmetric [ ] and equal-tailed ( ) 95% confidence intervals are superimposed on the figure. A vertical line has been drawn at $x = \bar{X}$.

Confidence pictures have bivariate analogues in which the relative height of $\hat{g}(\mathbf{x})$ is indicated by shading or colouring areas of the plane. To be explicit, let $n^{-1}\widehat{\boldsymbol{\Sigma}}$ (a 2×2 matrix) be an estimate of the variance matrix of the 2-vector $\widehat{\boldsymbol{\theta}}$, which estimates $\boldsymbol{\theta}$. A kernel density estimator applied to simulated values of $\mathbf{T}^* = n^{1/2}\widehat{\boldsymbol{\Sigma}}^{*-1/2}(\widehat{\boldsymbol{\theta}}^* - \widehat{\boldsymbol{\theta}})$ may be used to compute an estimator $\hat{g}$ of the density g of $\mathbf{T} = n^{1/2}\widehat{\boldsymbol{\Sigma}}^{-1/2}(\widehat{\boldsymbol{\theta}} - \boldsymbol{\theta}_0)$. A confidence picture may be constructed by shading the Euclidean plane $\mathbb{R}^2$ at the point $\widehat{\boldsymbol{\theta}} - n^{-1/2}\widehat{\boldsymbol{\Sigma}}^{1/2}\mathbf{x}$, in proportion to the value of $\hat{g}(\mathbf{x})$. Examples and illustrations of a closely related procedure (for inference on the surface of a sphere, rather than in the plane) are given by Fisher and Hall (1990). As in that work, confidence regions may be superimposed on the confidence picture to provide additional information.

Appendix IV

A Nonstandard Example: Quantile Error Estimation

IV.1 Introduction

In this appendix we discuss an example of a bootstrap estimate that fails to satisfy many of the properties that underpin the work throughout our monograph. In particular, the estimate fails to converge at a rate of $n^{-1/2}$, although it does not involve any explicit degree of smoothing (compare Sections 4.4 and 4.5); smoothing the bootstrap can significantly enhance the convergence rate of the estimate, although not to the extent of improving it to $n^{-1/2}$ (compare Section 4.1); and the estimate gives rise to an Edgeworth expansion where the polynomial in the term of size $n^{-1/2}$ is neither even nor odd (compare Sections 2.3 and 2.4). The example in question is that of the bootstrap estimate of mean squared error[2] of a sample quantile, and is introduced in Section IV.2 below. Sections IV.3 and IV.4 describe convergence rates and Edgeworth expansions, respectively.

The example serves to illustrate that even in an asymptotic sense, the properties that we have ascribed to the bootstrap in Chapters 3 and 4 are not universally true. They hold in a very important but nevertheless restricted class of circumstances. The range of problems to which bootstrap methods can be applied is so vast that it is impossible to give a general account which goes beyond first-order properties; the account in this monograph has concentrated on second- and higher-order properties.

IV.2 Definition of the Mean Squared Error Estimate

Let $\mathcal{X} = \{X_1, \ldots, X_n\}$ denote a random sample of size n drawn from a distribution with distribution function F, and write

$$\widehat{F}(x) = n^{-1} \sum_{i=1}^{n} I(X_i \leq x)$$

---

[2]In this context, many authors do not distinguish between bootstrap estimates of variance and mean squared error. Strictly speaking, the estimate treated here is of mean squared error, although the variance estimate has similar properties.

for the empirical distribution function of the sample. The bootstrap estimate of the pth quantile of F, $\xi_p = F^{-1}(p)$, is

$$\hat{\xi}_p = \widehat{F}^{-1}(p) = \inf\left\{x : \widehat{F}(x) \geq p\right\}$$

$$= X_{nr}, \tag{A.32}$$

where $X_{n1} \leq \ldots \leq X_{nn}$ denote the order statistics of $\mathcal{X}$ and $r = [np]$ is the largest integer not greater than np. The mean squared error of ξ_p is given by

$$\tau^2 = E\{(\hat{\xi}_p - \xi_p)^2\}$$

$$= \frac{n!}{(r-1)!\,(n-r)!} \int_{-\infty}^{\infty} (x - \xi_p)^2\, F(x)^{r-1}\,\{1 - F(x)\}^{n-r}\, dF(x)$$

$$= r\binom{n}{r} \int_0^1 \{F^{-1}(u) - F^{-1}(p)\}^2\, u^{r-1}(1-u)^{n-r}\, du, \tag{A.33}$$

of which the bootstrap estimate is

$$\hat{\tau}^2 = r\binom{n}{r} \int_0^1 \{\widehat{F}^{-1}(u) - \widehat{F}^{-1}(p)\}^2\, u^{r-1}(1-u)^{n-r}\, du$$

$$= \sum_{j=1}^{n} (X_{nj} - X_{nr})^2\, w_j, \tag{A.34}$$

where

$$w_j = r\binom{n}{r} \int_{(j-1)/n}^{j/n} u^{r-1}(1-u)^{n-r}\, du.$$

The bootstrap variance estimate differs in details, as follows. Bootstrap estimates of the mean and mean square of $\hat{\xi}_p$ are respectively

$$r\binom{n}{r} \int_0^1 \widehat{F}^{-1}(u)\, u^{r-1}(1-u)^{n-r}\, du = \sum_{j=1}^{n} X_{nj}\, w_j,$$

$$r\binom{n}{r} \int_0^1 \widehat{F}^{-1}(u)^2\, u^{r-1}(1-u)^{n-r}\, du = \sum_{j=1}^{n} X_{nj}^2\, w_j.$$

Therefore, the bootstrap variance estimate is

$$\sum_{j=1}^{n} X_{nj}^2\, w_j - \left(\sum_{j=1}^{n} X_{nj}\, w_j\right)^2.$$

IV.3 Convergence Rate of the Mean Squared Error Estimate

The variance and mean squared error of $\hat{\xi}_p$ are both asymptotic to

$$n^{-1} p(1-p) f(\xi_p)^{-2},$$

where $f = F'$ denotes the population density. See David (1981, p. 254ff). (We assume here that f is continuous at ξ_p and $f(\xi_p) > 0$.) In view of this result, estimation of either variance or mean squared error involves nonparametric estimation of f, explicitly or implicitly. We know from Section 4.4 that nonparametric density estimators typically converge at much slower rates than $n^{-1/2}$, due to the fact that they are restricted to using only local information. This is a feature of the problem, not just of particular estimator types; see, e.g., Hall (1989c). Therefore, it stands to reason that the bootstrap estimate $\hat{\tau}^2$ will have a relative error of larger order than $n^{-1/2}$. That is, $\hat{\tau}^2 \tau^{-2} - 1$ will be of larger order than $n^{-1/2}$.

It turns out that under mild regularity conditions, for example if $f(\xi_p) > 0$ and f has a bounded derivative in a neighbourhood of ξ_p, then $\hat{\tau}^2 \tau^{-2} - 1$ is of precise order $n^{-1/4}$ in probability. Indeed,

$$n^{1/4}(\hat{\tau}^2 \tau^{-2} - 1) \longrightarrow N(0, \eta^2)$$

in distribution, where $\eta^2 = 2\pi^{-1/2}\{p(1-p)\}^{1/2} f(\xi_p)^{-2}$. See Hall and Martin (1988b), and also Ghosh, Parr, Singh, and Babu (1984) and Babu (1986).

This rate of convergence may be improved by smoothing during the construction of $\hat{\tau}^2$, although the rate cannot be improved as far as $n^{-1/2}$ without making assumptions that border on those of a parametric model. To appreciate how the smoothing is carried out, recall from Section 4.4 that a nonparametric kernel estimator of $f = F'$ has the form

$$\hat{f}_h(x) = (nh)^{-1} \sum_{i=1}^{n} K\{(x - X_i)/h\},$$

where $\mathcal{X} = \{X_1, \ldots, X_n\}$ denotes the random sample drawn from the distribution with density f, K is the kernel function, and h is the bandwidth. By integrating $\hat{f}_h$ we obtain a smoothed estimator of F,

$$\widehat{F}_h(x) = \int_{-\infty}^{x} \hat{f}_h(y)\, dy = n^{-1} \sum_{i=1}^{n} \int_{-\infty}^{(x-X_i)/h} K(y)\, dy$$

$$= \int_{-\infty}^{\infty} h^{-1} K\{(x-y)/h\}\, d\widehat{F}(y),$$

where $\widehat{F}$ is the empirical distribution function of the sample $\mathcal{X}$. Recall that in constructing the bootstrap estimate $\hat{\tau}^2$ of τ^2, we simply eplaced F by $\widehat{F}$ in the definition (A.33) of τ^2. If instead we replace F by $\widehat{F}_h$ we obtain the smoothed bootstrap estimate

$$\hat{\tau}_h^2 = r \binom{n}{r} \int_0^1 \left\{ \widehat{F}_h^{-1}(u) - \widehat{F}_h^{-1}(p) \right\}^2 u^{r-1}(1-u)^{n-r}\, du$$

$$= r \binom{n}{r} \int_0^1 (x - \hat{\xi}_{p,h})^2\, \widehat{F}_h(x)^{r-1} \left\{ 1 - \widehat{F}_h(x) \right\}^{n-r} \hat{f}_h(x)\, dx\,,$$

where $\hat{\xi}_{p,h} = \widehat{F}_h^{-1}(p)$.

Take K to be an mth order kernel, where $m \geq 2$ is an integer. (An rth order kernel was defined in (4.69) in Chapter 4.) It may be proved that if the bandwidth h is taken to be of size $n^{-1/(2m+1)}$ then the relative error of $\hat{\tau}_h^2$ as an estimate of τ^2 is of order $n^{-m/(2m+1)}$. Indeed, if $h = C_1\, n^{-1/(2m+1)}$ then

$$n^{m/(2m+1)} \left(\hat{\tau}_h^2\, \tau^{-2} - 1 \right) \longrightarrow N(C_2, C_3)\,, \tag{A.35}$$

where C_2, $C_3 > 0$ are constants depending on C_1 and f. See Hall, DiCiccio, and Romano (1989), although note that the notation in that article is unsatisfactory because the same notation is used for both $[np]$ and the order of the kernel K. The regularity conditions for (A.35) demand that f have m derivatives.

Result (A.35) makes it clear that for sufficiently smooth distributions, the convergence rate of relative error for the quantile mean squared error estimate can be improved to $n^{-(1/2)+\epsilon}$, for any given $\epsilon > 0$, by smoothing the empirical distribution function. We noted in Section 4.1 that smoothing generally cannot improve the convergence rate of a bootstrap estimate, at least not in problems where that estimate can be expressed as a smooth function of a vector sample mean. The problem of estimating quantile variance is not of that form, and so does not fall within the compass of the remarks in Section 4.1.

IV.4 Edgeworth Expansions for the Studentized Bootstrap Quantile Estimate

Let $\hat{\xi}_p$ and $\hat{\tau}^2$ denote the bootstrap estimates of ξ_p and τ^2, defined in (A.32) and (A.34). Since $\hat{\xi}_p$ is asymptotically Normal $N(0, \tau^2)$ (David 1981, p. 255), and since $\hat{\tau}^2/\tau^2 \to 1$ in probability, $(\hat{\xi}_p - \xi_p)/\hat{\tau}$ has a limiting Standard Normal distribution. The rate of convergence to this limit may be described by an Edgeworth expansion, although the properties of terms in that expansion are very different from those described in Chapter 2. Of course, this is principally due to the fact that $\hat{\tau}^2 \tau^{-2} - 1$ is of order $n^{-1/4}$ rather than $n^{-1/2}$. Interestingly, the first term in the expansion is of size $n^{-1/2}$, not $n^{-1/4}$, but the polynomial in the coefficient of that term is neither even nor odd; in the cases encountered previously, the polynomial for the $n^{-1/2}$ term was always even.

To describe the result more explicitly, define

$$q(x) = \tfrac{1}{4} \{\pi p(1-p)\}^{-1/2} x(x^2 + 1 + 2^{3/2})$$

$$+ \tfrac{1}{6} \{p(1-p)\}^{-1/2} (1+p)(x^2 - 1)$$

$$- \left[\{p(1-p)\}^{-1/2} p + \{p(1-p)\}^{1/2} f'(\xi_p) f(\xi_p)^{-2} \right] x^2$$

$$- \{p(1-p)\}^{-1/2} \{\tfrac{1}{2}(1-p) + r - np\}.$$

Then

$$P\{(\hat{\xi}_p - \xi_p)/\hat{\tau} \le x\} = \Phi(x) + n^{-1/2} q(x) \phi(x) + O(n^{-3/4}) \qquad \text{(A.36)}$$

as $n \to \infty$; see Hall and Martin (1991). The polynomial q is of degree 3 and is neither an odd function nor an even function.

This result may be explained intuitively, as follows. The Edgeworth expansion of a Studentized quantile consists, essentially, of a "main" series of terms decreasing in powers of $n^{-1/2}$, arising from the numerator in the Studentized ratio, together with a "secondary" series arising from the denominator. On the present occasion the secondary series decreases in powers of $n^{-1/4}$ since the variance estimate has relative error $n^{-1/4}$. However, it may be shown that the first term in the secondary series vanishes. For both series, the jth term is even or odd according to whether j is odd or even, respectively. The term of order $n^{-1/2}$ in the combined series includes the first, even term of the main series and the second, odd term of the secondary series.

The fact that the polynomial q in formula (A.36) is not an even function means that the coverage error of a two-sided confidence interval for ξ_p, based on the Normal approximation to the distribution of $(\hat{\xi}_p - \xi_p)/\hat{\tau}$, is of size $n^{-1/2}$. This contrasts with the more conventional case treated in Section 2.3, where it was shown that when q is even, the coverage error of a two-sided Normal approximation interval is of size n^{-1}.

Hall and Sheather (1988) discussed Edgeworth expansions for the distribution of a Studentized quantile when the standard deviation estimate is based explicitly on a density estimator, of the type introduced by Siddiqui (1960) and Bloch and Gastwirth (1968).

Appendix V

A Non-Edgeworth View of the Bootstrap

Through much of this monograph we have described the performance of bootstrap methods in terms of their ability to reproduce Edgeworth expansions of theoretical distributions. The notions of second-order correctness and coverage accuracy, which lie at the heart of our assessment of different bootstrap methods, have been defined and derived using Edgeworth expansions. One could be excused for thinking that the Edgeworth view was the only one, or that it provided all the information we need about properties of the bootstrap in problems of distribution estimation. This is certainly not the case.

At least two points in our discussion of Edgeworth expansions give us cause to suspect that a reliance on those expansions conceals important aspects of the story. First, we noted in Section 2.2 that even in particularly simple cases, such as distributions of sums of independent random variables with all moments finite, Edgeworth expansions do not necessarily converge as infinite series. Thus, if we run the series to an extra term, we may actually reduce the accuracy of the series as an approximation to a distribution function. Second, we learned in Section 3.8 that explicit correction by Edgeworth expansion can produce abstruse results, owing to the fact that such expansions are not monotone functions. This can be a particularly serious problem with small samples, and it is often in just those cases that we place most importance on refinements of simple Normal approximations.

Beran (1982) and Bhattacharya and Qumsiyeh (1989) have compared bootstrap and Normal or Edgeworth approximations in terms of loss functions. To appreciate some of their results, let us define

$$\widehat{H}(x) = P(T^* \leq x \mid \mathcal{X})$$
$$= \Phi(x) + n^{-1/2}\, \hat{q}_1(x)\, \phi(x) + n^{-1}\, \hat{q}_2(x)\, \phi(x) + \ldots$$

to be the bootstrap approximation to the distribution function $H(x)$ of a Studentized statistic T, the latter function being given by

$$H(x) = P(T \leq x)$$
$$= \Phi(x) + n^{-1/2} q_1(x)\, \phi(x) + n^{-1} q_2(x)\, \phi(x) + \ldots. \quad \text{(A.37)}$$

Then

$$\tilde{H}(x) \; = \; \Phi(x) + n^{-1/2}\, \hat{q}_1(x)\, \phi(x)$$

is the Edgeworth approximation obtained by ignoring all but the term of size $n^{-1/2}$ in the estimated Edgeworth expansion. The respective L^p measures of loss for these bootstrap and Edgeworth approximations are

$$\ell_{\text{boot}}(x) \; = \; \big\{\, E|\widehat{H}(x) - H(x)|^p \,\big\}^{1/p},$$

$$\ell_{\text{Edge}}(x) \; = \; \big\{\, E|\tilde{H}(x) - H(x)|^p \,\big\}^{1/p}.$$

Bhattacharya and Qumsiyeh (1989) showed, amongst other things, that for any $1 \le p \le \infty$, and any x such that $q_2(x) \ne 0$,

$$\lim_{n\to\infty}\, \big\{\ell_{\text{boot}}(x)/\ell_{\text{Edge}}(x)\big\}$$

exists and is strictly less than 1. In this sense the bootstrap estimate $\widehat{H}$ outperforms the short Edgeworth approximation $\tilde{H}$ in any L^p metric.

Results of this nature help to quantify the performance of bootstrap estimates relative to short Edgeworth approximations, but they still take an essentially Edgeworth view of the bootstrap. For example, the proof of Bhattacharya and Qumsiyeh's theorem is based on approximating $\widehat{H}(x)$ by the Edgeworth expansion

$$\tilde{H}(x) \, + \, n^{-1}\, \hat{q}_2(x)\, \phi(x)\,.$$

Edgeworth expansions can be inaccurate when employed to estimate tail probabilities, and it is in just those cases that we usually wish to use the bootstrap to approximate distribution functions. To appreciate the inaccuracy, it is helpful to perform a few rough calculations concerning the order of the jth term $t_j(x) = n^{-j/2}\, q_j(x)\, \phi(x)$ in the expansion (A.37).

Recall from Chapter 2 that the polynomial q_j in (A.37) is of degree $3j - 1$. Assuming that the coefficients of q_j are increasing or decreasing no faster than geometrically, the term $t_j(x)$ is, for large x, approximately of size

$$n^{-j/2}\, C^j\, x^{3j-1}\, \phi(x) \; = \; (n^{-1/2}\, C\, x^3)^j\, x^{-1}\, \phi(x)$$

$$\approx \; (n^{-1/2}\, C\, x^3)^j\, \{1 - \Phi(x)\}\,,$$

where C is bounded away from zero and infinity. (Note that $x^{-1}\, \phi(x) \sim 1 - \Phi(x)$ as $x \to \infty$.) That is, the jth term in an expansion of the ratio

$$P(T > x)/\{1 - \Phi(x)\}$$

is approximately of size $(n^{-1/2}\, C\, x^3)^j$. It follows that if x is of larger order than $n^{1/6}$, the Edgeworth series will either not converge or not adequately describe the tail probability $P(T > x)$. When $n = 20$, $n^{1/6} = 1.648$, which is approximately the 5% point of a Standard Normal distribution. These admittedly crude calculations suggest that an Edgeworth view of the bootstrap could sometimes be misleading, even for moderate sample sizes.

A more careful analysis of this problem shows that there is, in fact, a substantial change in behaviour at $x \approx n^{1/6}$. It is at $n^{1/6}$ that approximation of tail probabilities by Edgeworth expansions becomes unsatisfactory. The new analysis requires the theory of large deviation probabilities, in which expansions of Edgeworth type are combined with others to provide a reasonably complete picture of the properties of $P(T > x)/\{1 - \Phi(x)\}$ as $x \to \infty$ and $n \to \infty$ *together*. The greatest amount of detail is available in the case of a sample mean, and so we shall investigate the ratio

$$P(S > x)/\{1 - \Phi(x)\},$$

where $S = n^{1/2}(\bar{X} - \mu)/\sigma$ and $\bar{X}$ is the mean of an n-sample drawn from a population with mean μ and variance σ^2. Without loss of generality we take $\mu = 0$ and $\sigma^2 = 1$. We shall show that Edgeworth approximations, even those that use the true polynomials q_j rather than the estimates $\hat{q}_j$, fail for x-values of size $n^{1/6}$ or larger; whereas bootstrap approximations are valid up to $x = o(n^{1/3})$.

Assume that the parent distribution has finite moment generating function in a neighbourhood of the origin, and that the characteristic function satisfies Cramér's condition,

$$\limsup_{|t|\to\infty} |E(e^{itX})| < 1.$$

Then it may be proved that for $0 \le x = o(n^{1/2})$ as $n \to \infty$, and for each integer $m \ge 1$,

$$P(S > x)/\{1 - \Phi(x)\}$$

$$= \exp\left\{n^{-1/2}\, x^3\, \lambda(x/n^{1/2})\right\}\left\{1 + \sum_{j=1}^{m} n^{-j/2}\, \psi(x) + R_m(x)\right\}, \quad \text{(A.38)}$$

where

$$R_m(x) = O\left[\{n^{-1/2}(x+1)\}^{m+1}\right], \quad \text{(A.39)}$$

$$\lambda(u) = \tfrac{1}{6}(EX^3) + \tfrac{1}{24}\left\{EX^4 - 3 - (EX^3)^2\right\}u + \dots \quad \text{(A.40)}$$

is a power series ("Cramér's series") in u whose coefficients are polynomials in population moments, and ψ_j is a bounded, continuous function depending on population moments. The series in (A.40) converges absolutely for $|u| \le u_0$, some $u_0 > 0$. Saulis (1973) gives a proof of (A.39).

For bounded x, the factor

$$1 + \sum_{j=1}^{m} n^{-j/2}\, \psi_j(x) + R_m(x)$$

in (A.38) represents essentially the Edgeworth expansion

$$P(S > x)/\{1 - \Phi(x)\}$$
$$= 1 - \sum_{j=1}^{m} n^{-j/2}\, p_j(x)\, \phi(x) \left\{1 - \Phi(x)\right\}^{-1} + O(n^{-(m+1)/2}), \quad (A.41)$$

with a modification to take care of the newly introduced term

$$\exp\left\{n^{-1/2}\, x^3\, \lambda(x/n^{1/2})\right\}$$

on the right-hand side of (A.38). The functions ψ_j are readily determined from the Edgeworth polynomials p_j and the coefficients of λ. This equivalence between the large deviation formula (A.38) and the Edgeworth expansion (A.41) is maintained for unbounded x, provided x increases less rapidly than $n^{1/6}$ as n increases. To appreciate why the change point is at $n^{1/6}$, observe from (A.40) that if $x = o(n^{1/2})$ then

$$n^{-1/2}\, x^3\, \lambda(x/n^{1/2}) = (n^{-1/6}\, x)^3\, (EX^3)\, \{1 + o(1)\}$$

as $n \to \infty$. Thus, if x is of size $n^{1/6}$ or greater then $n^{-1/2}\, x^3\, \lambda(x/n^{1/2})$ no longer converges to zero, and the term-by-term comparison that allows us to pass from (A.38) to (A.41) is no longer valid. For such an x, Edgeworth expansions do not adequately describe large deviation probabilities.

The beauty of bootstrap estimation is that it captures both of the components on the right-hand side of (A.38). Indeed, if $\hat{\lambda}$ and $\hat{\psi}_j$ denote versions of λ and ψ_j in which population moments are replaced by their sample counterparts, and $S^* = n^{1/2}(\bar{X}^* - \bar{X})/\hat{\sigma}$ denotes the version of S computed for a resample rather than a sample, then for $O \le x = o(n^{1/2})$,

$$P(S^* > x \mid \mathcal{X})/\{1 - \Phi(x)\}$$
$$= \exp\{n^{-1/2}\, x^3\, \hat{\lambda}(x/n^{1/2})\} \left\{1 + \sum_{j=1}^{m} n^{-j/2}\, \hat{\psi}_j(x) + \widehat{R}_m(x)\right\},$$

where $\widehat{R}_m(x) = O[\{n^{-1/2}(x + 1)\}^{m+1}]$ with probability one. See Hall (1990c). Now, $\hat{\lambda}$ is distant $O_p(n^{-1/2})$ from λ, and so

$$\left| n^{-1/2}\, x^3\, \hat{\lambda}(x/n^{1/2}) - n^{-1/2}\, x^3\, \lambda(x/n^{1/2}) \right| = O_p(n^{-1}\, x^3),$$

which equals $o_p(1)$ if $x = o(n^{1/3})$. Therefore, the bootstrap provides an accurate approximation to large deviation probabilities for values x as large as $o(n^{1/3})$, in the sense of having a relative error that equals $o_p(1)$. On the other hand, Edgeworth approximations (even those where the terms are known precisely, rather than estimated), fail for x larger than $o(n^{1/6})$. Further details are given in Hall (1990c).

These important properties are concealed by a view of the bootstrap that is based principally on Edgeworth expansion. Thus, while the present monograph describes an important aspect of theory for the bootstrap, it does not present a complete picture.

References

Abramovitch, L. and Singh, K. (1985). Edgeworth corrected pivotal statistics and the bootstrap. *Ann. Statist.* **13**, 116–132.

Abramowitz, M. and Stegun, I.A. (1972). *Handbook of Mathematical Functions*, 9th ed. Dover, New York.

Athreya, K.B. (1983). Strong law for the bootstrap. *Statist. Probab. Lett.* **1**, 147–150.

Athreya, K.B. (1987). Bootstrap of the mean in the infinite variance case. *Ann. Statist.* **15**, 724–731.

Babu, G.J. (1986). A note on bootstrapping the variance of the sample quantile. *Ann. Inst. Statist. Math.* **38**, 439–443.

Babu, G.J. and Bose, A. (1988). Bootstrap confidence intervals. *Statist. Probab. Lett.* **7**, 151–160.

Babu, G.J. and Singh, K. (1983). Inference on means using the bootstrap. *Ann. Statist.* **11**, 999–1003.

Babu, G.J. and Singh, K. (1984). On one term correction by Efron's bootstrap. *Sankhyā Ser. A* **46**, 219–232.

Babu, G.J. and Singh, K. (1985). Edgeworth expansions for sampling without replacement from finite populations. *J. Multivar. Anal.* **17**, 261–278.

Barnard, G.A. (1963). Contribution to discussion. *J. Roy. Statist. Soc. Ser. B* **25**, 294.

Barndorff-Nielsen, O.E. and Cox, D.R. (1979). Edgeworth and saddle-point approximations with statistical applications. (With discussion.) *J. Roy. Statist. Soc. Ser. B* **41**, 279–312.

Barndorff-Nielsen, O.E. and Cox, D.R. (1989). *Asymptotic Techniques for Use in Statistics*. Chapman and Hall, London.

Beran, R. (1982). Estimated sampling distributions: The bootstrap and competitors. *Ann. Statist.* **10**, 212–225.

Beran, R. (1984a). Bootstrap methods in statistics. *Jber. d. Dt. Math. Verein.* **86**, 24–30.

Beran, R. (1984b). Jackknife approximations to bootstrap estimates. *Ann. Statist.* **12**, 101–118.

Beran, R. (1987). Prepivoting to reduce level error of confidence sets. *Biometrika* **74**, 457–468.

Beran, R. (1988a). Balanced simultaneous confidence sets. *J. Amer. Statist. Assoc.* **83**, 670–686.

Beran, R. (1988b). Balanced simultaneous confidence sets. *J. Amer. Statist. Assoc.* **83**, 679–686.

Beran, R. (1988c). Prepivoting test statistics: A bootstrap view of asymptotic refinements. *J. Amer. Statist. Assoc.* **83**, 687–697.

Beran, R. and Millar, P. (1985). Asymptotic theory of confidence sets. *Proc. Berkeley Conf. in Honor of Jerzy Neyman and Jack Kiefer* (L.M. Le Cam and R.A. Olshen, eds.), vol. **2**, Wadsworth, Belmont, pp. 865–886.

Beran, R. and Millar, P.W. (1986). Confidence sets for a multivariate distribution. *Ann. Statist.* **14**, 431–443.

Beran, R. and Millar, P.W. (1987). Stochastic estimation and testing. *Ann. Statist.* **15**, 1131–1154.

Beran, R. and Srivastava, M.S. (1985). Bootstrap test and confidence regions for functions of a covariance matrix. *Ann. Statist.* **13**, 95–115.

Bhattacharya, R.N. (1987). Some aspects of Edgeworth expansions in statistics and probability. *New Perspectives in Theoretical and Applied Statistics* (M.L. Puri, J.P. Vilaplana, and W. Wertz, eds.), Wiley, New York, pp. 157–170.

Bhattacharya, R.N. and Ghosh, J.K. (1978). On the validity of the formal Edgeworth expansion. *Ann. Statist.* **6**, 434–451.

Bhattacharya, R.N. and Ghosh, J.K. (1989). On moment conditions for valid formal Edgeworth expansions. In: *Multivariate Statistics and Probability: Essays in Memory of P.R. Krishnaiah.* (C.R. Rao and M.M. Rao, eds.), Academic, New York, pp. 68–79.

Bhattacharya, R.N. and Qumsiyeh, M. (1989). Second order L^p-comparisons between the bootstrap and empirical Edgeworth expansion methodologies. *Ann. Statist.* **17**, 160–169.

Bhattacharya, R.N. and Rao, R.R. (1976). *Normal Approximation and Asymptotic Expansions.* Wiley, New York.

Bickel, P.J. (1974). Edgeworth expansions in nonparametric statistics. *Ann. Statist.* **2**, 1–20.

Bickel, P.J. and Freedman, D.A. (1980). On Edgeworth expansions and the bootstrap. Unpublished manuscript.

Bickel, P.J. and Freedman, D.A. (1981). Some asymptotic theory for the bootstrap. *Ann. Statist.* **9**, 1196-1217.

Bickel, P.J. and Freedman, D.A. (1983). Bootstrapping regression models with many parameters. *A Festschrift for Erich L. Lehmann* (P.J. Bickel, K.A. Doksum, and J.C. Hodges, Jr., eds.), Wadsworth, Belmont, pp. 28–48.

Bickel, P.J. and Yahav, J.A. (1988). Richardson extrapolation and the bootstrap. *J. Amer. Statist. Assoc.* **83**, 387-393.

Bloch, D.A. and Gastwirth, J.L. (1968). On a simple estimate of the reciprocal of the density function. *Ann. Math. Statist.* **39**, 1083-1085.

Bose, A. (1988). Edgeworth correction by bootstrap in autoregressions. *Ann. Statist.* **16**, 1709-1722.

Buckland, S.T. (1980). A modified analysis of the Jolly-Seber capture-recapture model. *Biometrics* **36**, 419–435.

Buckland, S.T. (1983). Monte Carlo methods for confidence interval estimation using the bootstrap technique. *Bull. Appl. Statist.* **10**, 194-212.

Buckland, S.T. (1984). Monte Carlo confidence intervals. *Biometrics* **40**, 811-817.

Buckland, S.T. (1985). Calculation of Monte Carlo confidence intervals. *Appl. Statist.* **34**, 296-301.

Bunke, O. and Riemer, S. (1983). A note on bootstrap and other empirical procedures for testing linear hypothesis without normality. *Math. Operat. Statist., Ser. Statist.* **14**, 517-526.

Chambers, J.M. (1967). On methods of asymptotic expansion for multivariate distributions. *Biometrika* **54**, 367-383.

Chebyshev, P.L. (1890). Sur deux théorèmes relatifs aux probabilités. *Acta Math.* **14**, 305-315.

Chen, Z. (1990). A resampling approach for bootstrap hypothesis testing. Unpublished manuscript.

Chibishov, D.M. (1972). An asymptotic expansion for the distribution of a statistic admitting an asymptotic expansion. *Theor. Probab. Appl.* **17**, 620-630.

Chibishov, D.M. (1973a). An asymptotic expansion for a class of estimators containing maximum likelihood estimators. *Theor. Probab. Appl.* **18**, 295-303.

Chibishov, D.M. (1973b). An asymptotic expansion for the distribution of sums of a special form with an application to minimum-contrast estimates. *Theor. Probab. Appl.* **18**, 649–661.

Chibishov, D.M. (1984). Asymptotic expansions and deficiencies of tests. *Proc. Internat. Congress Math., August 16-24 1983, Warsaw* (Z. Ciesielski and C. Olech, eds.), vol. **2**, North-Holland, Amsterdam, pp. 1063–1079.

Collomb, G. (1981). Estimation non-paramétrique de la régression: Revue bibliographique. *Internat. Statist. Rev.* **49**, 75–93.

Cornish, E.A. and Fisher, R.A. (1937). Moments and cumulants in the specification of distributions. *Internat. Statist. Rev.* **5**, 307–322.

Cox, D.R. and Hinkley, D.V. (1974). *Theoretical Statistics.* Chapman and Hall, London.

Cox, D.R. and Reid, N. (1987). Approximations to noncentral distributions. *Canad. J. Statist.* **15**, 105-114.

Cramér, H. (1928). On the composition of elementary errors. *Skand. Aktuarietidskr.* **11**, 13–74, 141–180.

Cramér, H. (1946). *Mathematical Methods of Statistics.* Princeton University Press, Princeton, NJ.

Cramér, H. (1970). *Random Variables and Probability Distributions*, 3rd ed. Cambridge University Press, Cambridge, UK.

David, H.A. (1981). *Order Statistics*, 2nd ed. Wiley, New York.

Davison, A.C. (1988). Discussion of papers by D.V. Hinkley and by T.J. DiCiccio and J.P. Romano. *J. Roy. Statist. Soc. Ser. B* **50**, 356–357.

Davison, A.C. and Hinkley, D.V. (1988). Saddlepoint approximations in resampling methods. *Biometrika* **75**, 417–431.

Davison, A.C., Hinkley, D.V., and Schechtman, E. (1986). Efficient bootstrap simulation. *Biometrika* **73**, 555–566.

Devroye, L. and Györfi, L. (1985). *Nonparametric Density Estimation: The L_1 View.* Wiley, New York.

De Wet, T. and Van Wyk, J.W.J. (1986). Bootstrap confidence intervals for regression coefficients when the residuals are dependent. *J. Statist. Comput. Simul.* **23**, 317–327.

Diaconis, P. and Efron, B. (1983). Computer-intensive methods in statistics. *Sci. Amer.* **248**, 116-130.

DiCiccio, T.J. and Romano, J.P. (1988). A review of bootstrap confidence intervals. (With discussion.) *J. Roy. Statist. Soc. Ser. B* **50**, 338-354.

DiCiccio, T.J. and Romano, J.P. (1989). The automatic percentile method: accurate confidence limits in parametric models. *Canadian J. Statist.* **17**, 155-169.

DiCiccio, T.J. and Tibshirani, R. (1987). Bootstrap confidence intervals and bootstrap approximations. *J. Amer. Statist. Assoc.* **82**, 163–170.

Do, K.-A. and Hall, P. (1991). On importance resampling for the bootstrap. *Biometrika* **78**, 161–167.

Draper, N.R. and Tierney, D.E. (1973). Exact formulas for additional terms in some important expansions. *Commun. Statist.* **1**, 495–524.

Ducharme, G.R., Jhun, M., Romano, J.P., and Truong, K.N. (1985). Bootstrap confidence cones for directional data. *Biometrika* **72**, 637–645.

Edgeworth, F.Y. (1896). The asymmetrical probability curve. *Philos. Mag., 5th Ser.* **41**, 90–99.

Edgeworth, F.Y. (1905). The law of error. *Proc. Cambridge Philos. Soc.* **20**, 36–65.

Edgeworth, F.Y. (1907). On the representation of a statistical frequency by a series. *J. Roy. Statist. Soc. Ser. A* **70**, 102–106.

Edgington, E.S. (1987). *Randomization Tests*, 2nd ed. Dekker, New York.

Efron, B. (1979). Bootstrap methods: Another look at the jackknife. *Ann. Statist.* **7**, 1–26.

Efron, B. (1981a). Nonparametric standard errors and confidence intervals. (With discussion.) *Canad. J. Statist.* **9**, 139–172.

Efron, B. (1981b). Nonparametric estimates of standard error: The jackknife, the bootstrap and other methods. *Biometrika* **68**, 589–599.

Efron, B. (1982). *The Jackknife, the Bootstrap and Other Resampling Plans.* SIAM, Philadelphia.

Efron, B. (1983). Estimating the error rate of a prediction rule: Improvement on cross-validation. *J. Amer. Statist. Assoc.* **78**, 316–331.

Efron, B. (1985). Bootstrap confidence intervals for a class of parametric problems. *Biometrika* **72**, 45–58.

Efron, B. (1987). Better bootstrap confidence intervals. (With discussion.) *J. Amer. Statist. Assoc.* **82**, 171–200.

Efron, B. (1988). Bootstrap confidence intervals: good or bad? (With discussion.) *Psychol. Bull.* **104**, 293–296.

Efron, B. (1990). More efficient bootstrap computations. *J. Amer. Statist. Assoc.* **85**, 79–89.

Efron, B. and Gong, G. (1981). Statistical theory and the computer. *Computer Science and Statistics: Proceedings of the 13th Symposium on the Interface* (W.F. Eddy, ed.), Springer-Verlag, New York, pp. 3–7.

Efron, B. and Gong, G. (1983). A leisurely look at the bootstrap, the jackknife, and cross-validation. *Amer. Statist.* **37**, 36–48.

Efron, B. and Tibshirani, R. (1986). Bootstrap methods for standard errors, confidence intervals, and other measures of statistical accuracy. (With discussion.) *Statist. Sci.* **1**, 54–77.

Esseen, C.-G. (1945). Fourier analysis of distribution functions. A mathematical study of the Laplace-Gaussian law. *Acta Math.* **77**, 1–125.

Faraway, J. and Jhun, M. (1990). Bootstrap choice of bandwidth for density estimation. *J. Amer. Statist. Assoc.* **85**, 1119-1122.

Finney, D.J. (1963). Some properties of a distribution specified by its cumulants. *Technometrics* **5**, 63–69.

Fisher, N.I. and Hall, P. (1989). Bootstrap confidence regions for directional data. *J. Amer. Statist. Assoc.* **84**, 996–1002.

Fisher, N.I. and Hall, P. (1990). On bootstrap hypothesis testing. *Austral. J. Statist.* **32**, 177–190.

Fisher, N.I. and Hall, P. (1991). Bootstrap algorithms for small samples. *J. Statist. Plann. Inf.* **27**, 157–169.

Fisher, R.A. (1928). Moments and product moments of sampling distributions. *Proc. London Math. Soc.* **30**, 199–238.

Fisher, R.A. and Cornish, E.A. (1960). The percentile points of distributions having known cumulants. *Technometrics* **2**, 209–226.

Freedman, D.A. (1981). Bootstrapping regression models. *Ann. Statist.* **9**, 1218–1228.

Freedman, D.A. and Peters, S.C. (1984). Bootstrapping a regression equation: Some empirical results. *J. Amer. Statist. Assoc.* **79**, 97–106.

Gasser, T., Sroka, L., and Jennen-Steinmetz, C. (1986). Residual variance and residual pattern in nonlinear regression. *Biometrika* **73**, 625–633.

Gayen, A.K. (1949). The distribution of "Student's" *t* in random samples of any size drawn from non-normal universes. *Biometrika* **36**, 353–369.

Geary, R.C. (1936). The distribution of "Student's" ratio for non-normal samples. *J. Roy. Statist. Soc. Supp.* **3**, 178–184.

Ghosh, M., Parr, W.C., Singh, K., and Babu, G.J. (1984). A note on bootstrapping the sample median. *Ann. Statist.* **12**, 1130–1135.

Gleason, J.R. (1988). Algorithms for balanced bootstrap simulations. *Amer. Statist.* **42**, 263–266.

Gnedenko, B.V. and Kolmogorov, A.N. (1954). *Limit Distributions for Sums of Independent Random Variables.* Addison-Wesley, Reading, MA.

Godambe, V.P. (1960). An optimum property of regular maximum likelihood estimation. *Ann. Math. Statist.* **31**, 1208–1211.

Godambe, V.P. (1985). The foundations of finite sample estimation in stochastic processes. *Biometrika* **72**, 419–428.

Godambe, V.P. and Thompson, M.W. (1984). Robust estimation through estimating equations. *Biometrika* **71**, 115–125.

Graham, R.L., Hinkley, D.V., John, P.W.M., and Shi, S. (1990). Balanced design of bootstrap simulations. *J. Roy. Statist. Soc. Ser. B* **52**, 185–202.

Gray, H.L. and Schucany, W.R. (1972). *The Generalized Jackknife Statistic.* Dekker, New York.

Hall, P. (1983). Inverting an Edgeworth expansion. *Ann. Statist.* **11**, 569–576.

Hall, P. (1986a). On the bootstrap and confidence intervals. *Ann. Statist.* **14**, 1431–1452.

Hall, P. (1986b). On the number of bootstrap simulations required to construct a confidence interval. *Ann. Statist.* **14**, 1453–1462.

Hall, P. (1987a). On the bootstrap and continuity correction. *J. Roy. Statist. Soc. Ser. B* **49**, 82–89.

Hall, P. (1987b). On the bootstrap and likelihood-based confidence regions. *Biometrika* **74**, 481–493.

Hall, P. (1987c). Edgeworth expansion for Student's t statistic under minimal moment conditions. *Ann. Probab.* **15**, 920–931.

Hall, P. (1988a). Theoretical comparison of bootstrap confidence intervals. (With discussion.) *Ann. Statist.* **16**, 927–985.

Hall, P. (1988b). On symmetric bootstrap confidence intervals. *J. Roy. Statist. Soc. Ser. B* **50**, 35–45.

Hall, P. (1988c). Rate of convergence in bootstrap approximations. *Ann. Probab.* **16**, 1665–1684.

Hall, P. (1988d). Unusual properties of bootstrap confidence intervals in regression problems. *Probab. Theory Rel. Fields* **81**, 247–273.

Hall, P. (1989a). On efficient bootstrap simulation. *Biometrika* **76**, 613–617.

Hall, P. (1989b). Antithetic resampling for the bootstrap. *Biometrika* **76**, 713–724.

Hall, P. (1989c). On convergence rates in nonparametric problems. *Internat. Statist. Rev.* **57**, 45–58.

Hall, P. (1990a). Performance of balanced bootstrap resampling in distribution function and quantile problems. *Probab. Theory Rel. Fields.* **85**, 239–260.

Hall, P. (1990b). Using the bootstrap to estimate mean squared error and select smoothing parameter in nonparametric problems. *J. Mult. Anal.* **32**, 177–203.

Hall, P. (1990c). On the relative performance of bootstrap and Edgeworth approximations of a distribution function. *J. Mult. Anal.* **35**, 108–129.

Hall, P. (1990d). Asymptotic properties of the bootstrap for heavy-tailed distributions. *Ann. Probab.* **18**, 1342–1360.

Hall, P. (1990e). Balanced importance resampling for the bootstrap. Unpublished manuscript.

Hall, P. (1990f). Effect of bias estimation on coverage accuracy of bootstrap confidence intervals for a probability density. *Ann. Statist.*, to appear.

Hall, P. (1990g). On bootstrap confidence intervals in nonparametric regression. Unpublished manuscript.

Hall, P. (1990h). On Edgeworth expansions and bootstrap confidence bands in nonparametric curve estimation. Unpublished manuscript.

Hall, P. (1991a). Bahadur representations for uniform resampling and importance resampling, with applications to asymptotic relative efficiency. *Ann. Statist.* **19**, 1062–1072.

Hall, P. (1991b). Edgeworth expansions for nonparametric density estimators, with applications. *Statistics* **22**, 215–232.

Hall, P. (1992). Transformations to remove skewness when constructing confidence intervals. *J. Roy. Statist. Soc. Ser. B*, to appear.

Hall, P., DiCiccio, T.J., and Romano, J.P. (1989). On smoothing and the bootstrap. *Ann. Statist.* **17**, 692–704.

Hall, P., Kay, J.W., and Titterington, D.M. (1990). Optimal difference-based estimation of variance in nonparametric regression. *Biometrika* **70**, 521–528.

Hall, P. and La Scala, B. (1990). Methodology and algorithms of empirical likelihood. *Internat. Statist. Rev.* **58**, 109–127.

Hall, P., Marron, J.S., and Park, B.U. (1990). Smoothed cross-validation. Unpublished manuscript.

Hall, P. and Martin, M.A. (1988a). On bootstrap resampling and iteration. *Biometrika* **75**, 661–671.

Hall, P. and Martin, M.A. (1988b). Exact convergence rate of bootstrap quantile variance estimator. *Probab. Theory Rel. Fields* **80**, 261–268.

Hall, P. and Martin, M.A. (1991). On the error incurred using the bootstrap variance estimate when constructing confidence intervals for quantiles. *J. Mult. Anal.* **38**, 70–81.

Hall, P., Martin, M.A., and Schucany, W.R. (1989). Better nonparametric bootstrap confidence intervals for the correlation coefficient. *J. Statist. Comput. Simul.* **33**, 161–172.

Hall, P. and Pittelkow, Y.E. (1990). Simultaneous bootstrap confidence bands in regression. *J. Statist. Comput. Simul.* **37**, 99–113.

Hall, P. and Sheather, S.J. (1988). On the distribution of a Studentized quantile. *J. Roy. Statist. Soc. Ser. B* **50**, 381–391.

Hall, P. and Titterington, D.M. (1989). The effect of simulation order on level accuracy and power of Monte Carlo tests. *J. Roy. Statist. Soc. Ser. B* **51**, 459–467.

Hall, P. and Wilson, S.R. (1991). Two guidelines for bootstrap hypothesis testing. *Biometrics* **47**, 757–762.

Hammersley, J.M. and Handscomb, D.C. (1964). *Monte Carlo Methods.* Methuen, London.

Hammersley, J.M. and Mauldon, J.G. (1956). General principles of antithetic variates. *Proc. Cambridge Philos. Soc.* **52**, 476–481.

Hammersley, J.M. and Morton, K.W. (1956). A new Monte Carlo technique: antithetic variates. *Proc. Cambridge Philos. Soc.* **52**, 449–475.

Härdle, W. (1989). Resampling for inference from curves. *Proc. 47th Session of the International Statistical Institute, Paris, 29 August-6 September 1989*, vol. **3**, International Statistical Institute, Paris, pp. 53–64.

Härdle, W. (1990). *Applied Nonparametric Regression*. Cambridge University Press, Cambridge, UK.

Härdle, W. and Bowman, A.W. (1988). Bootstrapping in nonparametric regression: Local adaptive smoothing and confidence bands. *J. Amer. Statist. Assoc.* **83**, 102–110.

Härdle, W. and Marron, J.S. (1991). Bootstrap simultaneous error bars for nonparametric regression. *Ann. Statist.* **19**, 778–796.

Härdle, W., Marron, J.S., and Wand, M.P. (1990). Bandwidth choice for density derivatives. *J. Roy. Statist. Soc. Ser. B* **52**, 223–232.

Hartigan, J.A. (1969). Using subsample values as typical values. *J. Amer. Statist. Assoc.* **64**, 1303–1317.

Hartigan, J.A. (1971). Error analysis by replaced samples. *J. Roy. Statist. Soc. Ser. B* **33**, 98–110.

Hartigan, J.A. (1975). Necessary and sufficient conditions for asymptotic joint normality of a statistic and its subsample values. *Ann. Statist.* **3**, 573–580.

Hartigan, J.A. (1986). Discussion of Efron and Tibshirani (1986). *Statist. Sci.* **1**, 75–77.

Hinkley, D.V. (1988). Bootstrap methods. (With discussion.) *J. Roy. Statist. Soc. Ser. B* **50**, 321–337.

Hinkley, D.V. (1989). Bootstrap significance tests. *Proc. 47th Session of the International Statistical Institute, Paris, 29 August-6 September 1989*, vol. **3**, International Statistical Institute, Paris, pp. 65–74.

Hinkley, D.V. and Shi, S. (1989). Importance sampling and the nested bootstrap. *Biometrika* **76**, 435–446.

Hinkley, D.V. and Wei, B.-C. (1984). Improvement of jackknife confidence limit methods. *Biometrika* **71**, 331–339.

Holst, L. (1972). Asymptotic normality and efficiency for certain goodness-of-fit tests. *Biometrika* **59**, 137–145.

Hope, A.C.A. (1968). A simplified Monte Carlo significance test procedure. *J. Roy. Statist. Soc. Ser. B* **30**, 582–598.

Hsu, P.L. (1945). The approximate distribution of the mean and variance of a sample of independent variables. *Ann. Math. Statist.* **16**, 1–29.

James, G.S. (1955). Cumulants of a transformed variate. *Biometrika* **42**, 529–531.

James, G.S. (1958). On moments and cumulants of systems of statistics. *Sankhyā* **20**, 1–30.

James, G.S. and Mayne, A.J. (1962). Cumulants of functions of random variables. *Sankhyā Ser. A* **24**, 47–54.

Johns, M.V. Jr (1988). Importance sampling for bootstrap confidence intervals. *J. Amer. Statist. Assoc.* **83**, 709–714.

Johnson, N.J. (1978). Modified *t*-tests and confidence intervals for asymmetrical populations. *J. Amer. Statist. Assoc.* **73**, 536–544.

Johnson, N.L. and Kotz, S. (1970). *Distributions in Statistics. Continuous Univariate Distributions*, vol. 1. Houghton Mifflin, Boston, MA.

Kaplan, E.L. (1952). Tensor notation and the sampling cumulants of *k*-statistics. *Biometrika* **39**, 319–323.

Kendall, M.G. and Stuart, A. (1977). *The Advanced Theory of Statistics*, vol. 1, 4th three-volume ed., Griffin, London.

Knight, K. (1989). On the bootstrap of the sample mean in the infinite variance case. *Ann. Statist.* **17**, 1168–1175.

Laird, N.M. and Louis, T.A. (1987). Empirical Bayes confidence intervals based on bootstrap samples. (With discussion.) *J. Amer. Statist. Assoc.* **82**, 739–757.

Lehmann, E.L. (1959). *Testing Statistical Hypotheses*. Wiley, New York.

Lehmann, E.L. (1986). *Testing Statistical Hypotheses*. 2nd ed. Wiley, New York.

Leonov, V.P. and Shiryaev, A.M. (1959). On a method of calculation of semi-invariants. *Theor. Probab. Appl.* **4**, 319–329.

Littlewood, D.E. (1950). *A University Algebra*. Heinemann, London.

Liu, R.Y. and Singh, K. (1987). On a partial correction by the bootstrap. *Ann. Statist.* **15**, 1713–1718.

Loh, W.-Y. (1987). Calibrating confidence coefficients. *J. Amer. Statist. Assoc.* **82**, 155–162.

Lunneborg, C.E. (1985). Estimating the correlation coefficient: The bootstrap approach. *Psychol. Bull.* **98**, 209–215.

Magnus, W., Oberhettinger, F., and Soni, R.P. (1966). *Formulas and Theorems for the Special Functions of Mathematical Physics*. Springer-Verlag, Berlin.

Mallows, C.L. and Tukey, J.W. (1982). An overview of techniques of data analysis, emphasizing its exploratory aspects. *Some Recent Advances in Statistics* (J. Tiago de Oliveira and B. Epstein, eds.). Academic Press, New York, pp. 111-172.

Mammen, E. (1991). *When Does Bootstrap Work: Asymptotic Results and Simulations*. Habilitationsschrift eingereicht bei der Fakultät für Mathematik der Ruprecht-Karls-Universität, Heidelberg.

Maritz, J.S. and Jarrett, R.G. (1978). A note on estimating the variance of the sample median. *J. Amer. Statist. Assoc.* **73**, 194–196.

Marriott, F.H.C. (1979). Barnard's Monte Carlo tests: How many simulations? *Appl. Statist.* **28**, 75-77.

Martin, M.A. (1989). On the bootstrap and confidence intervals. Unpublished PhD thesis, Australian National University.

Martin, M.A. (1990). On bootstrap iteration for coverage correction in confidence intervals. *J. Amer. Statist. Assoc.* **85**, 1105–1118.

McCullagh, P. (1984). Tensor notation and cumulants of polynomials. *Biometrika* **71**, 461–476.

McCullagh, P. (1987). *Tensor Methods in Statistics*. Chapman and Hall, London.

McKay, M.D., Beckman, R.M., and Conover, W.J. (1979). A comparison of three methods for selecting values of input variables in the analysis of output from a computer code. *Technometrics* **21**, 239–245.

Morey, M.J. and Schenck, L.M. (1984). Small sample behaviour of bootstrapped and jackknifed regression estimators. *ASA Proc. Bus. Econ. Statist. Sec.*, Amer. Statist. Assoc., Washington, DC, pp. 437–442.

Mosteller, F. and Tukey, J.W. (1977). *Data Analysis and Regression*. Addison-Wesley, Reading, MA.

Müller, H.-G. (1988). *Nonparametric Regression Analysis of Longitudinal Data*. Springer Lecture Notes in Statistics vol. **46**, Springer-Verlag, New York.

Müller, H.-G. and Stadtmüller, U. (1987). Estimation of heteroscedasticity in regression analysis. *Ann. Statist.* **15**, 610–625.

Müller, H.-G. and Stadtmüller, U. (1988). Detecting dependencies in smooth regression models. *Biometrika* **75**, 639–650.

Noreen, E.W. (1989). *Computer-Intensive Methods for Testing Hypotheses: an Introduction*. Wiley, New York.

Ogbonmwan, S.-M. and Wynn, H.P. (1986). Accelerated resampling codes with low discrepancy. Preprint, Department of Statistics and Actuarial Science, City University, London.

Ogbonmwan, S.-M. and Wynn, H.P. (1988). Resampling generated likelihoods. *Statistical Decision Theory and Related Topics* IV (S.S. Gupta and J.O. Berger, eds.), Springer-Verlag, New York, pp. 133–147.

Oldford, R.W. (1985). Bootstrapping by Monte Carlo versus approximating the estimator and bootstrapping exactly: Cost and performance. *Comm. Statist. Ser. B* **14**, 395–424.

Owen, A.B. (1988). Empirical likelihood ratio confidence intervals for a single functional. *Biometrika* **75**, 237–249.

Owen, A.B. (1990). Empirical likelihood confidence regions. *Ann. Statist.* **18**, 90–120.

Parr, W.C. (1983). A note on the jackknife, the bootstrap and the delta method estimators of bias and variance. *Biometrika* **70**, 719–722.

Pearson, E.S. and Adyanthaya, N.K. (1929). The distribution of frequency constants in small samples from non-normal symmetrical and skew populations. *Biometrika* **21**, 259–286.

Peters, S.C. and Freedman, D.A. (1984a). Bootstrapping an econometric model: Some empirical results. *J. Bus. Econ. Studies* **2**, 150–158.

Peters, S.C. and Freedman, D.A. (1984b). Some notes on the bootstrap in regression problems. *J. Bus. Econ. Studies* **2**, 406–409.

Petrov, V.V. (1975). *Sums of Independent Random Variables*. Springer-Verlag, Berlin.

Pfanzagl, J. (1979). Nonparametric minimum contrast estimators. *Selecta Statistica Canadiana* **5**, 105–140.

Phillips, P.C.B. (1977). A general theorem in the theory of asymptotic expansions as approximations to the finite sample distributions of econometric estimators. *Econometrica* **45**, 1517–1534.

Prakasa Rao, B.L.S. (1983). *Nonparametric Functional Estimation*. Academic Press, New York.

Quenouille, M.H. (1949). Approximate tests of correlation in time-series. *J. Roy. Statist. Assoc. Ser. B* **11**, 68–84.

Quenouille, M.H. (1956). Notes on bias in estimation. *Biometrika* **43**, 353–360.

Rasmussen, J.L. (1987). Estimating correlation coefficients: bootstrap and parametric approaches. *Psychol. Bull.* **101**, 136–139.

Reid, N. (1988). Saddlepoint methods and statistical inference. (With discussion.) *Statist. Sci.* **3**, 213–238.

Rice, J. (1984). Bandwidth choice for nonparametric regression. *Ann. Statist.* **12**, 1215–1230.

Roberts, F.S. (1984). *Applied Combinatorics*. Prentice-Hall, Englewood Cliffs.

Robinson, G.K. (1982). Confidence intervals and regions. *Encyclopedia of Statistical Sciences* (S. Kotz and N.L. Johnson, eds.), vol. **2**, Wiley, New York, pp. 120–127.

Robinson, J. (1986). Bootstrap and randomization confidence intervals. *Proceedings of the Pacific Statistical Congress, 20-24 May 1985, Auckland* (I.S. Francis, B.F.J. Manly, and F.C. Lam, eds.), North-Holland, Groningen, pp. 49–50.

Robinson, J. (1987). Nonparametric confidence intervals in regression: The bootstrap and randomization methods. *New Perspectives in Theoretical and Applied Statistics* (M.L. Puri, J.P. Vilaplana, and W. Wertz, eds.), Wiley, New York, pp. 243–256.

Robinson, J.A. (1983). Bootstrap confidence intervals in location-scale models with progressive censoring. *Technometrics* **25**, 179–187.

Romano, J.P. (1988). A bootstrap revival of some nonparametric distance tests. *J. Amer. Statist. Assoc.* **83**, 698–708.

Rosenblatt, M. (1971). Curve estimates. *Ann. Math. Statist.* **42**, 1815–1842.

Rothe, G. (1986). Some remarks on bootstrap techniques for constructing confidence intervals. *Statist. Hefte* **27**, 165–172.

Rubin, D.B. (1981). The Bayesian bootstrap. *Ann. Statist.* **9**, 130–134.

Sargan, J.D. (1975). Gram-Charlier approximations applied to t ratios of k-class estimators. *Econometrica* **43**, 327–346.

Sargan, J.D. (1976). Econometric estimators and the Edgeworth approximation. *Econometrica* **44**, 421–448.

Saulis, L.I. (1973). Limit theorems that take into account large deviations in the case when Yu.V. Linnik's condition is satisfied. (Russian) *Litovsk. Mat. Sb.* **13**, 173–196, 225–226.

Sazanov, V.V. (1981). *Normal Approximation—Some Recent Advances.* Springer-Verlag, New York.

Schemper, M. (1987a). Nonparametric estimation of variance, skewness and kurtosis of the distribution of a statistic by jackknife and bootstrap techniques. *Statist. Neerland.* **41**, 59–64.

Schemper, M. (1987b). On bootstrap confidence limits for possibly skew distributed statistics. *Comm. Statist. Ser. A* **16**, 1585–1590.

Schenker, N. (1985). Qualms about bootstrap confidence intervals. *J. Amer. Statist. Assoc.* **80**, 360–361.

Shao, J. (1988). On resampling methods for variance and bias estimation in linear models. *Ann. Statist.* **16**, 986–988.

Shorack, G.R. (1982). Bootstrapping robust regression. *Comm. Statist. Ser. A* **11**, 961–972.

Siddiqui, M.M. (1960). Distribution of quantiles in samples from a bivariate population. *J. Res. Nat. Bur. Stand. Sec. B* **64B**, 145–150.

Silverman, B.W. (1981). Using kernel density estimates to investigate multimodality. *J. Roy. Statist. Soc. Ser. B* **43**, 97–99.

Silverman, B.W. (1986). *Density Estimation for Statistics and Data Analysis.* Chapman and Hall, London.

Silverman, B.W. and Young, G.A. (1987). The bootstrap: To smooth or not to smooth? *Biometrika* **74**, 469–479.

Simon, J.L. (1969). *Basic Research Methods in Social Science.* Random House, New York.

Singh, K. (1981). On the asymptotic accuracy of Efron's bootstrap. *Ann. Statist.* **9**, 1187–1195.

Snijders, T.A.B. (1984). Antithetic variates for Monte Carlo estimation of probabilities. *Statist. Neerland.* **38**, 55-73.

Speed, T.P. (1983). Cumulants and partition lattices. *Austral. J. Statist.* **25**, 378–388.

Stein, M. (1987). Large sample properties of simulations using Latin hypercube sampling. *Technometrics* **29**, 143–151.

Stine, R.A. (1985). Bootstrap prediction intervals for regression. *J. Amer. Statist. Assoc.* **80**, 1026–1031.

Swanepoel, J.W.H., Van Wyk, J.W.J., and Venter, J.H. (1983). Fixed width confidence intervals based on bootstrap procedures. *Sequential Anal.* **2**, 289–310.

Tapia, R.A. and Thompson, J.R. (1978). *Nonparametric Probability Density Estimation*. Johns Hopkins University Press, Baltimore, MD.

Taylor, C.C. (1989). Bootstrap choice of the smoothing parameter in kernel density estimation. *Biometrika* **76**, 705–712.

Tibshirani, R. (1986). Bootstrap confidence intervals. *Computer Science and Statistics: Proceedings of the 18th Symposium on the Interface* (T.J. Boardman, ed.), Amer. Stat. Assoc., Washington, DC, pp. 267–273.

Tibshirani, R. (1988). Variance stabilization and the bootstrap. *Biometrika* **75**, 433–444.

Tukey, J.W. (1958). Bias and confidence in not-quite large samples. (Abstract) *Ann. Math. Statist.* **29**, 614.

Wallace, D.L. (1958). Asymptotic approximations to distributions. *Ann. Math. Statist.* **29**, 635–654.

Weber, N.C. (1984). On resampling techniques for regression models. *Statist. Probab. Lett.* **2**, 275–278.

Weber, N.C. (1986). On the jackknife and bootstrap techniques for regression models. *Proceedings of the Pacific Statistical Congress, 20-24 May 1985, Auckland* (I.S. Francis, B.F.J. Manly, and F.C. Lam, eds.), North-Holland, Groningen, pp. 51–55.

Whittaker, E.T. and Watson, G.N. (1927). *A Course of Modern Analysis*, 4th ed. Cambridge University Press, Cambridge, UK.

Withers, C.S. (1983). Expansions for the distribution and quantiles of a regular function of the empirical distribution with applications to nonparametric confidence intervals. *Ann. Statist.* **11**, 577–587.

Withers, C.S. (1984). Asymptotic expansions for distributions and quantiles with power series cumulants. *J. Roy. Statist. Soc. Ser. B* **46**, 389–396.

Woodroofe, M. and Jhun, M. (1988). Singh's theorem in the lattice case. *Statist. Probab. Lett.* **7**, 201-205.

Wu, C.F.J. (1986). Jackknife, bootstrap and other resampling methods in regression analysis. (With discussion.) *Ann. Statist.* **14**, 1261–1350.

Young, G.A. (1986). Conditioned data-based simulations: Some examples from geometrical statistics. *Internat. Statist. Rev.* **54**, 1–13.

Young, G.A. (1988). A note on bootstrapping the correlation coefficient. *Biometrika* **75**, 370–373.

Author Index

Subject Index

Springer Series in Statistics

(continued from p. ii)